Cambridge Astrophysics Series

The symbiotic stars

For my parents

THE SYMBIOTIC STARS

S.J.KENYON

Harvard-Smithsonian Center for Astrophysics

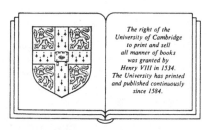

The right of the
University of Cambridge
to print and sell
all manner of books
was granted by
Henry VIII in 1534.
The University has printed
and published continuously
since 1584.

CAMBRIDGE UNIVERSITY PRESS

Cambridge

London   New York   New Rochelle

Melbourne   Sydney

CAMBRIDGE UNIVERSITY PRESS
Cambridge, New York, Melbourne, Madrid, Cape Town, Singapore, São Paulo, Delhi

Cambridge University Press
The Edinburgh Building, Cambridge CB2 8RU, UK

Published in the United States of America by Cambridge University Press, New York

www.cambridge.org
Information on this title: www.cambridge.org/9780521268073

First published 1986
This digitally printed version 2008

A catalogue record for this publication is available from the British Library

Library of Congress Cataloguing in Publication data

Kenyon, S. J.
The symbiotic stars.

Bibliography
Includes index
1. Stars, Symbiotic. I. Title.
QB843.S96K46   1986   523.8′4   85-9709

ISBN 978-0-521-26807-3 hardback
ISBN 978-0-521-09331-6 paperback

Contents

ded, the M-type spectrum returned to its former dominance and nor-
al Mira-like pulsations resumed.

While R Aqr was exhibiting its peculiar transformations, Plaskett
1928) was completing a detailed study of Z And. He identified two
istinct light sources in his spectra: a) the stellar source - associated
ith narrow emission lines of Fe II, Ti II, and other singly ionized
etals, and b) the nebular source - wider emission lines of He II,
III, and [O III]. He noted that the He I singlet to triplet intensity
atio was ~1 in Z And, while it was typically 1/3 in planetary nebulae.
lthough Plaskett did not notice any absorption features, Hogg (1932)
e-examined these plates on a suggestion by Merrill and identified TiO
ands.

In 1932, Merrill and Humason (1932) reported the discovery of three
ew "stars with combination spectra": CI Cyg, RW Hya, and AX Per.
On objective prism spectra, each of these objects had the TiO bands
normally observed in M4 giant stars and a superimposed He II λ4686
emission line. The Balmer lines also appeared as intense emission
features, and nebular lines from [O III] and [Ne III] were bright in CI
Cyg and AX Per. Although RW Hya was a known red variable
(Yamamoto 1924), CI Cyg and AX Per were new peculiar emission-
line stars. Lindsay (1932) and Greenstein (1937) subsequently reported
that both systems had undergone two 1-3 mag outbursts in the previous 30 years.

The discovery of the two new peculiar variables CI Cyg and AX Per
heightened interest in the "stars with combination spectra". Over the
next decade, roughly two dozen of these interesting variables were
identified by Merrill, Humason, and Burwell at Mt. Wilson Observatory, Swings and Struve at McDonald and Yerkes Observatories, Bloch
and Tcheng at Haute Provence Observatory, and Thackeray at
Radcliffe Observatory. An optical spectrum of the prototypical star
with combination spectrum, Z And, is shown in Figure 1.1. The
features Merrill associated with this class of objects (TiO absorption
bands, strong H I and He II emission lines, and a faint, blue continuum) are quite prominent on this spectrum; many systems were foun
to be conspicuously variable from month to month and sometin
from night to night, distinguishing them from other red varia
whose emission line spectra vary over much longer timescales.

Aside from its interesting optical spectrum, Z And displays
variety of photometric activity (Figure 1.2). Z And usually
found at V~10.5, with minima near V=11 every 757 days
and Webbink 1984). The occasional 2 mag outbursts are a

Preface

It has been over fifty years since P. Merrill and M. Humason of Mt.
Wilson Observatory reported the discovery of three M-type giant stars
with an unusually strong He II λ4686 emission line. Astronomers at
Harvard College Observatory found each of these "stars with combination spectra" was a long-period variable, and two of them, CI Cyg and
AX Per, had undergone a 3 mag nova-like eruption in the previous
thirty years. New objects were added to this group of peculiar variables over the next decade: some of these exhibited the regular radial
velocity variations expected of a binary system, while other systems
appeared to fluctuate randomly and were thought to be single stars. P.
Merrill eventually coined the term "symbiotic stars" to describe objects
whose spectra simultaneously display features associated with red giant
stars and planetary nebulae.

Symbiotic stars are now commonly accepted as binary systems, in
which a red giant star transfers material to its hotter companion.
Interactions between the components of a binary system are of special
interest in astrophysics today, and symbiotic stars present an exciting
laboratory in which to examine such basic physical processes as (i)
mass loss from red giants and the formation of planetary nebulae, (ii)
accretion onto compact stars and the evolution of nova-like eruptions,
and (iii) photoionization and radiative transfer within gaseous nebulae.
The physical conditions found in these systems are usually very
extreme, and they therefore present activity not easily observed in
other binaries.

Although observational studies continue to reveal interesting
phenomena in symbiotic stars, this seems an appropriate time to summarize our current knowledge of these binaries and to prepare for
future space-based observatories such as Space Telescope. Many

important problems remain unsolved, and I hope this monograph will stimulate new interest in symbiotic stars and assist in planning new observational and theoretical efforts to understand their interesting behavior.

Many colleagues have aided my preparation of this book, and I am sorry that there is insufficient space to acknowledge everyone. I owe a special debt to R.F. Webbink, who introduced me to symbiotic stars, supervised my Ph. D. research, and greatly improved the presentation that follows. I must also mention D.A. Allen, G.T. Bath, J.S. Gallagher, I. Iben, Jr., J.B. Kaler, J.E. Pringle, J.C. Raymond, S.G. Starrfield, and J.W. Truran for their comments and encouragement throughout this endeavor. I also thank C.M. Anderson, M. Kafatos, S. Kwok, A.G. Michalitsianos, E.R. Seaquist, A.R. Taylor, and P.A. Whitelock for providing numerous figures and diagrams. Most of the spectra reproduced in this volume were obtained at Kitt Peak National Observatory (operated by AURA, Inc., under contract with the National Science Foundation), and I thank G. Burbidge, S. Wolff, and the observatory staff for allocations of telescope time and technical assistance. Finally, J.A. Mattei (director of the American Association of Variable Star Observers) and F.M. Bateson (director of the Variable Star Section, Royal Astronomical Society of New Zealand) have my eternal gratitude for furnishing historical light curves of individual symbiotic stars. The work of the amateur astronomers across the world has been very important for improving my understanding of these systems, and I appreciate their largely unsung efforts.

1

Introduction

1.1   The Discovery of Symbiotic Stars

Most of the stars in the HD catalogue have si consisting of a bright continuum and a number of dar lines. These are the "normal" dwarf, giant and supergia fall in distinct bands in the HR diagram, and form the current understanding of stellar evolution. However, th many HD stars display bright emission lines in addition to in place of) the more usual absorption features. Fleming duced the first comprehensive list of these "stars with pecu and grouped them into various categories, including (i) n type stars, (iii) stars with bright hydrogen lines, and (iv) variables. A small class of red variables was especially although their spectral characteristics appeared identical other red long-period variables, their range of variability se for their spectral type. Cannon later isolated another gr stars with bright lines of H I and He II; these included t variables Z And and SY Mus.

One of the first systematic spectroscopic studies of emission-line star was Merrill's (1919) investigation of the long-period variable R Aqr. He found conspicuously bri nebular lines about one month before a predicted maximum of the spectrum appeared to be that of a normal Md var strong TiO absorption bands and sharp, bright H I lines. B sion lines of He II, C III, and N III appeared in 1926 and strong until 1933 (Merrill 1936). During these years, the an he Mira-like brightness variations diminished, although the e oscillations remained constant. Once the high ionization

Preface

It has been over fifty years since P. Merrill and M. Humason of Mt. Wilson Observatory reported the discovery of three M-type giant stars with an unusually strong He II λ4686 emission line. Astronomers at Harvard College Observatory found each of these "stars with combination spectra" was a long-period variable, and two of them, CI Cyg and AX Per, had undergone a 3 mag nova-like eruption in the previous thirty years. New objects were added to this group of peculiar variables over the next decade: some of these exhibited the regular radial velocity variations expected of a binary system, while other systems appeared to fluctuate randomly and were thought to be single stars. P. Merrill eventually coined the term "symbiotic stars" to describe objects whose spectra simultaneously display features associated with red giant stars and planetary nebulae.

Symbiotic stars are now commonly accepted as binary systems, in which a red giant star transfers material to its hotter companion. Interactions between the components of a binary system are of special interest in astrophysics today, and symbiotic stars present an exciting laboratory in which to examine such basic physical processes as (i) mass loss from red giants and the formation of planetary nebulae, (ii) accretion onto compact stars and the evolution of nova-like eruptions, and (iii) photoionization and radiative transfer within gaseous nebulae. The physical conditions found in these systems are usually very extreme, and they therefore present activity not easily observed in other binaries.

Although observational studies continue to reveal interesting phenomena in symbiotic stars, this seems an appropriate time to summarize our current knowledge of these binaries and to prepare for future space-based observatories such as Space Telescope. Many important problems remain unsolved, and I hope this monograph will stimulate new interest in symbiotic stars and assist in planning new observational and theoretical efforts to understand their interesting behavior.

Many colleagues have aided my preparation of this book, and I am sorry that there is insufficient space to acknowledge everyone. I owe a special debt to R.F. Webbink, who introduced me to symbiotic stars, supervised my Ph. D. research, and greatly improved the presentation that follows. I must also mention D.A. Allen, G.T. Bath, J.S. Gallagher, I. Iben, Jr., J.B. Kaler, J.E. Pringle, J.C. Raymond, S.G. Starrfield, and J.W. Truran for their comments and encouragement throughout this endeavor. I also thank C.M. Anderson, M. Kafatos, S. Kwok, A.G. Michalitsianos, E.R. Seaquist, A.R. Taylor, and P.A. Whitelock for providing numerous figures and diagrams. Most of the spectra reproduced in this volume were obtained at Kitt Peak National Observatory (operated by AURA, Inc., under contract with the National Science Foundation), and I thank G. Burbidge, S. Wolff, and the observatory staff for allocations of telescope time and technical assistance. Finally, J.A. Mattei (director of the American Association of Variable Star Observers) and F.M. Bateson (director of the Variable Star Section, Royal Astronomical Society of New Zealand) have my eternal gratitude for furnishing historical light curves of individual symbiotic stars. The work of the amateur astronomers across the world has been very important for improving my understanding of these systems, and I appreciate their largely unsung efforts.

1

Introduction

1.1   The Discovery of Symbiotic Stars

Most of the stars in the HD catalogue have simple spectra, consisting of a bright continuum and a number of dark absorption lines. These are the "normal" dwarf, giant and supergiant stars that fall in distinct bands in the HR diagram, and form the basis for our current understanding of stellar evolution. However, the spectra of many HD stars display bright emission lines in addition to (or perhaps in place of) the more usual absorption features. Fleming (1912) produced the first comprehensive list of these "stars with peculiar spectra" and grouped them into various categories, including (i) novae, (ii) O-type stars, (iii) stars with bright hydrogen lines, and (iv) long-period variables. A small class of red variables was especially interesting: although their spectral characteristics appeared identical to those of other red long-period variables, their range of variability seemed small for their spectral type. Cannon later isolated another group of red stars with bright lines of H I and He II; these included the irregular variables Z And and SY Mus.

One of the first systematic spectroscopic studies of a peculiar emission-line star was Merrill's (1919) investigation of the enigmatic long-period variable R Aqr. He found conspicuously bright [O III] nebular lines about one month before a predicted maximum. The rest of the spectrum appeared to be that of a normal Md variable, with strong TiO absorption bands and sharp, bright H I lines. Bright emission lines of He II, C III, and N III appeared in 1926 and remained strong until 1933 (Merrill 1936). During these years, the amplitude of the Mira-like brightness variations diminished, although the period of the oscillations remained constant. Once the high ionization lines had faded, the M-type spectrum returned to its former dominance and normal Mira-like pulsations resumed.

While R Aqr was exhibiting its peculiar transformations, Plaskett (1928) was completing a detailed study of Z And. He identified two distinct light sources in his spectra: a) the stellar source - associated with narrow emission lines of Fe II, Ti II, and other singly ionized metals, and b) the nebular source - wider emission lines of He II, N III, and [O III]. He noted that the He I singlet to triplet intensity ratio was ~ 1 in Z And, while it was typically 1/3 in planetary nebulae. Although Plaskett did not notice any absorption features, Hogg (1932) re-examined these plates on a suggestion by Merrill and identified TiO bands.

In 1932, Merrill and Humason (1932) reported the discovery of three new "stars with combination spectra": CI Cyg, RW Hya, and AX Per. On objective prism spectra, each of these objects had the TiO bands normally observed in M4 giant stars and a superimposed He II $\lambda 4686$ emission line. The Balmer lines also appeared as intense emission features, and nebular lines from [O III] and [Ne III] were bright in CI Cyg and AX Per. Although RW Hya was a known red variable (Yamamoto 1924), CI Cyg and AX Per were new peculiar emission-line stars. Lindsay (1932) and Greenstein (1937) subsequently reported that both systems had undergone two 1-3 mag outbursts in the previous 30 years.

The discovery of the two new peculiar variables CI Cyg and AX Per heightened interest in the "stars with combination spectra". Over the next decade, roughly two dozen of these interesting variables were identified by Merrill, Humason, and Burwell at Mt. Wilson Observatory, Swings and Struve at McDonald and Yerkes Observatories, Bloch and Tcheng at Haute Provence Observatory, and Thackeray at Radcliffe Observatory. An optical spectrum of the prototypical star with combination spectrum, Z And, is shown in Figure 1.1. The features Merrill associated with this class of objects (TiO absorption bands, strong H I and He II emission lines, and a faint, blue continuum) are quite prominent on this spectrum; many systems were found to be conspicuously variable from month to month and sometimes from night to night, distinguishing them from other red variables whose emission line spectra vary over much longer timescales.

Aside from its interesting optical spectrum, Z And displays a wide variety of photometric activity (Figure 1.2). Z And usually can be found at $V \sim 10.5$, with minima near $V = 11$ every 757 days (Kenyon and Webbink 1984). The occasional 2 mag outbursts are accompanied by remarkable spectral changes, as the blue component brightens to dominate the late-type absorption bands. Most of the emission lines vanish as well, although the bright H I and He I lines usually develop P Cygni absorption components. This behavior in outburst led observers such as Merrill, Swings, and Struve to suspect the binary nature of all symbiotic systems, although verifying binary motion proved to be a difficult task. Merrill termed Z And and related systems *symbiotic stars* in 1941 (cf. Merrill 1958), since two seemingly hostile stellar components, a cool red giant and a small hot companion, seemed to live in general harmony (although with occasional disagreements!!).

1.2   The Definition of "Symbiotic Star"

Before proceeding with a discussion of the observed properties of and the theoretical models for symbiotic stars, it is necessary to provide an operational definition for this group of peculiar variables. The definition of a class of variable stars historically depends on choosing one or more "prototypes", and then discovering other stars that resemble the prototypes. Merrill's original definition recognized Z And, BF Cyg, CI Cyg, RW Hya, and AX Per as type examples of the class of

Figure 1.1 - Optical spectrum of the symbiotic star Z And in a quiescent state obtained by the author on JD 2445212.
The prominent TiO bands redward of $\lambda5000$ and strong emission lines from H I (e.g. $\lambda4861$ and $\lambda4340$), He I ($\lambda5876$), and He II ($\lambda4686$) are characteristic of quiescent symbiotic stars.

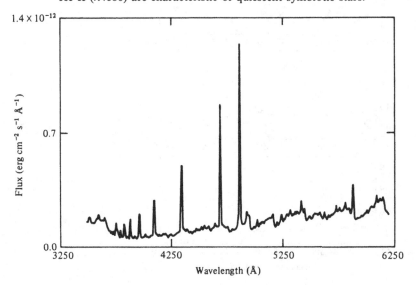

4   *Introduction*

Figure 1.2 - Visual light curve of Z And.
The data plotted are 20 day means of observations collected by the
AAVSO. Small amplitude variations ($\Delta V \sim 0.5$ mag) are characteristic
of quiescent phases, and 2-3 mag eruptions occur every 10-15 years.

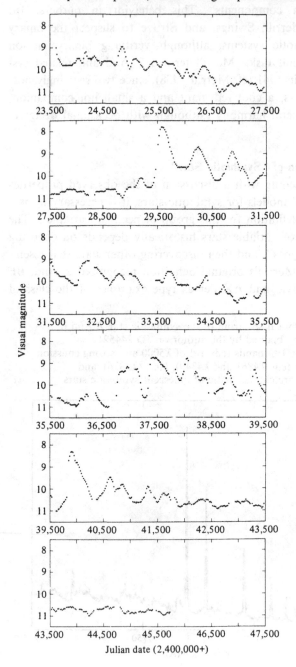

symbiotic stars, leading to the inclusion of other objects such as Y CrA, V443 Her, AR Pav, and AG Peg into the exclusive group. More recent attempts to "improve" Merrill's classification scheme have added a number of quantitative requirements, including:

(i)   irregular optical variability (Boyarchuk 1969),

(ii)   the presence of emission lines from ions with an ionization potential exceeding 55 eV (Allen 1979), and

(iii)   the object cannot be simply classified as something else (Nussbaumer 1982).

While each of these restrictions may have some merit, it can be said that every known symbiotic star has, at one time or another, violated *all* the classification criteria invented in the past 50 years. In some instances, the symbiotic nature of a given object has become obvious only after years of painstaking effort. A case in point is the well-known system AG Peg. Cannon and Fleming originally classified AG Peg (known then as BD + 11°4673) as a P Cygni star, since its optical spectrum displayed a blue continuum and H I emission lines with blue-shifted absorption components. Over the next 20 years, the spectrum evolved from a B-type star into an O-type star, and then into a Wolf Rayet star with TiO absorption bands (Merrill 1959 and references therein).

Given the mischievous nature of most variable stars, it seems best to rely on type examples of the symbiotic phenomenon, rather than develop a rigid set of classification criteria. I have chosen two proto-typical systems, CI Cyg and V1016 Cyg, whose observed properties best represent those found among the objects generally called symbiotic stars or stars with combination spectra (as discussed in the following chapters and the Appendix). These properties may be briefly summar-ized as:

1) the absorption features of a *late-type giant star* (a G, K or M star which is not obviously a main sequence or a supergiant star); these include Ca I, Ca II, Na I, Fe I, H_2O, CN, CO, TiO, and VO, among others;

2) bright H I and He I *emission lines* and either

(a) additional *bright lines* of ions with an ionization potential of at least 20 eV (e.g., [O III]) and an equivalent width exceed-ing 1 Å, or

(b) an A or F-type continuum with additional *absorption lines* from H I, He I, and singly ionized metals.

Knowledgeable readers will recognize that (2b) represents the outburst state of a typical symbiotic star - since it may be possible to catch a given system in a semi-permanent outburst state (TX CVn or BL Tel, perhaps?), this condition is included for completeness. It is important to note that optical and infrared observations provide necessary but perhaps not sufficient information for classifying potential symbiotic stars. EG And is an obvious example of a system with rather weak optical emission lines and very intense emission features in the satellite ultraviolet.

1.3   The Space Distribution of Symbiotic Stars

The general confusion concerning the nature of symbiotic stars may lie in the manner in which they have been discovered. The earliest systems were noted as either (i) peculiar long-period variables (R Aqr) or (ii) stars with peculiar spectra (Z And). More recent identifications have relied on objective prism surveys coupled with infrared photometry, and these objects tend to be spectroscopically different from those discovered in the 1930's and 1940's. A number of systems were found accidentally as a result of 2-7 mag nova-like outbursts (T CrB and V1016 Cyg), and were only later realized to be symbiotic. There has yet to be a systematic search for symbiotic star systems in both hemispheres, although Allen (1979) has examined fields in the vicinity of the galactic center rather completely.

Roughly 125 symbiotic stars are known in our galaxy, and another 25 objects are suspected of having symbiotic tendencies. Their angular distribution on the sky exhibits a fairly strong concentration towards the galactic plane (Figure 1.3), in a manner similar to that of field red giants, galactic novae, planetary nebulae, and other members of the intermediate or old disk population (Boyarchuk 1975; Payne-Gaposchkin 1957). The lack of symbiotic stars in the direction of the galactic anti-center probably reflects an overall decrease in the stellar density and in the star formation rate (Mihalas and Binney 1981). The concentration of systems towards the galactic center is likely to be more extreme than indicated by Figure 1.3 due to the effects of interstellar reddening.

Very little quantitative kinematic information is available for symbiotic stars. None have been associated with any open or globular star cluster, although the suspected symbiotic star V1148 Sgr lies near the globular cluster NGC 6553. Reliable trigonometric parallaxes are non-existent, and the known proper motions are very small (Table A.6). Spectroscopic parallaxes have been derived by Allen (1980) from infrared photometry at K (2.2 μm) and an absolute K magnitude-spectral type relation for M-type giants. While these distances appear reasonable, Allen suggested that the galactic distribution might be better understood if the cool components in symbiotic stars were somewhat more luminous than normal field giants (i.e., $M_V \sim -2$ instead of the more normal $M_V \sim -0.5$). Allen's distances imply a mean z_o of ≈ 550 pc (where the space density of symbiotic stars perpendicular to the galactic plane is given by $\rho(z) \approx e^{-z/z_o}$; Duerbeck 1984), which is somewhat of an overestimate since interstellar reddening biases the sample against systems with low galactic latitude. This value for z_o is consistent with an old disk population, as opposed to a young disk or halo population.

Radial velocities have been measured in some systems, although great care must be taken when interpreting the measured values. Most velocities have been derived from the emission lines, and may therefore reflect motions of gas streams rather than the system as a whole. Wallerstein (1981) compiled a list of relatively reliable radial velocities, and calculated a velocity dispersion of 63 ± 14 km s^{-1}. While this is consistent with that expected from an old disk population, it should be noted that velocities are not available for the vast majority of known symbiotic stars. The systems that have been measured are generally the brightest (and perhaps, therefore, the closest), and the velocity measurements have not been acquired in a systematic way. A new sur-

Figure 1.3 - Galactic distribution of symbiotic stars in $-60° < b^{II} < 60°$ and $-180° < l^{II} < 180°$.

The concentration of symbiotic systems towards the galactic plane is similar to that seen in planetary nebulae and galactic novae, and suggests these objects belong to an old disk population.

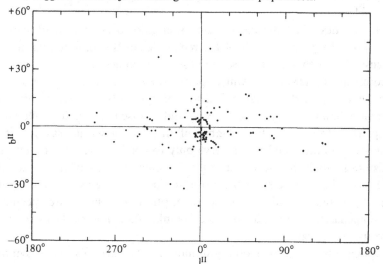

vey of a large sample of symbiotic stars, especially those in the direction of the galactic center, is needed to measure more accurately the velocity dispersion among symbiotic systems.

Many new extragalactic symbiotic stars have been discovered in the past few years; five are now known in the Magellanic Clouds and another was identified in the Draco dwarf spheroidal galaxy. Four of the LMC and SMC objects contain carbon stars, while the fifth is enveloped by a cool dust shell. Since only 2 of ≈ 125 galactic symbiotics contain carbon stars (UV Aur and UKS Ce-1), the systems in the Clouds are not easily related to those in the Milky Way. Blanco, et al. (1978) have noted that field carbon stars are much more common in the Clouds than in our own galaxy, and it is reassuring that carbon star symbiotics are prevalent as well. This fact implies that symbiosis is not particular to a given type of late-type giant, and can infect any such star regardless of its chemical composition (Allen 1983). The discovery of a carbon star symbiotic in the Draco dwarf spheroidal galaxy is interesting, since dwarf spheroidals are usually thought to contain only very old stars.

Since the LMC objects do not resemble the majority of known symbiotic systems, they cannot be used to calibrate the distances of their galactic counterparts. A sensitive search for symbiotics in M31 and M33 might resolve this problem, providing a sufficient number of systems could be located. A symbiotic binary in M31 would have $V \sim 22$ (24), providing $M_V \sim -2$ (0) is a typical visual magnitude for the cool component. Thus, symbiotic stars in M31 and M33 are within reach of ground-based telescopes, although suitable candidate stars have yet to be identified.

Having described the most basic observational properties of symbiotic stars, the primary goals of the remainder of this monograph are to discuss the behavior of these interesting variables and to develop a theoretical framework in which present-day and future observations can be interpreted. A large fraction of symbiotic behavior forms a common bond among these objects, and is discussed in the following chapters. However, a fair number of these variables display features that are "unique", and which serve only to confuse the overall picture. This book would be incomplete without some mention of this material, so I have chosen to discuss the more (in)famous systems individually in the Appendix. Aside from a description of the more peculiar phenomena, the Appendix provides tabular data not easily integrated into the main text.

Optical observations naturally dominated the early history of symbiotic stars, but recent progress in the field has been accomplished with the aid of high-speed digital detectors that span the entire electromagnetic spectrum. These new instruments provide *quantitative* information that can be compared to theoretical models, and thus the casual, but exhaustive, qualitative descriptions of spectra that so dominated the 1950's and 1960's are no longer warranted. Throughout this work, I shall therefore try to emphasize the astrophysical problems raised by the observations, and to call attention to new data that are needed to test the basic theories that have been developed to explain the diverse phenomena.

2

The Quiescent Phase - Theory

2.1   Introduction

The composite nature of a typical symbiotic star is shown effectively in Figure 1.1, which exhibits the spectroscopic features commonly associated with these interesting variables, namely (i) a bright red continuum, (ii) strong TiO absorption bands, (iii) prominent high ionization emission lines, and (iv) a weak blue continuum. Berman (1932) and Hogg (1934) were the first to suggest that these phenomena are most naturally explained by a binary star. Both authors suggested that the behavior of an "average" star with combination spectrum could be understood if an M-type giant had a faint O- or B-type companion star similar to that known to exist in o Ceti. This faint star would give rise to the weak blue continuum, and surrounding nebulosity would provide an obvious emission-line region. The small-scale fluctuations observed in these lines and in the continuum are a result of binary motion, while the larger flares are caused by instabilities in the hot source itself.

A few years later, Kuiper (1940) proposed an alternative binary model, suggesting that a system composed of a normal main sequence star and a Roche lobe-filling companion might explain such peculiar stars as Z And and β Lyr. Matter lost by the giant falls onto its fainter companion, giving rise to a hot emission region. In this case, flares and other types of random variability are a result of instabilities in the mass-losing star, rather than in the hot, compact star. This explanation gained some support with the detection of gas streams from the giant components in AX Mon and 17 Lep (Cowley 1964 and references therein). Such motions were not directly observed in symbiotic systems, but periodic radial velocity (and photometric) variations indicative of binary motion were observed in some objects by Merrill and others.

The single star alternatives to the binary models were developed in response to the failure to find periodic variations in a number of symbiotic objects. Menzel (1946; also Sobolev 1960) envisioned that a combination spectrum might be produced in an extremely evolved long period variable by its condensed stellar core (emission lines) and cool, tenuous atmosphere (absorption lines). Other efforts (e.g. Aller 1954; Bruce 1955,1956; Gauzit 1955a,b) sought to eliminate the hot companion star from the binary model in favor of a very active coronal or chromospheric region surrounding an otherwise normal red giant. Hack and Selvelli (1982 and references therein) have recently interpreted their observations of CH Cyg in terms of such a model, while Oliversen, et al. (1980) proposed that a large starspot (and the associated magnetic activity) on the surface of a K giant might explain the behavior of AG Dra. Stencel and Ionson (1979) sought to quantify these hypotheses by suggesting that a strong magnetic field might act to inhibit the mass loss commonly observed in M giants. They concluded that a 1000 G field could constrain the outward flow of material, and might energize a low temperature corona. However, the radiative losses of such a corona are sufficient to explain only the low excitation optical spectra of symbiotic stars such as EG And.

Wood (1973) expanded on the original suggestions by Aller and Gauzit that the strong emission lines might be produced in a shock-heated chromosphere or corona. He investigated the dynamical properties of the atmosphere of an asymptotic branch star, and showed that the development of relaxation oscillations might energize an emission-line region. Wood suggested that the low-amplitude photometric variations and the eruptive behavior observed in symbiotics are naturally explained by pulsations in an evolved red giant, while the blue continuum and strong emission lines might be produced in optically thin gas above the shocked region.

2.2   Synthetic Spectra for Symbiotic Stars

We shall see in the next chapter that high quality spectrophotometric data covering nearly the entire electromagnetic spectrum are now available for many symbiotic systems. Thus, it is no longer sufficient merely to postulate an active coronal region or a hot, binary companion star, as such models may be tested directly by the observations. As an example, coronal models for red giant stars predict that X-ray emission should be correlated with the flux from the He II $\lambda 1640$

emission line, and observations of ≈ 20 normal red giants show good agreement with theory (Hartmann, Dupree, and Raymond 1981). These results should be applicable if symbiotics are simply red giants with unusually active coronal regions, but the X-ray fluxes predicted from observations of He II $\lambda 1640$ emission lines in R Aqr and AG Dra are 1-2 orders of magnitude *above* the observed flux from the *Einstein* X-ray satellite. T CrB and V1017 Sgr, on the other hand, show no indication for substantial He II emission, and yet were also detected by *Einstein*. Another possible coronal diagnostic is the N V $\lambda 1240$/C IV $\lambda 1548$ line ratio, which is $\approx 1/4$ to $1/2$ in a typical corona (Hartmann, Dupree, and Raymond 1982). While a ratio of $1/2$ does not require a coronal model, observations of CI Cyg and EG And yield N V/C IV $\approx 1/15$ to $1/20$ (Stencel, et al. 1982; Stencel and Sahade 1980). Such low ratios would seem to preclude a coronal model for the emission lines in these objects, and lend support to the more traditional binary interpretation.

The first attempt to compare model symbiotic stars with observations was made by Boyarchuk (1969, 1975, and references therein). He proposed that a symbiotic star was composed of three sources: (i) a giant of spectral type G-M, (ii) a small hot star with an effective temperature, T_h, of $\sim 10^5$ K, and (iii) an ionized nebula with an electron temperature, T_e, of 17,000 K and an electron density, n_e, exceeding 10^6 cm^{-3}. The hot star was assumed to radiate as a blackbody, while case B recombination was adopted for the nebula (i.e., a nebula optically thick to photons with $\lambda < 912$ Å and optically thin to photons with $\lambda > 912$ Å). Boyarchuk's theoretical energy distributions satisfactorily reproduced spectroscopic observations over $\lambda\lambda 3200$-8000 for a number of systems, including Z And and AG Peg. He found that the secular variations of Z And could be understood if the hot component varied in effective temperature, but maintained a roughly constant bolometric luminosity.

The launch of various ultraviolet satellites has permitted new tests of the binary model, as high quality spectroscopic observations are now available down to ~ 1200 Å. Kenyon and Webbink (1984) expanded on Boyarchuk's original analysis, and considered two types of hot components: (i) a hot, compact star similar to the central star of a planetary nebula, and (ii) an accretion disk surrounding a low mass white dwarf or main sequence star. They combined energy distributions for late-type giant stars (Johnson 1966; O'Connell 1973; Wu, et al. 1980) with theoretical distributions for the hot component and the ionized nebula (including the strong H-Balmer and He II emission lines) to construct synthetic spectra from 0.1 to 3.5 μm. Examples of these synthetic spectra are shown in Figure 2.1 for a binary consisting of an M2 III giant, an ionized nebula with $T_e = 10^4$ K, and a white dwarf with $R = 0.01\ R_\odot$ at the indicated effective temperatures. At low effective temperatures, the white dwarf spectrum is very weak compared to that of its giant companion. As the white dwarf's temperature increases, the UV spectrum becomes increasingly dominated by nebular continuum radiation. At an effective temperature of 100,000 K, the He II λ1640 line is very strong relative to the UV continuum, and there is a weak Balmer jump at λ3646. However, the optical lines are fairly weak and Hβ has an equivalent width of only 5 Å. The UV spectrum is quite remarkable at an effective temperature of 200,000 K: λ1640 is extremely intense, and the jumps at λ2052 and λ3646 are quite sharp. The optical spectrum, except for the bright emission lines, is virtually identical to a normal M-type giant. In this

Figure 2.1 - Synthetic spectra for a white dwarf ($R = 0.01\ R_\odot$) at the indicated effective temperatures in combination with an M2 III giant and a gaseous nebula (Kenyon and Webbink 1984). Both H I and He II emission lines are very prominent at high effective temperatures in these models, although the blue optical continuum is never very strong. A distance of 10 pc to the source is assumed for this and all subsequent synthetic spectra.

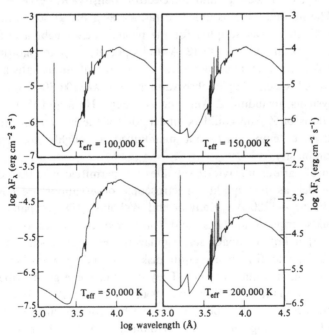

spectrum, Hβ has an equivalent width of roughly 50 Å, which is comparable to that observed in some symbiotic stars.

Examples of spectra for binaries containing an accreting main sequence star are shown in Figure 2.2. At low accretion rates, the optical spectrum is indistinguishable from a normal M2 III giant star, and the UV continuum is very weak. Hard photons emitted by an optically thick boundary layer at the inner edge of the disk are responsible for the nebular emission lines, which are nearly invisible at these low rates (10^{-7} $M_\odot$ yr^{-1}). Such a binary would not be classified as a symbiotic at all, but rather as a normal Me variable (cf. Merrill 1940). For somewhat larger rates ($\approx 10^{-6}$ $M_\odot$ yr^{-1}), the nebular and Me-type Balmer lines will have roughly comparable intensities. The accreting secondary does not announce itself in the optical until $\dot{M} > 10^{-6}$ $M_\odot$ yr^{-1}, when the Balmer lines are fairly strong and the spectrum appears composite. The blue veiling and the emission lines grow in strength as the accretion rate increases. For accretion rates approaching the Eddington limit (10^{-3} $M_\odot$ yr^{-1}), the continuum is totally dominated by the disk, and there is little evidence for a late-

Figure 2.2 - Synthetic spectra for an accreting main sequence star ($M = 1$ $M_\odot$, 1 $R_\odot$) at a binary inclination of 30° in combination with an M2 III giant and a gaseous nebula (Kenyon and Webbink 1984).

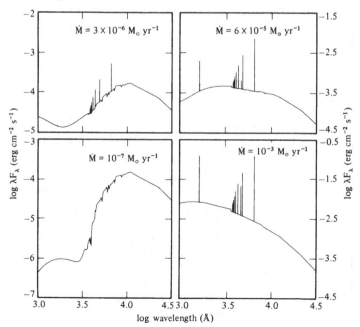

type giant. Note that the giant must fill its tidal lobe if such mass loss rates ($\dot{M} > 10^{-5} M_\odot$ yr^{-1}) are to be maintained for an extended period of time.

The situation is quite different if the M-type giant has an accreting 1 $M_\odot$ white dwarf companion (Figure 2.3). Weak H-Balmer and He II lines are visible in the optical even at very low accretion rates, and they strengthen markedly as the accretion rate (and emission from the optically thick boundary layer) increases. The Balmer jump is a much more pronounced emission feature in the white dwarf accretor. This is a result of the deeper potential well which produces many more ionizing photons at a given accretion luminosity. At moderate accretion rates ($10^{-8} M_\odot$ yr^{-1}), the UV spectrum is similar to the black-body models described earlier, especially if these systems suffer some interstellar reddening. Even at the highest allowed rates ($10^{-5} M_\odot$ yr^{-1}; the Eddington limit), the optical absorption features of the giant are still rather strong. Nebular emission is extremely intense at this accretion rate, and the approximation of optically thin emission lines is no longer valid (the computed equivalent width of Hβ is nearly 1200 Å!).

Figure 2.3 - Synthetic spectra for an accreting white dwarf ($M = 1 M_\odot$, $R = 1 R_\odot$) at an inclination of 30° in combination with an M2 III giant and a gaseous nebula (Kenyon and Webbink 1984).

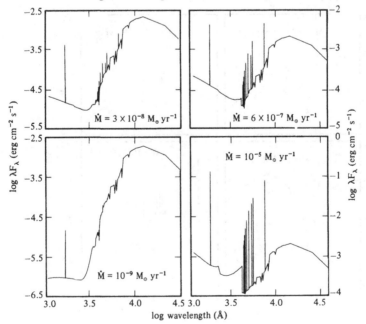

It is important to realize that while accretion may not be the dominant energy source for the hot component in a symbiotic binary, mass transfer may still be necessary for the hot stellar sources described in Figure 2.1. Calculations discussed by Paczynski (1971; also Iben 1982) suggest that such hot white dwarfs (or subdwarfs) remain luminous for $10^3 - 10^4$ yr, and then evolve into normal (cold) white dwarfs. Material from the giant is therefore needed to fuel the hot source if a binary containing a hot white dwarf is to remain symbiotic for the $10^5 - 10^6$ yr lifetime of its red giant companion star. For example, a luminosity of $\sim 100\ L_\odot$ can be maintained if the hydrogen-burning shell of an accreting white dwarf is replenished at a rate of $\sim 10^{-9}\ M_\odot\ \mathrm{yr}^{-1}$ (Iben 1982). This is comparable to the minimum rate needed to produce a symbiotic optical spectrum in an accreting white dwarf, and results in an accretion luminosity of $\sim 5\ L_\odot$ for $R \sim 0.01\ R_\odot$. Such a small contribution to the overall energy output does not significantly affect the flux distribution of the hot stellar sources shown in Figure 2.1, but does impose some constraints on the nature of the mass-losing star, as noted below.

2.3 Observational Diagnostics

Symbiotic stars were originally discovered as M giants with abnormally strong emission lines of hydrogen and helium. This simple definition for a symbiotic star has been applied to the *optical spectra* of the model binaries described above, with the results listed in Table 2.1. A number of hot components do not emit enough high energy photons to ionize the surrounding nebula. These systems have been classified as normal giants, although an ultraviolet satellite might reveal them to be binaries. A wide variety of hot components generate a symbiotic optical spectrum, including hot stellar sources, accreting white dwarfs, and accreting main sequence stars. Although the optical spectra of

Table 2.1. *Optical classification of model binaries*[1]

Hot component	Normal giants	Symbiotic stars
Hot dwarf (no accretion)	$T_h < 50,000$ K $L_h < 25 L_\odot$	$T_h > 50,000$ K $L_h > 25 L_\odot$
Accreting main sequence star	$\dot{M} < 10^{-6}$	$\dot{M} > 10^{-6}$
Accreting white dwarf star	$\dot{M} < 10^{-9}$	$\dot{M} > 10^{-9}$

[1] adapted from Kenyon and Webbink (1984)

these model binaries are qualitatively similar, there are obvious differences in their UV continua and optical emission lines. Quantitative diagnostic criteria are therefore needed to distinguish various models, and we now discuss a few such diagnostics.

2.3.1   *The Ultraviolet Continuum*

Although two symbiotic stars were detected by the UV photometers on *OAO-2* and *TD-1* (Thompson, et al. 1978; Gallagher, et al. 1979), it is the *International Ultraviolet Explorer* (*IUE*) that brought the study of symbiotic stars into the space age. In order to compare these important observations with theoretical models, it is necessary to provide a measure of the UV continuum of each synthetic symbiotic star. For simplicity, the UV continuum of each model is characterized by continuum magnitudes, $m_\lambda = -2.5 \log F_\lambda - 21.1$, at 1300, 1700, 2200, and 2600 Å, which are used to construct two reddening free color indices:

$$C_1 = (m_{1300} - m_{1700}) - 0.46(m_{2200} - m_{2600}) \tag{2.1}$$

$$C_2 = 0.55(m_{1700} - m_{2600}) - 0.45(m_{1300} - m_{1700}). \tag{2.2}$$

The first of these, C_1, measures the intrinsic *slope* of the far-UV continuum, and should be relatively insensitive to variations in the extinction curve or to calibration errors in ultraviolet flux measurements. The second color index, C_2, measures the *continuity* of the UV continuum across the 2200 Å extinction bump. This UV color is more vulnerable to variations in the extinction curve, since the extinction at 1700 Å may not be well-correlated with that at 2600 Å (Savage and Mathis 1979; Meyer and Savage 1981).

The C_1 and C_2 color indices are unique functions of either the accretion rate or effective temperature for each type of model symbiotic star. This is shown effectively in Figure 2.4, which plots C_1 vs. C_2 for a complete set of models. The classification scheme presented in Table 2.1 suggests that each class of model occupies its own region in the color-color plot:

    1) main sequence accretors: $C_1 > -0.5$, $-0.25 < C_2 < 0.25$;

    2) white dwarf accretors: $C_1 < -0.25$, $-0.6 < C_2 < -0.2$;

    3) hot stellar sources: $C_1 < -0.5$, $C_2 > -0.4$.

While these regions overlap somewhat, it appears that the hot components of symbiotic stars can be differentiated solely on the basis of their C_1 and C_2 colors.

Observational diagnostics

19

2.3.2 Optical Emission Lines

Emission lines are very prominent in symbiotic optical spectra, and the strongest features are usually transitions in H I and He II. The Balmer decrements suggest that case B recombination is appropriate for the majority of these objects (Boyarchuk 1975), so the predicted flux in Hβ and He II λ4686 may be computed easily from the numbers of H and He$^+$ ionizing photons, N_γ(H) and N_γ(He$^+$). If the nebula is radiation bounded, these fluxes are given by (assuming $T_e = 20{,}000$ K):

$$F(H\beta) = 3.7 \times 10^{-14}\, N_\gamma(H)/d^2 \tag{2.3}$$

$$F(\text{He II } \lambda 4686) = 6.3 \times 10^{-14}\, N_\gamma(\text{He}^+)/d^2 \tag{2.4}$$

Figure 2.4 - Color-color relation between C_1 and C_2 for main sequence accretors (solid line), white dwarf accretors (dashed line), and hot stellar sources (dot-dashed line). The branches of the trajectory for hot stellar sources refer, at high source temperature, to nebular electron temperatures of (a) $T_e = 10{,}000$ K and (b) $T_e = 20{,}000$ K; at low temperatures, the branches define sequences at (c) constant luminosity (log $L=4$) and (d) constant gravity (log $g=4$). The arrows in the upper right-hand corner of the diagram indicate the displacements in the position of an object produced by increases in the brightness by $\delta m = -0.2$ in each of the UV continuum points. The darkest portions refer to the locus of models that give rise to identifiably symbiotic optical spectra (Table 2.1; Kenyon and Webbink 1984).

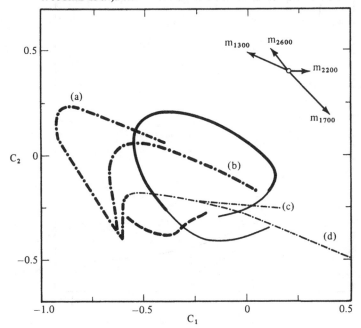

While numerical values for $F(H\beta)$ and $F(He\ II\ \lambda4686)$ can be predicted by the binary models if the distance, d, is known, it is more useful to predict the flux ratio as:

$$F(H\beta)/F(He\ II\ \lambda4686) = 0.6\ N_\gamma(H)/N_\gamma(He^+). \qquad (2.5)$$

The ratio, $N_\gamma(H)/N_\gamma(He^+)$, has been plotted for a blackbody ionization source in Figure 2.5, and provides a useful temperature indicator for $T_h \geq 50,000$ K.

An alternate temperature diagnostic utilizes the He I $\lambda4471$ line, in addition to the strong H I and He II lines (Iijima 1981). Defining

$$K = 2.22F(He\ II\ \lambda4686)/[4.16F(H\beta) + 9.94F(He\ I\ \lambda4471)], \quad (2.6)$$

the temperature of the hot star (in units of 10^4 K) may be estimated from:

$$T_4 = 19.38\ \sqrt{K} + 5.13. \qquad (2.7)$$

Figure 2.5 - The behavior in the ratio of the number of H-ionizing photons to the number of He^+-ionizing photons as a function of effective temperature, assuming a radiation-bounded nebula and a blackbody ionization source.

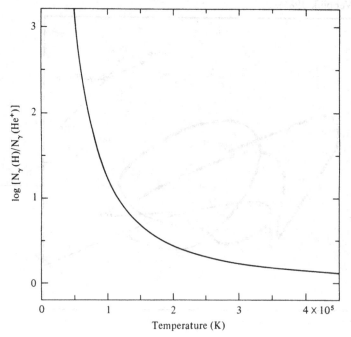

This expression is valid for $7 < T_4 < 20$, and should yield results comparable to those obtained with Figure 2.5 and equation (2.5). The main drawbacks with this temperature estimator are that (i) He I $\lambda 4471$ is usually very weak $[F(\text{He I } \lambda 4471)/F(\text{H}\beta) \sim 0.05]$, and (ii) K is somewhat sensitive to interstellar reddening.

A disadvantage of both temperature estimators is that the lower members of the H I Balmer lines may be optically thick (cf. Chapter 3), in which case the effective temperature of the hot star is overestimated. If Hβ has been reduced in intensity due to optical depth effects, the He I $\lambda 5876$/He II $\lambda 4686$ intensity ratio provides a somewhat more sensitive temperature estimate (Ferland 1979). Unfortunately, the intensity of the $\lambda 5876$ line is sensitive to the electron density and electron temperature within the nebula (Cox and Daltabuit 1971), which may not be known with good accuracy (cf. Chapter 3).

The $\lambda 4686$ line is visible in most symbiotic systems, but is absent or weak in a few others. Kaler (1978) has shown that either the [O III] $\lambda 5007$ or the [Ne III] $\lambda 3868$ line can be used as a temperature discriminant in planetary nebulae with relatively low effective temperatures. If $I(\lambda 5007)$ and $I(\lambda 3868)$ measure the ratio of the intensities of these two lines with respect to that of Hβ, T_h may be computed from:

$$\log T_h = 4.426 + 4.827 \times 10^{-2} I(\lambda 5007) - 1.374 \times 10^{-3} I^2(\lambda 5007) \qquad (2.8)$$

or

$$\log T_h = 4.490 + 0.4484 I(\lambda 3868) - 0.1548 I^2(\lambda 3868). \qquad (2.9)$$

These relations have been derived for a large sample of planetary nebulae, and are believed to be accurate to $\pm 10\text{-}20\%$ for $4.4 < \log T_h < 4.8$ (Kaler 1978).

2.3.3 The Radio Continuum

The temperature diagnostics derived from various emission lines are unable to determine the physical nature of the hot component, but do yield information regarding the flux distribution shortwards of the Lyman edge. Radio continuum emission provides a more direct probe of the hot source, as the radio flux density should be well-correlated with the nature of the ionizing star. Continuum emission is expected to be thermal bremsstrahlung in the radio-frequency regime, with the optical depth through the nebula at a frequency ν given by (Osterbrock 1974):

$$\tau_\nu = 1.1 \times 10^{-8} (\nu/5 \text{ GHz})^{-2.1} E. \qquad (2.10)$$

$E = \int n_e n_i ds$ is the emission measure in cm^{-6} pc. For densities and nebular radii appropriate to most symbiotic stars (Chapter 3), $E \approx 10^9 - 10^{12}$ cm^{-6} pc, so $\tau \approx 10 - 10^3$ at 5 GHz (6 cm). Since the nebula is optically thick, the observed intensity is roughly:

$$S_\nu \approx 0.2(\nu/5 \text{ GHz})^2 (R_\nu/d)^2 \text{ } \mu\text{Jy}, \qquad (2.11)$$

where R_ν is the "photospheric" radius of the nebula in AU and d is the distance in kpc. The current sensitivity of advanced radio telescopes such as the *VLA* is approximately 1 mJy, and thus symbiotic stars are expected to be detectable only if they have large ionized regions ($R_{5 \text{ GHz}} \gtrsim 100$ AU, for a distance of 1 kpc).

In view of the strong dependence of radio intensity on the nebular radius, symbiotic stars with small, dense nebulae are not expected to be radio sources. Systems with lobe-filling giants and extended accretion disks are unlikely to have ionized nebulae beyond the dimensions of the binary orbit, implying $R_{5 \text{ GHz}} \approx 1$ AU. The estimated radio flux for such a system, $\approx 0.2 - 1.0$ μJy for $d \approx 1 - 3$ kpc, is considerably below the *VLA* detection limit. On the other hand, systems containing red giants losing mass via a stellar wind are predicted to be luminous radio sources. Adopting $n \propto r^{-2}$ and normalizing the density to that of a typical red giant atmosphere (e.g., Watanabe and Kodaira 1978), $R_{5 \text{ GHz}} \approx 100$ AU for the hot stellar sources discussed earlier ($T_h = 10^5$ K, $R = 0.1$ $R_\odot$). Such a nebula is optically thick at $\nu \sim 15$ GHz and should have $S_\nu = 2$ mJy at $\nu = 5$ GHz, well within the range of the *VLA*.

2.4   Summary

The simple model developed in the preceding pages proposes that symbiotic stars are binaries composed of (i) a late-type giant star, (ii) a hot component, and (iii) an ionized nebula. The hot component might be a hot stellar source similar to the central star of a planetary nebula (e.g., Boyarchuk 1975), or an accreting white dwarf or main sequence star (e.g., Kenyon and Webbink 1984). Although radio and optical observations can provide a crude measure of the temperature and luminosity of the hot object, the physical nature of this component is determined best by satellite ultraviolet data. Tests of this picture will be discussed in Chapter 3.

While the observational diagnostics described in this Chapter probe the hot component directly, it is important to understand that the physical properties of the cool giant ultimately determine the observable characteristics of an individual symbiotic star. For the 1-3 yr orbital periods that are typical of these systems (Chapter 3), a normal red giant star with a radius, R_g, $\sim 50\,R_\odot$ is easily contained within its tidal lobe, while a bright asymptotic branch giant or supergiant with $R_g \sim 200\,R_\odot$ is likely to fill its tidal lobe (Allen 1973). Since mass transfer seems essential for *all* binary models (Sec 2.2), the physical size of the giant determines *how* this material is lost and to what extent it interacts with the hot component. Lobe-filling giants apparently lose mass at rates exceeding $10^{-6}\,M_\odot\,\mathrm{yr}^{-1}$ (Webbink 1979), and this is roughly the lower limit required to form a symbiotic system with an accreting main sequence star. Detailed calculations suggest that white dwarfs or subdwarfs expand to giant dimensions when subjected to large accretion rates over a long period of time (e.g., Iben 1982 and references therein), and therefore would not be detected as symbiotic stars. Red giants that underfill their tidal lobes can only lose mass in a stellar wind, and observations of single stars suggest wind-driven mass loss rates of $\sim 10^{-7}\,M_\odot\,\mathrm{yr}^{-1}$ for normal red giants and $\sim 10^{-5}\,M_\odot\,\mathrm{yr}^{-1}$ for Mira variables and supergiants (Zuckerman 1980). Calculations described by Tutukov and Yungel'son (1982) and Livio and Warner (1984) show that perhaps 1-5% of this material could be gained by a hot companion star, implying an accretion rate $< 10^{-7}\,M_\odot\,\mathrm{yr}^{-1}$. This can energize a binary containing a white dwarf or hot subdwarf, but is insufficient for a main sequence star, as noted above. Thus, symbiotic stars might be divided into two categories based on the nature of the hot component and the giant component: (i) systems containing a lobe-filling giant and a low mass main sequence star, and (ii) systems containing a white dwarf or subdwarf and a red giant losing mass in a stellar wind. This hypothesis can only be tested by observations, and we now turn to a discussion of data acquired during quiescent phases.

3

The Quiescent Phase - Observations

3.1 Introduction

Initial spectroscopic observations of symbiotic stars revealed them to be peculiar beasts which somehow combine the absorption features of a red giant with the emission lines of a planetary nebula (Figure 1.1). Analysis of their photometric behavior led to additional confusion, as long, apparently quiescent, intervals would be rudely interrupted by 2-3 mag outbursts (Figure 1.2). While the eruptions have captured the imagination of many researchers, a few groups recognized the value of acquiring photometric and spectroscopic data during the lengthy quiescent phases. These historical data, briefly summarized in the Appendix, clarify the long-term behavior of these systems, and provide strong support for the binary models proposed by Berman, Hogg, and Kuiper.

The field of symbiotic stars has grown tremendously in the past twenty years, and data covering a significant portion of the electromagnetic spectrum have been acquired for a number of systems. This information allows definitive tests of the basic models developed in the 1930's and 1940's, and generally supports the basic binary picture discussed in Chapter 2. New results from infrared, optical, and radio photometry demonstrate that symbiotic stars are binary systems ($P \sim 200$ days to > 20 years) composed of a late-type giant star and a hot component surrounded by an ionized nebula. Detailed spectroscopic observations have resulted in refined portraits of individual systems, which show a wonderful variety of accreting hot components and mass-losing red giants.

3.2   Photometry

While most discussions of symbiotic stars highlight their spectroscopic peculiarities, much can be learned from photometry at a variety of well-chosen wavelengths. Broad-band visual observations (especially the wealth of data collected by amateurs worldwide) have been the primary source of symbiotic binary periods, while infrared JHKL measurements provided an impetus to divide the systems into two fundamental classes. Finally, radio "photometry" gave the first indication of the sizes of the gaseous nebulae surrounding symbiotic binaries, and may ultimately provide such fundamental parameters as the binary separation and the luminosity of the hot star.

3.2.1   Optical Photometry

One of the most useful collections of symbiotic star observations is the voluminous amount of optical photometry. Pioneering discussions showed these systems to be semi-regular variables with amplitudes of a few tenths of a magnitude and periods of 600-900 days (Townley, Cannon, and Campbell 1928; Greenstein 1937; Mayall 1937). Although these variations seemed quite peculiar when compared to those experienced by other long-period red variables, the amplitude of a given variation did tend to increase towards short wavelengths. Analysis of a few systems led to the conclusion that both components of a symbiotic system were variable with roughly comparable periods (Payne-Gaposchkin 1946).

These periodic photometric variations generally were not associated with binary motion until the late 1960's, when Hoffleit (1968) noted the mean light curve of CI Cyg resembled that of an eclipsing binary, and Belyakina (1968, 1970) remarked that 0.3 mag variations in AG Peg's blue continuum followed the 800 day spectroscopic period discovered by Merrill. Belyakina's spectroscopic observations demonstrated that light maximum coincides with a maximum in the observed number of Fe II and [Fe II] emission lines. According to the radial velocity data, these maxima occur during inferior conjunction, when the hot component in AG Peg passes in front of its red giant companion. Belyakina interpreted the observations with a model in which ultraviolet radiation from the hot star heats up the facing hemisphere of the giant. This hemisphere then radiates more energy (at a higher temperature) than the opposite hemisphere, and binary motion results in a modulation of the flux detected at earth (Figure 3.1).

The good agreement between the photometric and spectroscopic periods of AG Peg suggests binary periods might be determined best by optical photometry. Hoffleit (1970) and Pucinskas (1970) established photometric periods for a number of symbiotic stars, including BF Cyg and AX Per. More recently, Meinunger (1979, 1981) established new periods for AG Dra and AG Peg. He found a 554 day

Figure 3.1 - The reflection effect in symbiotic stars. The top panel shows the configuration of a hypothetical symbiotic binary at five orbital phases, with the arrow pointing in the direction of the observer. The small hot component heats up the facing hemisphere of the cool giant, causing the giant to radiate more energy when its hot side faces the earth. Optical light curves for AX Per and AG Peg are displayed in the center and bottom panels, respectively, and these variations are believed to be a result of the reflection effect. Spectroscopic observations suggest the deep minimum in AX Per is caused by an eclipse of the hot component by the giant, which does not occur in AG Peg.

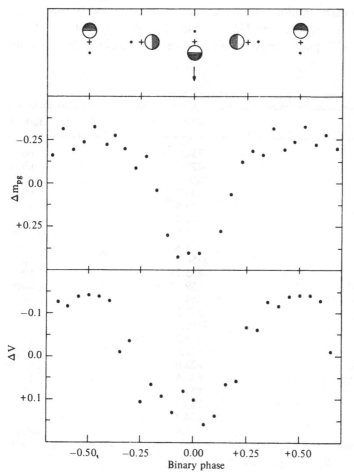

variation in AG Dra, with an amplitude of nearly 1 mag in U and 0.2 mag in B and V. In a subsequent paper, he refined the period of AG Peg to 827 days using patrol plates from Sonneberg Observatory.

Period-search techniques have now been applied to photometric data for a fairly large sample of symbiotic stars, and reliable periods have been derived for approximately 20 systems (cf. Table 3.1). A good example of these recent successes is provided by an original star with combination spectrum, AX Persei. Kenyon (1982) analyzed the photographic and photovisual data compiled by Mjalkovskij (1977), and found a best period of 681.6 days. This is very nearly the same period determined by Payne-Gaposchkin (1946), who suggested the Harvard light curve could only be explained by fluctuations in both the hot and cool components. The folded photographic light curve (Figure 3.1) resembles closely the light curve of AG Peg, and might therefore be a result of binary motion. Oliversen and Anderson (1983) later determined that minima in the Hβ and He II λ4686 emission line fluxes coincided with photometric minima. The spectroscopic data is most

Table 3.1. *Binary periods for symbiotic stars*

Symbiotic star	Period	Method[1]
Z And	756.9	ph
EG And	470:	ph
R Aqr	16060::	ph
UV Aur	395.2	ph
TX CVn	70.8:	sp
V748 Cen	564.8	ph
T CrB	227.5	ph,sp
BF Cyg	757.3:	ph
CH Cyg	5750::	sp
CI Cyg	855.25	ph
V1329 Cyg	959	ph,sp
AG Dra	554	ph
RW Hya	372.5:	ph
BX Mon	1380:	ph
SY Mus	627.0	ph
AR Pav	604.6	ph,sp
AG Peg	827.0	ph,sp
AX Per	681.6	ph
V2601 Sgr	850:	ph
V2756 Sgr	243:	ph
CL Sco	624.7:	ph
BL Tel	778.6	ph,sp

[1] ph - photometry, sp - spectroscopy

easily understood if the hot component is *eclipsed* during photometric minima. Thus, Payne-Gaposchkin's original interpretation of AX Per's light curve was basically correct, as both components do in fact vary: (i) the hot component is eclipsed, while (ii) the cool component presents warm and cool hemispheres to our line-of-sight.

At least two symbiotic stars, R Aqr and RR Tel, are known to contain Mira variables as a result of optical photometry (Mattei and Allen 1979; Kenyon and Bateson 1984). The pulsational periods of these two objects (~ 387 and 380 days, respectively) are comparable to those of other Mira systems, although their amplitudes are considerably smaller than expected for a normal Mira. Their mean light curves are flat-bottomed (cf. the Appendix), suggesting that most of the light at minimum is produced by the hot companion star. The hot star in R Aqr usually has $V \sim 11$, with occasional excursions up to $V \sim 10$. The hot component in RR Tel underwent a large outburst in 1944 (Chapter 5; Appendix), and has not yet returned to minimum.

While broad-band visual observations have been of great importance · in following the long-term variations in symbiotic stars, intermediate filter photometry disclosed the nature of rapid variations (e.g., minutes) in the hot components of a few symbiotic stars. This "flickering" was observed in CH Cyg by Wallerstein (1968) and Cester (1968) during a bright phase. Subsequent observations verified the flickering, and confirmed that the amplitude of a flicker increases with decreasing wavelength. Both T CrB and RS Oph were later found to flicker, as well, although a survey discussed by Walker (1977) suggests that the majority of symbiotic stars are constant to 1% over timescales of 20 minutes to 2 hours. The flickering in T CrB is most noticeable in the ultraviolet (25-30% variations), and is invisible in the red (less than 1% variations). RS Oph varies by 15-20% in white light; the wavelength dependence of the flickering has not been observed due to the star's intrinsic faintness.

The rapid oscillations in brightness observed in T CrB, CH Cyg, and RS Oph are characteristic of the much shorter period cataclysmic binaries ($P < 1$-2 days). In these systems, a low mass main sequence star fills its tidal lobe and transfers matter, via an accretion disk, to a white dwarf companion. A "hot spot" is formed at the outer edge of the disk (where the stream of material from the mass-losing star impacts the disk), and it is currently thought that small variations and/or inhomogeneities in the flow result in a "pulsation" or flickering of the hot spot (and the disk itself). Since flickering is observed in nearly all cataclysmic variables with high mass transfer rates, it is commonly considered a "signature" of accretion onto a white dwarf (e.g., Robinson 1976). By analogy with the cataclysmics, then, the flickering in symbiotic stars might also be the result of accretion.

Although the interpretation of flickering in cataclysmic binaries is very secure, it is somewhat naive to assume that flickering in other binary systems is the result of the same phenomenon. While the interaction of the stream of matter lost by the lobe-filling star with the accretion disk is not a well-understood process, the temperature of the hot spot must depend on the position of the spot in the disk (which determines its luminosity) as well as its surface area. If M_1 and $\dot{M}$ represent the mass of the central star (in $M_\odot$) and the mass accretion rate (in $10^{-5} M_\odot$ yr^{-1}), this temperature may be estimated to be (Bath, et al. 1974; Tylenda 1977):

$$T_S \approx 78,700 \ M_1^{1/3} \dot{M}^{2/9} R_S^{-7/9} \ \text{K}, \tag{3.1}$$

where R_S is the radius (in $R_\odot$) at which the spot occurs in the disk. For cataclysmic systems, $R_S = 0.25 - 0.4$ A, where A is the binary separation, and thus the spot temperature is $\sim$ 20,000 K for $\dot{M} \sim 10^{-10} M_\odot$ yr^{-1}. It seems unlikely that R_S/A can be significantly smaller in a symbiotic star, so if $M_1 = 1 \ M_\odot$ and $A = 1$ AU (Table A.5), T_S is given by:

$$T_S \approx 2500 - 4000 \ (\dot{M}/10^{-5} \ M_\odot \ \text{yr}^{-1})^{2/9} \ \text{K}. \tag{3.2}$$

Thus, the *maximum* temperature of the "hot" spot in a symbiotic star is roughly 2500-4000 K, and substantial optical flickering from this region in a symbiotic binary containing an accreting dwarf star is not expected (although variations might be observed in the red or infrared).

Since the temperature of disk material increases with decreasing radius, optical flickering of the disk itself might be observed in some systems. In general, the large sizes of the disks in these binaries should lead to longer flickering timescales than in cataclysmic variables: simple calculations of the optical radius of a standard disk model (Shakura and Sunyaev 1973; Lynden-Bell and Pringle 1974) suggest that variations observed over the course of minutes in cataclysmic binaries are likely to require hours in an average symbiotic binary. Night to night variations have been observed in the emission lines of BF Cyg and a few other symbiotic stars (Swings 1970), which could be a result of inhomogeneities in a presumed accretion flow. Shorter timescale variations (e.g., minutes) are possible at the inner edges of disks in systems containing accreting white dwarf stars; there is, as yet, no direct evidence for white dwarfs in T CrB, CH Cyg, or RS Oph.

Aside from measurements of flickering, occasional observations with narrow and intermediate band filters have been successful in monitoring the overall energy distributions in several systems. Kaler and collaborators have discussed variations in the continua of EG And (Kaler and Hickey 1983) and CH Cyg (Kaler, Kenyon, and Hickey 1983). The blue continuum of CH Cyg increased by almost a factor of three during a five month period in 1981 (Figure 3.2), while contemporaneous JHKL photometry shows that the system maintained a roughly constant near-infrared flux. The photometric energy distributions are in excellent agreement with spectrophotometric observations that show the absorption features of an M giant (TiO) and an F supergiant (H I, Ca II, and Ti II). In this case, spectrophotometry revealed the absorption spectra of both binary components, while optical and infrared photometry demonstrated the variable nature of the hot component. Such complementary observations should prove to be instructive for other symbiotic stars.

3.2.2   Infrared Photometry

Infrared observations of a symbiotic star (R Aqr) were first obtained by Pettit and Nicholson (1933, and references therein) in the 1920's. While these data were necessarily crude, they showed that

Figure 3.2 - The energy distribution of CH Cyg at four epochs. The error bars at the bottom of the Figure indicate the widths of the interference filters used to obtain the photometry (adapted from Kaler, Kenyon, and Hickey 1983).

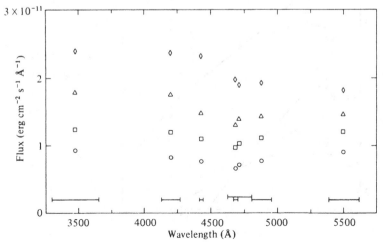

while R Aqr might be peculiar in the optical, its infrared properties were nearly identical to those of normal Mira variables. Other symbiotic stars could not be observed with their instrument, but Pettit and Nicholson demonstrated that the giant components of these systems could be profitably studied in the infrared where veiling from the hot component is minimized.

Nearly all symbiotic stars have now been observed with broad-band infrared photometry, and a few examples of the resulting energy distributions are displayed in Figure 3.3. Most systems have IR continua which resemble 2500-3500 K blackbodies, with small excesses of emission superimposed at 10 and 20 μm (e.g., Stein, et al. 1969; Gillett, Merrill, and Stein 1971; Swings and Allen 1972). This is typical of most M giants and supergiants; the small excesses are usually ascribed to blackbody emission from silicate grains at temperatures of 100-300 K

Figure 3.3 - Broadband infrared energy distributions for four symbiotic stars. The systems T CrB and CH Cyg have distributions similar to those of normal M giants, and are therefore termed *S-type* (stellar) symbiotics. The flux distributions of HM Sge and H2-38 are significantly redder than that of an average M star; these are known as *D-type* (dusty) systems, since they appear to be affected by circumstellar dust emission and/or absorption (Webster and Allen 1975).

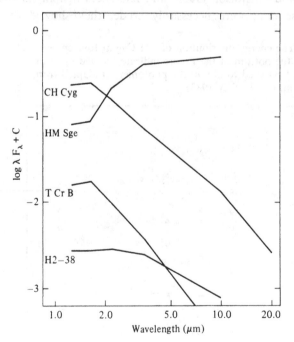

(Merrill and Ridgeway 1979). Indeed, the H-K and J-K colors of these *S-type* (stellar; Webster and Allen 1975) symbiotics are consistent with the colors of mean giants and supergiants (Frogel, et al. 1978) once they have been corrected for a modest amount of interstellar reddening ($A_K = 0.1 - 0.2$ mag; Figure 3.4). Recent photometry has suggested these systems may be prone to small-amplitude variations ($\Delta K < 0.3$ mag; Taranova and Yudin 1982, 1983). These fluctuations may be intrinsic to the giant component (i.e. pulsations), or may be a result of orbital motion (as in the case of the optical variations discussed earlier in this Chapter). Infrared observations over the course of a few binary cycles will be needed to determine the nature of these oscillations.

Although most symbiotics resemble normal M stars in the IR, a few systems ($\sim 20\%$) have IR continua that peak near 3.5 μm (cf. Figure 3.3). This corresponds to a color temperature of ~ 1000 K, which is much too low for an ordinary stellar photosphere. These *D-type* (dusty) systems are far-removed from the majority of symbiotics in a color-color diagram and appear to have IR colors that are nearly those expected from a combination of an M-type star and a 1000 K black-

Figure 3.4 - Infrared J-K vs. H-K plot for the known symbiotic stars. Observations of S-type systems are plotted as filled circles, while crosses indicate the positions of D-type systems in our own galaxy. Two objects in the Magellanic Clouds are denoted by open circles. The locus of K-M giants is shown by the dashed line (Frogel, et al. 1978).

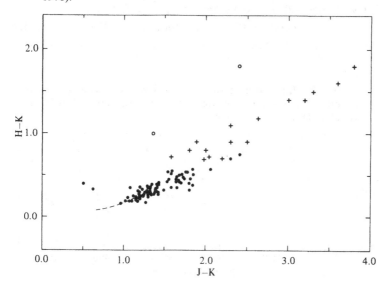

body (Figure 3.4; Webster and Allen 1975; Allen 1982). In the most extreme cases (e.g., HM Sge), the presence of the M-type star is not easily deduced from JHK photometry. However, nearly all the D-type systems have the strong H_2O and CO absorption bands that are commonly found in late-type giants (Allen 1982, and references therein), confirming that M-type stars are indeed present in these systems.

As with optical photometry, systematic IR observations have revealed long-period variations in many systems (Figure 3.5). The amplitudes of these fluctuations tend to be ~ 1 mag, with periods ranging from 200 to 600 days (Table 3.2; Feast, Robertson, and Catchpole 1977; Feast, et al. 1983a,b; Whitelock, et al. 1983a,b). These amplitudes and periods are typical of those found in Mira variables, and are generally observed in D-type systems with prominent H_2O absorption bands. The distribution of Table 3.2 seems to favor longer periods than that of normal field Miras, suggesting that symbiotics prefer to contain very evolved late-type giant stars.

3.2.3   Radio Observations

While radio observations of active galactic nuclei and dense molecular clouds appear to dominate today's journals, several stellar objects are strong radio sources as well. It has been over a decade since Antares B was detected by Hjellming and Wade (1971); this was followed by the detection of the first symbiotic star, RR Tel, in 1973 (Wright and Allen 1978). The contemporaneous discoveries of radio emission from V1016 Cyg (Purton, Feldman, and Marsh 1973) and V1329 Cyg (Altenhoff and Wendker 1973) were followed by other surveys (Bath and Wallerstein 1976; Altenhoff, et al. 1976) that provided no new sources. An extensive investigation by Wright and Allen

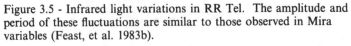

Figure 3.5 - Infrared light variations in RR Tel. The amplitude and period of these fluctuations are similar to those observed in Mira variables (Feast, et al. 1983b).

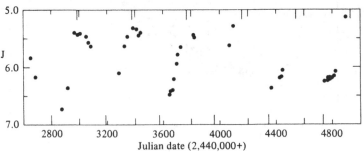

(1978) was more fruitful, and recorded nine radio symbiotics. While these early surveys were limited in sensitivity and resolution, they showed that some symbiotic stars possess large ionized nebulae with radii > 100 AU. RX Pup was found to be particularly interesting, as its radio flux density varied by a factor of 2 in 24 hr (Seaquist 1977)!

The advent of the Very Large Array (VLA) has opened up many exciting possibilities for the study of radio symbiotic stars. A recent survey by Seaquist, Taylor, and Button (1984) represents an increase in sensitivity by a factor of 10 over previous attempts, with an extraordinary increase in resolving power. They examined 59 symbiotics with $\delta > -25°$, and their 15 new detections more than doubled the number of known radio symbiotics (Table 3.3). Most of the new sources are quiescent S-type symbiotics; before this survey it was generally thought that only D-type systems (and perhaps S-types in outburst such as AG Peg) were strong radio emitters (e.g., Kwok 1982).

The properties of the radio symbiotics are best understood in terms of the binary models discussed in Chapter 2. We showed there that the vast majority of these sources should contain a red giant losing mass in a stellar wind. Radio continuum emission results from the portion of this wind that is ionized by the hot component. In general, the radio spectrum should be a power law (Wright and Barlow 1975),

$$S_\nu \propto \nu^\alpha, \tag{3.3}$$

where S_ν is the flux density at a frequency ν. Typical values of α for optically thin and optically thick constant density nebulae are -0.1 and $+2.0$, respectively. It is probably more realistic to suppose that the nebula has a density profile given by $n \propto r^{-2}$, appropriate to the equation of continuity at constant outflow velocity. The spectral

Table 3.2. *Pulsational periods for Mira-like components in symbiotics*

Symbiotic star	Period (days)
R Aqr	387
BI Cru	280
V1016 Cyg	470
RX Pup	580
HM Sge	> 400
RR Tel	380
He 2-34	370
He 2-38	433
He 2-106	400

index, α, should then be $+0.6$ for a completely ionized optically thick nebula, with the mass loss rate given by (Wright and Barlow 1975; Panagia and Felli 1975):

$$\dot{M} \sim 1.5 \times 10^{-6} (v/10 \text{ km s}^{-1})(S_6 \text{ cm}/1 \text{ Jy})^{3/4}$$

$$(d/1 \text{ kpc})^{3/2} \ M_\odot \text{ yr}^{-1}. \tag{3.4}$$

The values of α observed in radio symbiotics are listed in Table 3.3, and the results do not cluster around any of the expected values. The preferred value is $\alpha \sim +1.0 \pm 0.2$, significantly higher than $\alpha = +0.6$. Note that if a portion of the nebula is optically thin, α will tend to be smaller than $+0.6$, rather than larger as observed. A few systems do have $\alpha \sim 0.6$, and derived values for their mass loss rates also appear in the Table. These values for $\dot{M}$ are comparable to those observed in single red giants and Mira variables (e.g., Zuckerman 1980).

The simple theory discussed by Wright and Barlow (1975; also Panagia and Felli 1975) assumes that the entire nebula lost by the giant is ionized by the hot companion. This is probably a good assumption for systems with very luminous hot components, but is perhaps inap-

Table 3.3. *Radio symbiotic stars*

Symbiotic Star	F_6 cm (mJy)	Spectral index	$\dot{M}(M_\odot \text{ yr}^{-1})$
Z And	1.2	0.62	9.8×10^{-8}
R Aqr	10 (var)	0.6	5.9×10^{-8}
BF Cyg	2.1	0.98	-
CH Cyg	1.4	1.18	-
RW Hya	0.4	-	-
RX Pup	16 (var)	0.0	-
V2416 Sgr	3.5	0.92	-
V455 Sco	2.4	-	-
RT Ser	1.3	0.91	-
AS 210	1.8	-	-
AS 245	0.9 (var)	-	-
AS 296	0.6	-	-
H 1-36	45	1.0	-
H 2-38	13	-	-
He 2-106	40	0.65	1.3×10^{-6}
He 2-171	2.1 (var)	0.65	6×10^{-7}
He 2-173	0.5	-	-
He 2-176	14	-	-
He 2-390	0.7	-	-
SS 38	13	-	-
SS 96	2.6 (var)	1.05	-
SS 122	1.6	-	-

propriate for the fainter objects. Taylor and Seaquist (1984) have considered a two dimensional theory based on the shape of the ionization front created by the hot star, which successfully explains both the large values of α and the high frequency turnover observed in some sources. The shape of this front is solely a function of a dimensionless parameter, X, which depends on the photon luminosity of the hot source, L_{ph}, the binary separation, A, the mass loss rate of the cool star, $\dot{M}_g$, and the wind velocity of the red giant star, v $(X = [4\pi A L_{ph}/\alpha_B] [\mu m_H v/\dot{M}_g]^2)$. At small $X(X < 1/3)$, the ionized portion of the wind is entirely surrounded by the neutral region, as shown in Figure 3.6. The ionized nebula becomes a cone-shaped region as X

Figure 3.6 - The shape of the ionization front in a symbiotic binary as a function of the ionization parameter X (defined in the main text). The H II region (unshaded in the Figure) is extensive for a luminous hot component (large X), while small, elliptical nebulae are characteristic of faint hot components (small X; Taylor and Seaquist 1984).

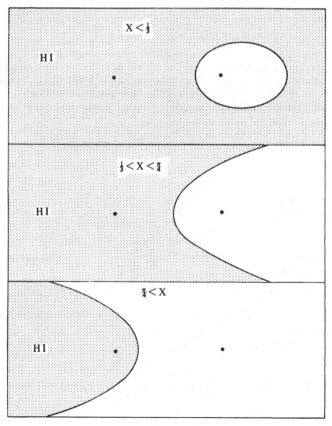

approaches $\pi/4$, although it is still primarily ionization bounded. Once $X=\pi/4$, the wind is completely ionized along $\theta=\pi/2$. Most of the nebula is ionized for $X>\pi/4$, as only a cone-shaped region shielded by dense material surrounding the cool giant remains neutral.

The behavior of the radio spectrum depends on the ionization parameter, X, and the viewing angle of the observer with respect to the axis connecting the binary components, θ, as illustrated in Figure 3.7. Naturally, the spectral index, α, approaches -0.1 at very large frequencies in all cases, since the shape of the ionized region is irrelevant when the nebula is optically thin. As the optical depth through the nebula approaches 1, α must increase, since an optically thick nebula emits fewer radio photons. The value of α in the optically thick regime is limited by X: (i) for $X>1$, most of the nebula is ionized and α reaches the canonical value of $+0.6$, while (ii) for $X<1$, most of the nebula is neutral and α can be as high as $+2.0$ (the upper limit for an optically thick nebula). Note that the dependence of α on the viewing angle implies the radio spectrum should vary in phase with the binary orbit.

Figure 3.7 - The behavior of the spectral index, α, as a function of the viewing angle, θ, and the normalized frequency (with $\nu=1$ corresponding to a frequency at which the optical depth through the nebula equals unity). The nebula is optically thin at large frequencies, and $\alpha \sim -0.1$. The nebula becomes optically thick at lower frequencies, and α increases dramatically (Taylor and Seaquist 1984).

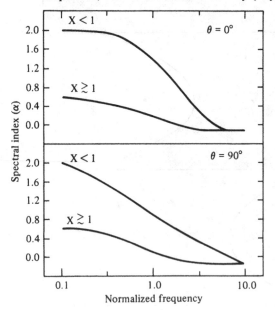

As noted above, early radio surveys detected many D-type systems and a few S-type objects, implying that radio emission somehow preferred symbiotics with evolved Mira components (cf. Webster and Allen 1975). Even with the increased sensitivity of the *VLA*, the propensity for D-type systems to be strong radio sources remains very obvious, as shown in Figure 3.8. Most of the strong sources ($S_v \geq 10$ mJy) are D-type systems, and 70% of all D-type objects observed in the radio have positive detections. Conversely, 73% of the S-type systems surveyed fall below the *VLA* detection limit, and only 2 of these (R Aqr and AG Peg) have intensities exceeding 5 mJy. It is interesting to note that AG Peg is in the midst of a major outburst (which, in truth, disqualifies it from this discussion), while R Aqr contains a Mira variable (and perhaps should be considered a D-type system, although it does not show a significant 3 μm excess).

The dominance of D-type symbiotics among radio sources is easily understood within the context of the binary model discussed above. Systems containing lobe-filling giants are expected to possess small ionized nebulae surrounding the accreting secondary, and thus should not emit much energy at radio wavelengths ($S \sim 2$ μJy was estimated for

Figure 3.8 - Distribution of radio symbiotic stars as a function of flux density and infrared type. The strongest sources tend to be D-type systems containing Mira variables (shaded region), and ~70% of all D-type objects observed with the VLA have been detected at radio wavelengths. Many S-type systems (unshaded regions) have been detected with the VLA, and these tend to be weak radio sources.

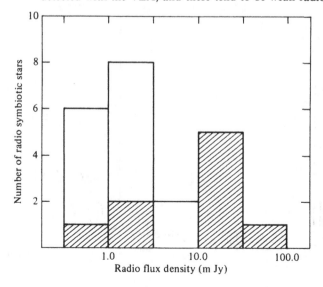

these sources in Chapter 2). Thus, some fraction of S-type systems fall below the *VLA* detection limit. The remaining symbiotics contain giants losing mass in a stellar wind, and therefore have extended nebulae. Mira variables losing mass at very high rates should have the most extended nebulae, and therefore the largest radio emission, assuming comparable hot components. The four systems with $S_v > 10$ mJy all contain Mira variables

3.3   Spectroscopy

Since symbiotic stars were originally discovered as spectroscopically peculiar objects, a great deal of information has been collected concerning the variability of the continuum and the strong emission lines. The ultraviolet/optical spectrum of a typical symbiotic star, CI Cyg, is shown in Figure 3.9. Strong absorption bands of TiO are quite prominent in CI Cyg, while weaker lines of Ca I, Mg I, and Fe I have also been identified. A weak continuum from the hot component is visible shortwards of ~ 4000 Å, and is obviously reddened by interstellar extinction. The strong Balmer jump at $\lambda 3646$ suggests some of the blue continuum is produced by an ionized nebula; radiation below $\lambda \sim 2000$ Å is likely to be emitted by the hot component itself. The

Figure 3.9 - The combined optical/ultraviolet spectrum of CI Cygni. Strong TiO bands are visible for $\lambda > 6000$ Å, and emission lines from H I, He II, C III], and C IV are also quite prominent.

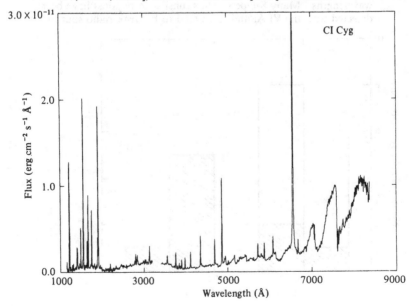

most striking features of this spectrum are the intense emission lines: H I and He II dominate the optical, while He II, C III], C IV, Si IV, and N V are very strong in the ultraviolet.

Before c. 1970, only optical spectra of symbiotic stars were available, and these were recorded on photographic plates which could not be calibrated to any great accuracy. Modern instrumentation has opened up the complete electromagnetic spectrum, and high quality spectroscopic data have been generated from X-ray to radio frequencies. Rather than summarize the results in individual wavelength domains, I have chosen to present a discussion of the major spectroscopic features, and begin with a discussion of the spectrum of the cool component.

3.3.1   The Late-Type Absorption Spectrum

A major element of a symbiotic star is the late-type absorption spectrum visible in the red and the near-infrared. A few examples of the various types of cool components present in symbiotics are shown in Figures 3.10-3.11. The strongest absorption features tend to be narrow lines of Ca I, Na I, and Ca II, and wide bands of CO, H_2O, and TiO. Lines of Fe I and other neutral metals are also quite common (Roman 1953; Kenyon, et al. 1982), and La I has been detected in UV Aur (Sanford 1945). Several weak absorption bands, including CN and CaCl, have been suspected in some symbiotic stars, but have yet to be confirmed (Roman 1953; Rybski 1973). Although these features are expected in late-type stars, an accurate spectral classification of the absorption spectrum is complicated by the variable blue continuum, which veils many absorption features shortwards of ~ 5000 Å. This problem was first encountered by Merrill (1933) when he classified AX Per as gM3 using TiO bands and gK0 using neutral metal lines in the blue. Spectral types ranging from G5 III to K5 III and K0 Ib have been proposed for the late-type star in AG Dra (Mirzoyan and Bartaya 1960; Belyakina 1969; Huang 1982), and similar difficulties occur in other systems.

The nature of the late-type giant star is an important constraint on binary models, as noted by Kenyon and Gallagher (1983). In order for disk accretion to occur, especially at the high rates required for main sequence accretors, the M primary must fill its tidal lobe. Given the 2-4 yr period of a typical symbiotic binary, this M star must then be an evolved red giant near the tip of the red giant branch or an asymptotic branch giant. This is a potential problem for current models, as optical classification of symbiotic primaries suggests they are rather normal M giants (e.g., Boyarchuk 1975). It therefore seems likely that symbiotic primaries are (i) systematically misclassified at too low a luminosity, or (ii) there is a major deficiency in our understanding of mass transfer processes in binaries containing highly evolved stars.

Preliminary attempts to reclassify symbiotic star primaries on the basis of near-infrared spectra have been discussed by Allen (1980a) and Kenyon and Gallagher (1983). Strong CO and H_2O bands have now been detected in the vast majority of symbiotics, and these serve as powerful probes of the late-type component. The 2.3 μm CO band appears to be especially important as the depth of this band scales with

Figure 3.10 - Optical spectra of the symbiotic stars UV Aurigae (lower panel) and BX Mononcerotis (upper panel). The majority of these systems resemble BX Mon, with strong TiO bands and weak absorption lines of neutral metals. A few systems (primarily those discovered in the Magellanic Clouds) are similar in many respects to the spectrum of UV Aur, which shows the strong CN bands visible in carbon stars.

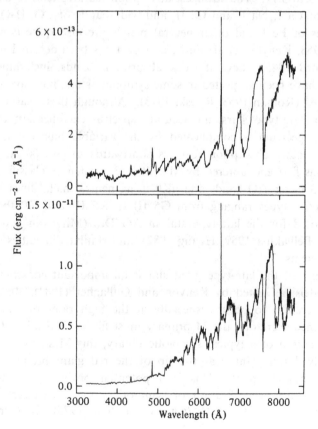

the spectral type and luminosity class in normal field giants and super-giants (Figure 3.12; Baldwin, Frogel, and Persson 1973; Kenyon and Gallagher 1983). If the late-type stars in symbiotics are all class III giants, then it is very straightforward to measure their spectral types (and therefore their effective temperatures). Allen adopted this viewpoint, and obtained spectral types for a large fraction of southern symbiotic stars. His results suggest symbiotics contain red giants which, on average, are significantly cooler than field red giants.

An alternative interpretation of Allen's results is that symbiotic stars contain giants which are more luminous than normal field giants, since the CO band becomes stronger as luminosity increases (Figure 3.12).

Figure 3.11 - Infrared spectra of the symbiotic stars R Aquarii and CH Cygni. CH Cyg is a typical *S-type* symbiotic, with a strong CO absorption band at 2.3 μm, while R Aqr shows the strong H_2O bands viewed in systems containing evolved Mira variables (Kenyon and Gallagher 1983; Whitelock, et al. 1983a)

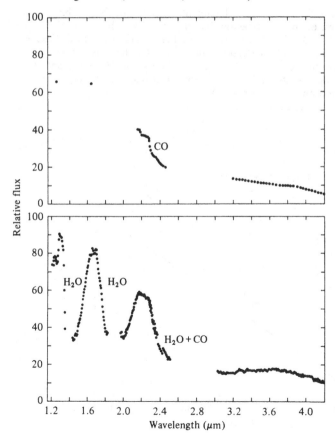

Kenyon and Gallagher (1983) showed that Z And, T CrB, and CI Cyg have signifcantly stronger CO bands than would be expected on the basis of their optical spectral types. If these spectral types provide a reliable measurement of the cool star's effective temperature, Kenyon and Gallagher's results suggest that bright giants or supergiants may be present in some symbiotic systems. Thus, optical spectra may *not* provide an accurate estimate for the luminosity of the late-type giants in symbiotic stars.

This method can be extended to higher resolution, as modern instrumentation is capable of resolving the CO band into its component spectral lines. Such data are potentially quite useful for determining the evolutionary status of these giants, as the individual lines are expected to be sensitive to variations in temperature and luminosity.

TiO bands are usually prominent on optical spectra of late-type stars, and Sharpless (1956) has discussed the importance of these

Figure 3.12 - Schematic behavior of the depth of the CO band as a function of spectral type and luminosity class. The depth of this band is weak in M0 III giants, and strengthens markedly with decreasing temperature along the giant branch. The CO band is not very temperature sensitive for bright supergiants, but at a given spectral type, supergiants tend to have deeper CO bands than do giants (Kenyon and Gallagher 1983).

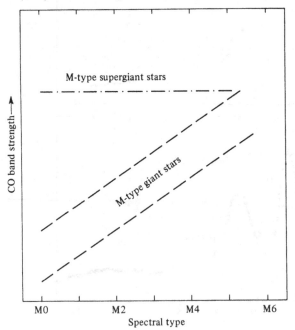

features as indicators of the spectral type of K-M stars. Strong bands at $\lambda\lambda7054$, 7744, 7821 are visible in M0-M3 giant stars, and additional bands appear in M4 and later-type giants. Andrillat (1982) and Andrillat and Houziaux (1982) discussed observations in the near-infrared, and attempted to classify symbiotic absorption spectra on the basis of Sharpless' classification scheme. In most cases, these spectral types agree very well with those derived previously by Merrill, Swings and Struve, and Tcheng and Bloch; new spectral types were derived for RX Pup (M5-M6 III) and TX CVn (M1 III). Unfortunately, the published results of this analysis have been limited to the well-known systems.

The classification criteria discussed by Sharpless can be accurately quantified, since the individual absorption bands are themselves temperature sensitive. O'Connell (1973) and Ramsay (1981) have presented calibrations of the $\lambda7054$ band (using low resolution spectra) and the $\lambda8860$ band (using high resolution spectra), respectively. The initial results of a program to classify the cool components of symbiotic stars using the strong $\lambda7054$ TiO band are presented in Figure 3.13 and Table 3.4. Following O'Connell, an index, I_{TiO}, is defined to measure the depth of the band at $\lambda7054$ relative to an interpolated continuum point at the same wavelength:

$$I_{TiO} = -2.5 \log (F_{7100}/F_{7050} + [F_{7400} - F_{7050}]/7), \qquad (3.5)$$

where F_λ represents the flux in a 20 Å bandpass centered at λ. The TiO index has been measured for a small sample of M-type stars, and plotted as a function of spectral type in Figure 3.13. The correlation of I_{TiO} with spectral type is sufficiently good, and generally independent of luminosity class, to allow an estimate of spectral types for a few symbiotic stars based on a measurement of I_{TiO}. These spectral types are in good agreement with those suggested in the literature by Merrill and others (as noted in the Appendix), and confirm that a bright giant or supergiant is present in CI Cyg and perhaps other systems. Obviously, the evolutionary status of the late-type components in these binaries remains unclear; however, a combination of red optical spectra and low resolution infrared spectra should eventually resolve this problem.

3.1.2 *The Blue Continuum*

One of the more notable features of a typical symbiotic star is a weak blue continuum which tends to dominate the spectrum at wavelengths less than about 5000 Å (Figure 3.9). Early attempts to determine the nature of this continuum were hampered by the atmospheric cutoff at 3200 Å and the lack of modern photoelectric detectors. In spite of these restrictions, Boyarchuk (1969, 1975, and references therein) demonstrated that calibrated photographic spectra of symbiotic stars could be understood as a combination of radiation from (i) an M-type giant, (ii) a hot subdwarf with $T_h \sim 10^5$ K, and (iii) an ionized nebula with $T_e = 15,000 - 20,000$ K. This combination successfully reproduced the observed continua of most symbiotic stars over $\lambda\lambda$3200-8000, and stood as the basic physical model for over 15 years.

The advent of satellite ultraviolet and X-ray observations has gen-

Figure 3.13 - The dependence of the depth of the λ7054 TiO band as a function of spectral type for class III giants (filled circles), class II giants (crosses), and class I supergiants (open circles). Although the band-depth seems rather independent of luminosity class, cooler giants tend to have significantly stronger TiO bands than their warmer counterparts.

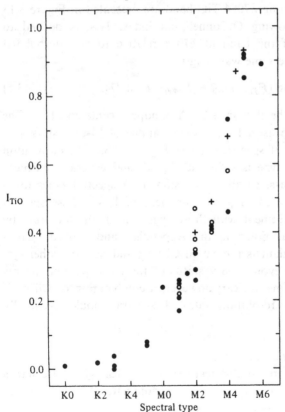

erally led to a greater appreciation of the hot components in symbiotic binaries. Initial ultraviolet observations were made with the *ANS*, *OAO-2 and TD-1* satellites, and consisted of broad-band measurements at selected UV wavelengths roughly evenly spaced between 1000 Å and 3500 Å. AG Peg and SY Mus were the only objects detected with these instruments, and SY Mus was invisible in 2 of 4 ANS bandpasses (Thompson, et al. 1978). OAO-2 photometry of AG Peg revealed a very bright source that resembled a WN star reddened by $E_{B-V} = 0.20$ (Gallagher, et al. 1979). These observations supported the model first proposed for AG Peg by Boyarchuk (1967), and confirmed the binary nature of at least one symbiotic star.

The launch of the sensitive *IUE* satellite made possible observations of many more systems, and has permitted direct observations of an extended underlying continuum attributable directly to the hot component. The effectiveness of *IUE* was demonstrated when Keyes and Plavec (1980a,b) reported initial observations of AG Peg (Figure 3.14), and refined the conclusions developed by Gallagher, et al. (1979). They found the continuum could be fit by a model stellar atmosphere (log g = 4.5) with an effective temperature of 30,000 K reddened by $E_{B-V} = 0.12$ or by a 100,000 K blackbody (also reddened by $E_{B-V} = 0.12$) with superimposed Balmer continuum radiation. Although the low temperature solution cannot explain the He II emission (both λ1640 and λ4686 are present as strong emission lines), Wolf

Table 3.4. *Classification of the M giants in symbiotic stars*

Symbiotic star	I_{TiO}	Spectral type
Z And	0.47	M3
EG And	0.38	M2
TX CVn	0.12	K5-M1
T CrB	0.70	M4
BF Cyg	0.71	M4
CI Cyg	0.72	M4
AG Dra	0.01	K0-K3
YY Her	0.53	M3-M4
V443 Her	0.77	M4-M5
RW Hya	0.30	M2
BX Mon	0.72	M4
RS Oph	0.21	M1
AG Peg	0.31	M2
AX Per	0.71	M4
AS 289	0.54	M3-M4
AS 296	0.71	M4

Rayet stars with strong He II lines also have low derived effective temperatures (which is a result of high electron temperatures in the outflowing gas; Willis and Wilson 1978).

While the majority of symbiotic stars have UV continua resembling that of AG Peg, a few have continua which cannot be fit simply by a reddened model stellar atmosphere combined with Balmer nebular radiation. CI Cyg is an example of a system with a predominantly flat continuum at short wavelengths that rises slowly for $\lambda > 2500$ Å. A reddened A star nominally provides an acceptable fit to the observed continuum, but such a star cannot possibly produce the high energy photons needed to explain the emission-line spectrum (Slovak 1982; Stencel, et al. 1982). An alternative explanation for the UV continua in stars such as CI Cyg is an accretion disk surrounding a low mass star (cf. Chapter 2). Adopting reasonable estimates for the distance and interstellar extinction (and correcting for the effects of the disk inclination with respect to the line-of-sight), the observed UV luminosity in these symbiotic stars implies accretion rates of $10^{-7} M_\odot \text{ yr}^{-1} (10^{-5} M_\odot \text{ yr}^{-1})$ if the accreting star is a white dwarf (main sequence star). These rates are far in excess of those typically associated with other interacting binaries such as cataclysmic variables

Figure 3.14 - The ultraviolet spectrum of AG Peg as observed with the *IUE* satellite. The UV continuum rises sharply for $\lambda < 2000$ Å, and, when de-reddened by $E_{B-V} \sim 0.1$, is consistent with either a 30,000 K model atmosphere or a 100,000 K blackbody and superposed Balmer continuum emission, as explained in the text.

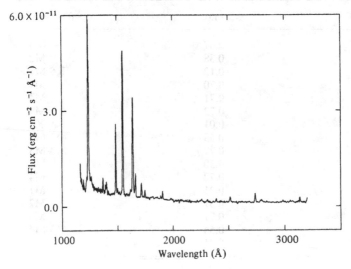

and low mass X-ray binaries (in which the accretion rates are estimated to lie between 10^{-8} and 10^{-10} $M_\odot$ yr^{-1}; Patterson 1984).

A significant fraction of symbiotic stars have now been observed with the *IUE* satellite, and many of the low resolution spectra are sufficiently well-exposed to display a distinct continuum. Recalling the ultraviolet diagnostics derived in Chapter 2, four continuum magnitudes (m_{1300}, m_{1700}, m_{2200}, and m_{2600}) have been measured for these objects and used to construct (C_1, C_2) for each set of spectra. Representative error ellipses have been plotted on the color-color diagram in Figure 3.15, and it is apparent that many systems lie near only one of the model sequences. These objects have unique solutions for the nature of the hot component within the framework of the basic theory. These solutions, listed in Table 3.5, consist of a pair of parameters: a mass transfer rate, $\dot{M}$ (for main sequence and white dwarf accretors), or an effective temperature, T_h (for hot stellar sources), obtained by fitting the (C_1, C_2) color indices to model colors; and a color excess, E_{B-V}, derived by fitting the individual colors ($m_{1300} - m_{1700}$, $m_{1700} - m_{2600}$, and $m_{2200} - m_{2600}$; cf. Kenyon and Webbink 1984).

It is interesting that main sequence accretors provide the best-fit solution for 5 symbiotic stars (CI Cyg, YY Her, AR Pav, AX Per, and CL Sco), and acceptable solutions for two other objects (Z And and Y

Figure 3.15 - Observed (C_1, C_2) indices for a sample of symbiotic stars. The model trajectories have been repeated from Figure 2.4, while the observations have been plotted as 1 σ error ellipses. The ellipses tend to cluster around the model tracks, and suggest symbiotic stars possess either accreting main sequence stars or hot stellar sources (Kenyon and Webbink 1984).

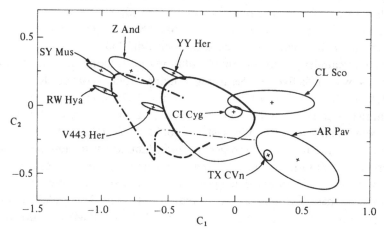

Table 3.5. *Classification of the hot components in symbiotic stars*

System	Type[1]	$\log \dot{M}$ $(M_\odot \, \mathrm{yr}^{-1})$	$\log T_h$ (K)	E_{B-V}
Z And	hs	-	5.17	0.27
	ms	-5.43	-	0.29
EG And	hs	-	4.84	0.09
UV Aur	hs	-	4.30	0.30
TX CVn	hs	-	3.99	0.03
Y CrA	hs	-	5.18	0.56
	ms	-4.98	-	0.39
BF Cyg	hs	-	4.63	0.49
CI Cyg	ms	-4.40	-	0.45
AG Dra[2]	hs	-	5.04	0.01
AG Dra[3]	hs	-	5.07	0.03
YY Her	ms	-5.20	-	0.13
V443 Her	hs	-	4.90	0.31
RW Hya	hs	-	4.95	0.01
SY Mus	hs	-	5.07	0.38
AR Pav	ms	-3.92	-	0.26
AG Peg	hs	-	4.76	0.15
AX Per	ms	-4.30	-	0.29
CL Sco	ms	-4.56	-	0.52

[1] hs - hot stellar source; ms - accreting main sequence star
[2] quiescent data
[3] outburst data

CrA; these two systems also have solutions for hot stellar sources). The derived accretion rates of $\sim 10^{-5} \, M_\odot \, \mathrm{yr}^{-1}$ require the giants of these systems to fill their tidal lobes; this constraint is satisfied for CI Cyg and AR Pav (Thackeray and Hutchings 1974; Kenyon and Gallagher 1983). The nature of the giants in Z And, Y CrA, YY Her, AX Per, and CL Sco is uncertain: additional observations are needed to determine if these giants do indeed fill their tidal lobes.

The remainder of the adopted solutions correspond to hot stellar sources with effective temperatures ranging from $\sim 10{,}000$ K up to $\sim 120{,}000$ K. No good candidates for white dwarf accretors have been identified among the systems examined to date. However, the hot stellar components in symbiotic systems could be accreting at rates low enough for the intrinsic hydrogen burning luminosity to dominate the accretion luminosity. Indeed, some accretion appears necessary to maintain the high observed temperatures, as noted in Chapter 2. The giants in these systems are not required to fill their tidal lobes, although some (e.g., AG Dra; Kenyon and Webbink 1984) may do so.

R Aqr is one of two quiescent symbiotics containing a Mira variable to be observed with *IUE*, and these observations did not reveal the hot source expected from optical spectroscopy (Johnson 1980; Michalitsianos, et al. 1980). The UV continuum of the D-type system H1-36 is also quite weak, although a few emission lines were reported by Allen (1983). Johnson suggested that silicate particles ejected by the Mira might result in circumbinary extinction without an associated 2200 Å absorption feature. If this is true, the (C_1, C_2) colors would not provide a reddening-free representation of the UV continuum. Ultraviolet observations of D-type systems in outburst tend to confirm this notion, and future ultraviolet continuum observations of quiescent D-type systems therefore may be difficult to interpret (Kenyon and Webbink 1984).

The high effective temperatures derived for some symbiotic stars imply significant fluxes at energies < 100 eV, and it is thus natural to assume these objects would be strong soft X-ray sources. Unfortunately, only five quiescent systems were bright enough to be detected by the *Einstein Observatory*, and one of these was the neutron star symbiotic V2116 Oph (Doty, Hoffman, and Lewin 1981). Two recurrent novae, T CrB and V1017 Sgr, were observed at low count rates by the *Einstein* Image Proportional Counter (IPC; Cordova and Mason 1984). Neither of these objects has a particularly intense optical emission-line spectrum, and T CrB does not appear to have a very hot ultraviolet continuum source (Cassatella, et al. 1982). Kenyon and Webbink (1984) suggested that the behavior of T CrB in the UV might be explained by an optically thin, unsteady accretion disk surrounding a main sequence star. If this is true, then X-ray emission from T CrB (and by analogy from V1017 Sgr as well) might be produced in the optically thin boundary layer at the inner edge of the disk.

Discounting the atypical system V2116 Oph, the brightest X-ray symbiotic star is AG Dra. Anderson, et al. (1982; also Anderson, Cassinelli and Sanders 1981) measured a flux of 0.27 count s^{-1} with the IPC, and could fit the spectrum by adopting either (i) a nebular model with $T_e \sim 10^6$ K, or (ii) a blackbody model with $T_h = 150,000 - 350,000$ K. While a blackbody with $T_h \sim 250,000$ K provides the best fit to the IPC data, a somewhat lower temperature is not excluded and is in better agreement with that derived above from a fit to the ultraviolet continuum and the optical emission lines. AG Dra is virtually unreddened, and is a prime candidate for observations in the EUV ($\lambda\lambda 100-1000$) - such observations would provide a useful check on the effective temperature derived from X-ray and UV data.

The lack of additional *Einstein* detections is disheartening, since objects such as Z And, RW Hya, and SY Mus should be copious X-ray emitters if their derived effective temperatures are correct. One must keep in mind, however, that X-rays may be absorbed by interstellar *and* circumbinary gas. Both Z And and SY Mus have substantial amounts of interstellar reddening ($E_{B-V} \sim 0.3$; $N_H \sim 1.5 \times 10^{20}$ cm^{-2}), and soft X-rays are likely to be effectively absorbed. Although RW Hya suffers negligible interstellar reddening ($E_{B-V} < 0.01$), it does possess a dense circumstellar nebula with $n_e \sim 10^8$ cm^{-3} (Kafatos, Michalitsianos, and Hobbs 1980). This implies a column density of hydrogen ions, N_i, $\sim 5 \times 10^{21}$ cm^{-2} for a nebular radius of 1 AU (cf. Table 3.7), which is significantly greater than the interstellar column densities of hydrogen atoms towards Z And and SY Mus. Thus soft X-rays may be absorbed by the circumbinary gas in symbiotic systems, and never reach earthly detectors.

Attempts to model the blue continua of symbiotic stars must account for variations in the flux distribution as a function of time (e.g., Sahade, Brandi, and Fontenla 1984; Stencel 1984). Stencel, et al. (1982) monitored CI Cyg during 1978-79, and found minima in the UV continuum flux and in the fluxes of various high ionization emission lines (e.g., He II $\lambda 1640$) which coincided with the optical minimum reported by Belyakina (1979,1984). These data provide valuable evidence that the cool giant in CI Cyg eclipses the hot component, as proposed by Hoffleit, Pucinskas, and Belyakina. Similar observations of AR Pav were discussed by Hutchings, et al. (1983); again, these show that the hot component is eclipsed during primary minimum. In both cases, the *IUE* spectrophotometry served to complement contemporaneous optical photometry and spectroscopy, verifying the nature of the orbital variations.

Another example of UV variations is provided by the southern symbiotic system SY Mus. When first observed by *IUE*, this system showed a very weak UV continuum and many strong high ionization emission lines (Figure 3.16). A fairly strong continuum appeared nine months later, and many of the emission lines had increased in intensity. The UV continuum and emission line fluxes remained at a high level for a few months, only to return to the level of September 1980 in May 1984 (Michalitsianos and Kafatos 1982; Kenyon, et al. 1985). The behavior observed with *IUE* parallels the optical light curve presented by Kenyon and Bateson (1984; Figure A.19), and suggests the hot component is eclipsed during primary minimum as in CI Cyg and AR Pav. It is interesting that the maximum in the UV emission line fluxes coincides with the short secondary maximum in the visual light curve. As usual, additional observations are required to interpret this phenomenon.

Fluctuations in the ultraviolet continua of other symbiotic stars have yet to be convincingly associated with binary motion. The variability observed in Z And, EG And, and AX Per qualitatively resembles the behavior reported in CI Cyg, SY Mus, and AR Pav, but additional observations are needed. The variations in the spectrum of T CrB (Cassatella, et al. 1982) appear to be unrelated to the orbital phase or to activity observed at optical wavelengths. Similarly rapid, and seemingly random, oscillations in the near-ultraviolet continuum of CH Cyg (Hack and Selvelli 1982) demonstrate that some variations are likely to be the result of activity in the hot component itself, rather than simple binary motion. Small changes in the mass accretion rate or in the photospheric radius might be responsible for the large-amplitude variability observed in these latter systems.

3.3.3   The Emission Lines

Bright emission lines are a defining feature of symbiotic satrs, and they are indicative of an intense radiation field emitted by the hot component. Lines of H I and He I are usually visible on optical spectra of all systems, and tend to be accompanied by Mg II, C III], and C IV emission in the satellite ultraviolet. Additional He II and N V lines are prominent in higher excitation systems, indicating the presence of a somewhat hotter binary companion to the red giant star. The most extreme symbiotics show intense lines from [Ne V], [Fe VII], and the mysterious λ6830 feature. The formation of Fe VII requires photons with energies exceeding 100 eV, and thus these latter objects appear to contain very hot components.

It is interesting that the symbiotics presenting the strongest [Fe VII] lines tend to be D-type objects with weak optical continua, a fact first noted by Webster and Allen (1975). The optical spectra of these systems closely resemble planetary nebulae (e.g., [O III] λ5007 > Hβ), except for the presence of strong emission bands at λλ6830,7088. These bands, which have yet to be convincingly associated with any atomic transition as noted in Chapter 5, *are observed only in symbiotic stars* and are strongest in D-type systems (Allen 1980b). Both bands are absent among S-type systems unambiguously identified as containing accreting main sequence stars (e.g., CI Cyg, AR Pav, and AX Per), but are definitely visible on spectra of objects believed to contain hot stellar sources (e.g., AG Dra). Since [Ne V] and [Fe VII] emission lines

Figure 3.16 - Time variations in the ultraviolet spectrum of SY Muscae. The disappearance of the blue continuum and many strong emission lines coincides with an optical minimum, and is the result of an eclipse of the hot component by its red giant companion (Michalitsianos and Kafatos 1984).

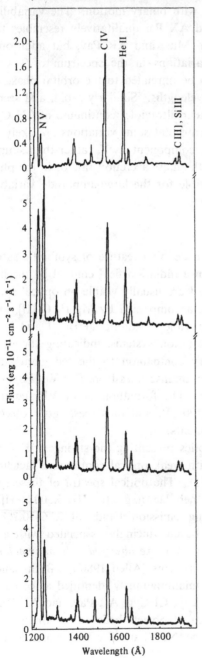

are common to both types of objects (and [Fe X] has been reported in CI Cyg and AX Per), it appears that λ6830 emission is associated with compact accreting objects such as white dwarfs or subdwarfs (Allen, 1980b).

Since λ6830 emission appears to be characteristic of hot stellar sources, it is tempting to classify symbiotics on the basis of the λ6830 band strength. Roughly 85% of all systems with [Fe VII] emission display the λ6830 band (Allen 1984), and could therefore be identified as systems containing hot stellar sources. The remaining objects which have noticeable [Fe VII] emission, but no trace of λ6830, are likely candidates for accreting main sequence stars. It is interesting to note that the predicted ratio of white dwarf to main sequence accretors is $\sim 10:1$, in reasonable agreement with this classification scheme (Chapter 6).

3.3.3.1 Recombination Lines. Strong recombination lines of H I, He I, and He II are observed in most symbiotic systems, and it is tempting to follow the traditional nebular analysis and estimate (i) the amount of interstellar reddening, and (ii) the effective temperature of the hot source. However, Blair, et al. (1983) and Kenyon (1983) have noted that the lower members of the Balmer series rarely conform to case B recombination with a standard interstellar extinction law. Large derived values for Hα/Hβ ($\sim 4 - 5$) point to nebulae that are optically thick at Hα ($\tau \sim 1 - 2$), implying electron densities, n_e, of $\sim 10^8$ cm^{-3} (Drake and Ulrich 1980). These densities are far in excess of those found in typical planetary nebulae and H II regions ($n_e < 10^4$ cm^{-3}; Osterbrock 1974), but are comparable to those derived from independent analyses of ultraviolet forbidden and intercombination lines, as discussed below.

Kenyon (1983) has suggested that the He II λ1640/λ4686 ratio provides a better measure of interstellar extinction than do the H I Balmer lines. The He II line ratios should not be grossly distorted by optical depth effects, although great care must be taken to secure contemporaneous optical and ultraviolet spectra. Kenyon's results are in good agreement with independent estimates of the reddening based on (i) ultraviolet measurements of the λ2200 interstellar absorption feature (Table 3.5), and (ii) optical measurements of H I column density through the galaxy (Burstein and Heiles 1982).

Although Hα appears to be optically thick, He II λ4686/Hβ still provides a useful estimate for the temperature of the central star, as noted in Chapter 2 (cf. Iijima 1981). Observed values for λ4686/Hβ are listed in Table 3.6, along with temperatures derived using equation (2.5) and Figure 2.5. These temperatures exceed those computed from the ultraviolet continuum indices, as might be expected in an object in which Hβ has been reduced in intensity through optical depth effects. Other complications involved in interpreting the λ4686/Hβ ratio include (i) collisional ionization associated with accretion processes (Kenyon and Webbink 1984) and (ii) collisional excitation of He⁺ in an extended atmosphere or outflowing wind (Willis and Wilson 1978). These effects all tend to result in an enhanced λ4686 flux, and the temperatures listed in the Table therefore may be regarded as overestimates.

Since the H I Balmer lines are optically thick, it might be profitable to observe the Paschen or Brackett series, in which the optical depth should be much smaller (Drake and Ulrich 1980). Thronson and Harvey (1981) secured high resolution spectra of HM Sge in the infrared, and noted many strong Brackett emission lines. Assuming case B recombination, the line ratios imply an optical extinction of 12 ± 1 mag for a standard interstellar extinction curve. This is significantly larger than the reddening derived from optical emission lines and the ultraviolet continuum, $A_V \sim 2-3$ mag, as discussed in the Appendix.

A similar effect is present in the quiescent D-type system H1-36 (Allen 1983). Strong H_2O and CO absorption bands dominate the near-infrared spectrum of H1-36, and this is inconsistent with the

Table 3.6. *He II Zanstra temperatures for symbiotic* stars[1]

Symbiotic	λ4686/Hβ	T_h (K)
Z And	0.54	180,000
BF Cyg	0.02	75,000
CI Cyg	0.39	155,000
V1016 Cyg	0.38	160,000
V1329 Cyg	0.37	160,000
AG Dra	0.45	170,000
YY Her	0.22	125,000
V443 Her	0.14	110,000
RS Oph	<0.08	<95,000
AG Peg	0.55	180,000
AX Per	0.24	135,000
HM Sge	0.32	145,000
CL Sco	0.07	90,000
AS 289	0.32	145,000
BD-21° 3873	0.26	135,000
M1-2	0.03	80,000

[1] data from Blair, et al. (1983) and the author

expectation of extensive emission from a 1000 K dust shell. A relatively normal near-IR flux distribution can be obtained by dereddening the observations by $A_K \sim 1.5$ mag, although the reddening deduced from the optical H I lines is much smaller ($A_K \sim 0.1 - 0.2$ mag). If this is true of all D-type systems, then dust *emission* may not be important in symbiotic stars.

Allen (1983) suggested that most of the obscuration in H1-36 might be local to the giant, with the hot component and its associated nebular region lying outside the dust cocoon. This explanation is inadequate for HM Sge, as the H I Balmer and Brackett lines are presumably formed in the same region. An unusual extinction law could be responsible for the behavior of the hydrogen lines, especially if the dust is composed primarily of graphite grains which have an extreme infrared extinction law (Thronson and Harvey 1981). However, it is difficult to understand how an oxygen-rich Mira variable with strong H_2O absorption bands produces vast amounts of graphite grains, so the origin of the extra extinction in HM Sge (and perhaps other symbiotic systems) remains uncertain.

3.3.3.2 Other Permitted Lines. Fairly strong emission from Mg II, Fe II, Ti II, and other singly ionized metals is present in many symbiotics, and Fe II and [Fe II] are probably the most prevalent transitions in spectra of RR Tel (Thackeray 1977). The Mg II doublet at $\lambda\lambda2796,2803$ is an indicator of a late-type stellar chromosphere, and the width of the Mg II emission core appears to be correlated with the absolute visual magnitude, M_V (Stencel and Mullan 1980). Prominent Mg II lines have been observed in a number of systems, including R Aqr and RW Hya, and this might prove to be a useful probe of the late-type component in symbiotic stars. Unfortunately, interstellar Mg II is an efficient absorber, and thus detailed measurements of the doublet are hampered by interstellar absorption features if the distance to the emission object exceeds 1 kpc (Michalitsianos, et al. 1982). The equivalent width-M_V relation has been applied to R Aqr, and results in an M_V (-2.5) that far exceeds the M_V derived from an expansion parallax for R Aqr ($M_V > -1$; Johnson 1980). This result suggests that the Mg II line is affected by radiation from the hot binary companion, or that additional Mg II emission is produced in the gaseous nebula surrounding R Aqr.

Narrow Fe II and [Fe II] lines also provide a measure of conditions in the giant's atmosphere, as their relative line strengths imply $n_e \sim 10^{11}$ cm^{-3} in some systems (Viotti 1976). Belyakina (1968, 1970) noted that the number of Fe II lines varied periodically on spectra of AG Peg, and was able to correlate this variation with radial velocity measurements presented by Merrill (1959, and references therein). Similar fluctuations in the Fe II lines have not been reported in other symbiotic stars, although BF Cyg and CH Cyg are likely candidates for such phenomena. Since Mg II and Fe II are formed under similar circumstances, it might be possible to derive orbital periods in some systems by following the intensity of the Mg II doublet discussed above.

3.3.3.3 Forbidden and Intercombination Lines.

Optical [O III] ($\lambda\lambda4363,4959,5007$) and [Ne III] ($\lambda\lambda3869,3968$) lines are visible in the spectra of many symbiotic stars, and these provide a useful probe of the effective temperature of the ionizing star when He II $\lambda4686$ is weak. This condition is satisfied for R Aqr and BF Cyg; applying equations (2.8) and (2.9) results in $T_h \sim 60,000$ K for R Aqr (Kaler 1981; Kenyon 1983) and $T_h \sim 35,000$ K for BF Cyg (Kenyon 1983). The solution for BF Cyg is comparable to the $T_h \sim 40,000$ K derived from (C_1, C_2) color indices (Table 3.5) and significantly less than the $T_h \sim 75,000$ K calculated from the He II $\lambda4686$ flux (Table 3.6).

The optical [O III] line fluxes are very useful when combined with satellite ultraviolet observations of the O III] $\lambda\lambda1661,1667$ doublet, as the $\lambda5007/(\lambda1661+\lambda1667)$ intensity ratio is sensitive to variations in n_e and T_e within an ionized nebula (Nussbaumer 1982). Other temperature and density diagnostics are accessible with ultraviolet spectra, including C III], N III], N IV], and [Ne IV]. These transitions are usually strong in symbiotic stars, and results from various investigations have been combined in Table 3.7. The electron densities listed in the Table are generally consistent with those estimated from the H I Balmer line ratios: most S-type systems have $n_e \sim 10^8$ cm^{-3}, while the majority of D-type objects have $n_e \sim 10^6$ cm^{-3}. A range of densities is typical of an individual system, suggesting that the nebulae are not homogeneous.

The electron temperatures derived in these studies tend to cluster near 10,000-20,000 K independent of the nature of the late-type giant, as might be expected in a photoionized nebula with solar abundances (Osterbrock 1974). Z And is the lone exception to this conclusion, and Nussbaumer (1982) has noted that a lower value for T_e would be derived for Z And if its abundances were slightly non-solar.

The values derived for n_e and T_e assume that all lines of a given multiplet are emitted at the same location within a given nebula.

While this assumption cannot be tested for the majority of symbiotic stars, an indication of the complications involved can be inferred from the optical [O III] lines in AG Peg. The two nebular lines ($\lambda\lambda4959$, 5007) have similar profiles, consisting of two distinct peaks (with the red component stronger than the blue) separated by ≈120 km s^{-1}. These features are relatively constant in velocity and have shown no large changes in their profiles. One interpretation for this behavior is that the [O III] $\lambda\lambda4959$, 5007 emission originates in a shell of material expanding away from the hot central star at ≈60 km s^{-1} (Hutchings, Cowley, and Redman 1975, and references therein).

The behavior of the $\lambda4363$ auroral line is quite different from its nebular counterparts. This line is composed of two unequal components which vary in velocity with a period of 820-830 days. It is curious that the relative phasing and velocity amplitude of the two components are very different, implying they are produced in distinct regions in the AG Peg nebula (Hutchings, Cowley, and Redman 1975). Evidently, three distinct [O III]-emitting regions are present in AG Peg, and this effect must not be overlooked when interpreting the [O III] lines of other symbiotic stars. There is evidence for similar behavior in CI Cyg, as the $\lambda4363/\lambda5007$ flux ratio remained constant throughout eclipse, while the $(\lambda1661+\lambda1667)/\lambda5007$ flux ratio markedly decreased (Oliversen and Anderson 1983; Stencel, et al. 1982). Simultaneous optical and ultraviolet observations of other eclipsing symbiotics (such as

Table 3.7. *Nebular densities and temperatures based on ultraviolet spectra*

Symbiotic	n_e (cm^{-3})	T_e(K)
Z And	2×10^{10}	80,000
R Aqr	10^6	15,000
BF Cyg	10^{5-8}	40,000
CI Cyg	10^9	10,000
V1016 Cyg	3×10^6	15,000
V1329 Cyg	3×10^6	12,500
AG Dra	3×10^9	-
V443 Her	10^9	15,000
RW Hya	10^{8-9}	15,000
SY Mus	10^{10}	15,000
AG Peg	10^{7-11}	15,000
AX Per	2×10^8	12,500
RX Pup	10^{9-11}	15,000
HM Sge	10^{6-7}	15,000
RR Tel	10^{6-8}	15,000
HD 330036	10^6	15,000

SY Mus and AX Per) might shed some light on this interesting phenomenon.

A possible explanation for the behavior of the emission lines is that some of them are not formed close to the hot component. Friedjung, Stencel and Viotti (1983) examined high dispersion *IUE* spectra, and noted a systematic redshift of the high ionization resonance lines with respect to the intercombination lines. They suggested that photon scattering in an optically thick wind lost from a corona or a chromosphere surrounding the cool star might be responsible for this effect. Their additional suggestion that such a region might be energized by excess rotation of the cool giant should be accepted with caution, since it is not obvious that the giant can be spun up by tidal forces. The ratio of the synchronization timescale for a star with a convective envelope (Zahn 1966) to the nuclear evolutionary timescale for the red giant (e.g., Becker 1979) is:

$$\tau_{sync}/\tau_{nuc} \sim (r^{-6}/2)(M_{env}/10^{-2}\,M_\odot)^{-1}(L/10^4 L_\odot), \tag{3.6}$$

where r $(= R_L)$ is the fractional radius of the giant, M_{env} is the mass of its convective envelope, and L is its luminosity. For reasonable choices of M_{env} (0.5 $M_\odot$) and L (2500 $L_\odot$), $\tau_{sync}/\tau_{nuc} \sim 0.2$ if the giant fills half its tidal lobe (i.e., $P \approx 3$ years), and significant synchronization might occur if dynamical processes associated with mass loss from a nearly lobe-filling giant do not affect the evolution of the cool star. As the radius of the giant approaches the tidal limit, such dynamical processes dominate the nuclear evolution of the giant, and although τ_{sync} shortens considerably ($\approx 10^4$ yr) the dynamic evolution of the giant may be more rapid than any other tidal process (e.g., Webbink 1979).

The electron densities derived from ultraviolet data are very important, as they may be combined with fluxes from strong recombination lines (such as Hβ) to estimate nebular radii as a function of distance. Assuming a constant density nebula and case B recombination, the nebular radius is approximately:

$$R_{neb} \sim 0.3 \; d^{2/3} n_9^{-2/3} f^{1/3}(H\beta) \; AU, \tag{3.7}$$

where d is the distance in kpc, n_9 is the electron density in units of 10^9 cm^{-3}, and $f(H\beta)$ is the reddening corrected Hβ flux in units of 10^{-12} erg cm^{-2} s^{-1} (Kenyon 1983). Although case B recombination is not precisely applicable to symbiotic stars and symbiotics do not possess constant density nebulae, as noted above, this expression should provide a reasonable estimate for the radius of the nebula.

Equation (3.7) is a complicated function of density, distance, and nebular flux, but considerable progress can be made by adopting values for the distance (1 kpc) and Hβ flux (10^{-11} erg cm^{-2} s^{-1}) that are typical of S and D-type systems (Blair, et al. 1983; Kenyon 1983). Thus, R_{neb} ($\sim n_e^{-2/3}$) is solely a function of the electron density within the nebula. This implies that the nebula of a typical D-type system ($n_e \sim 10^6$ cm^{-3}) should have a radius approximately 100 times larger than that of an average S-type object ($n_e \sim 10^9$ cm^{-3}; Table 3.7), which is independently confirmed by radio observations discussed in Sec 3.2.3 (equation [2.11]).

3.3.3.4 Line Profiles. While emission line fluxes provide information concerning the physical structure of the gaseous nebula and the ionizing source, high resolution observations of these lines yield some measure of dynamic motions within the gas. Most recent optical work has concentrated on the strong Hα line, although profiles of Hβ, He I λ5876, and He II λ4686 are available in some instances (e.g., Oliversen and Anderson 1982). Typical Hα profiles consist of a strong centrally-reversed peak and broad wings, indicative of velocities approaching 500 km s^{-1} (Figure 3.17). Symmetric profiles are rarely seen, and the central reversal is barely visible in some cases (e.g., BF Cyg). The twin emission peaks in T CrB appear to be formed in a differentially rotating accretion disk, but the large values of Hα/Hβ (~ 5) point to self-absorption as the most likely cause for the profiles displayed in Figure 3.17.

Cyclic evolution of the Hα profile is visible in all well-studied symbiotic stars, and this behavior could be associated with binary motion. The variations observed in AX Per suggest that certain profile shapes prefer specific orbital phases (Figure 3.18), but these data have yet to yield quantitative information regarding mass motions within the binary system. Similar profile fluctuations have been noted in EG And (Figure A.2), and are likely to be present in other systems.

The rapid profile variations observed in some symbiotics are probably associated with dynamical activity within the binary, rather than a result of orbital motion. Anderson, Oliversen, and Nordsieck (1980) noted significant changes in CH Cyg's Hα profile on one week timescales, and suggested that inhomogeneities in an accretion disk surrounding the hot component in this system could explain their observations. Similar behavior in the shell-like absorption features of CH Cyg has been reported by Faraggiana and Hack (1980) and Wallerstein (1983), who interpreted their results in terms of a slowly expanding spherical shell of material ejected by the hot (or cool) object. It is tempting to associate rapid line profile variations with the flickering observed by Wallerstein (1968) and Cester (1968), and discussed earlier in this Chapter. Detailed monitoring of various strong emission lines using narrow-band interference filters coupled with simultaneous continuum observations might reveal the nature of the flickering in CH Cyg and other symbiotic star systems.

Profiles of the strong ultraviolet emission lines are generally symmetric, and composed of a narrow core and broad wings. Some structure is observed in a few systems, and the individual components appear to vary in intensity over the course of months (Figure 3.19). Since the resolution of *IUE* is only ~ 20 km s^{-1}, it is difficult to associate this activity with binary motion. Inhomogeneities in material (i) accreted by the hot component, or (ii) ejected by the cool component could be the cause of this behavior, although contemporaneous optical and ultraviolet observations may be needed to understand the data in detail.

Figure 3.17 - Hα profiles for a sample of symbiotic stars. These lines typically have velocity widths of a few hundred km s^{-1}, although larger velocities have been observed in some systems. The central reversal is apparently constant in velocity, while the blue peak shows variations in intensity and position (C.M. Anderson, private communication).

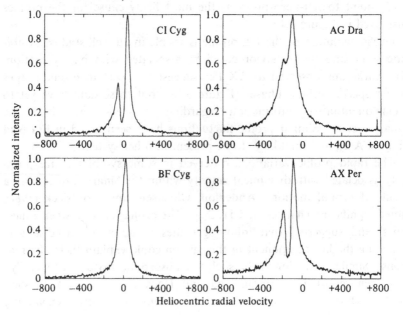

Figure 3.18 - Variations in the Hα profile in AX Per as a function of orbital phase. The shape of the line profile may be correlated with specific orbital phases, which are shown at right (C.M. Anderson, private communication).

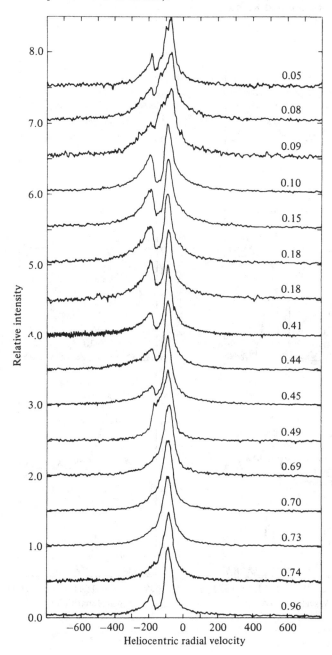

Figure 3.19 - Time variations of the strong C IV (left panel) and He II (right panel) emission lines in RX Pup. These lines are usually unresolved in symbiotic stars, but RX Pup shows multiple emission components that vary dramatically as a function of time (Michalitsianos and Kafatos, private communication).

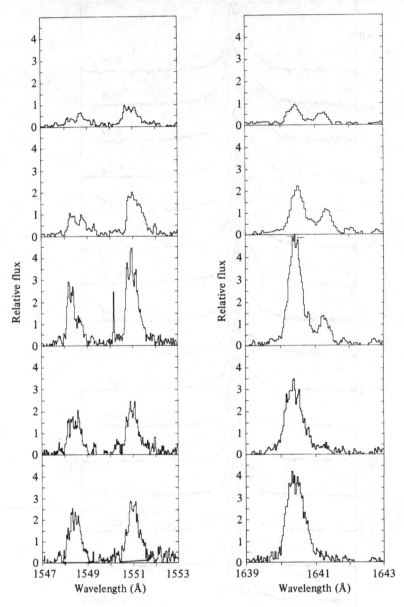

3.3.3.5 Radio Line Emission. Since symbiotic stars display a wealth of emission lines in the infrared, optical, and ultraviolet, it seems natural that they should have strong line emission in the radio regime. One possible source of line emission is, of course, hydrogen recombination, and radio recombination lines have been detected in H II regions (e.g., Osterbrock 1974). However, the extreme electron densities observed in symbiotics make it very unlikely that many of the lines commonly detected in H II regions would ever be visible in a symbiotic system. The ability to detect a given line depends on the ratio of the brightness temperature at line center (T_L) to that of the continuum (T_C). In thermodynamic equilibrium, this ratio is simply:

$$T_L/T_C = [1 - e^{-(\tau_L + \tau_C)}]/[1 - e^{-\tau_C}] - 1, \tag{3.8}$$

where τ_L and τ_C are the optical depths in the line and the continuum, respectively. For conditions appropriate to most symbiotic systems, the optical depth in Hα is of order unity (Boyarchuk 1969, 1975), so $\tau_L \ll 1$ for radio lines. However, the optical depth in the continuum is expected to be quite large at low frequencies (Osterbrock 1974). For an inverse-square density profile, the optical depth equals 1 at a frequency of roughly 10-20 GHz for most systems (Seaquist, Taylor, and Button 1984). Lines at frequencies less than 10-20 GHz are therefore not expected to be detected in symbiotic stars, and radio recombination lines have not been reported in any symbiotic system.

A promising alternative to line emission from hydrogen recombination is *maser* emission from hyperfine transitions in the excited states of various molecular species. Three types of molecular masers, OH, H_2O, and SiO have been identified in late-type stars, and these yield valuable information regarding the structure of the circumstellar nebulae surrounding evolved stars (Snyder 1980). An OH maser ($\nu = 1612.23$, 1665.40 and 1667.36 MHz) probes the outer layers of the nebula ($r > 10^{16}$ cm), while H_2O ($\nu = 22.2$ GHz) and SiO ($\nu = 43.1$, 86.2 and 129.4 GHz) masers probe regions much closer to the mass-losing star ($r \sim 10^{14-15}$ cm). The intensities of the various maser lines all show fluctuations which are associated with the pulsational period of the underlying star (Cahn and Elitzur 1979; Jewell, et al. 1979), suggesting that the masers are radiatively pumped by infrared photons emitted by the pulsating star.

Maser emission has been searched for in a number of symbiotic stars, and the results have been discouraging (Brocka 1979; Cohen and Ghigo 1980). Only R Aqr has been unambiguously identified as an SiO maser source; the remaining systems only have upper limits. Both the 43 and 86 GHz transitions of SiO have been seen in R Aqr (Lepine, LeSqueren, and Scalise 1978; Engels 1979; Zuckerman 1979; Ghigo and Cohen 1980), and the source is variable by at least a factor of 5-6 at 43 GHz. While this is somewhat larger than the observed amplitude in the infrared ($\Delta K \sim 1$ mag), the ratio L_{SiO}/L_{IR} in R Aqr seems normal when compared to that of other maser sources (Engels). In view of the complex optical radial velocity curve for R Aqr, Zuckerman suggested that the maser might be used to track the orbit of the M-type star. This advice should be accepted with caution, since Wolff and Carlson (1982) reported variations in the SiO line profile in various Mira variables.

3.3.4   Spectroscopic Orbits

As noted in Chapter 1, Berman (1932) and Hogg (1934) were the first to suggest that the symbiotic phenomenon is most naturally explained by a binary star. The red continuum and absorption features can then be associated with a late-type giant, while a variable hot component provides the blue continuum and strong emission lines. Although this model simply accounts for the spectroscopic appearance of a symbiotic star (and is probably the only model that can naturally explain the flux distribution in Figure 3.9), it has taken a great deal of effort to demonstrate convincingly *binary motion* in a fair number of systems. This is a natural consequence of the longer periods expected in a binary containing a red giant star. For the 1-2 AU separation needed to contain the giant within its Roche lobe, the orbital period is ~ 1-2 years. This binary, on average, will be found to have a radial velocity amplitude of 15-25 km s^{-1}, which is a challenge, but not an insurmountable one, to measure over a few binary periods. Unfortunately, seemingly random variations in the structure of the strong absorption and emission lines, as well as in the level of the blue continuum, greatly complicate symbiotic radial velocity curves. It is a tribute to the persistence and dedication of Merrill and Thackeray (among others) that fairly reliable spectroscopic orbits are available for *any* symbiotic stars.

Of the 133 known symbiotic stars (Table A.1), only T CrB, AR Pav, and AG Peg have what might be termed reliable spectroscopic orbits (and recent photometric accounts cast some doubt on the orbit of AG Peg; see the Appendix). Merrill (1959 and references therein) spent nearly 40 years observing AG Peg, and estimated a period of 800 days. Cowley and Stencel (1973) and Hutchings, Cowley, and Redman (1975; also Hutchings and Redman 1972) refined this estimate to 820-

830 days, and estimated a mass ratio of 3:1 (giant : hot star). Meinunger's (1981) photometric period of 827 days is nicely bracketed by these latter estimates, apparently confirming the orbital elements derived by Hutchings, Cowley and Redman.

Thackeray's (1959 and references therein; see also Thackeray and Hutchings 1974) determination of the orbit of AR Pav was simplified by periodic eclipses of the hot component by the M giant every 605 days. For reasonable values of the orbital inclination, complementary photometric and spectroscopic solutions yield a mass ratio of nearly 1, with the F-type primary slightly more massive than its M companion star (Andrews 1974; Menzies, et al. 1982). As with AG Peg, optical photometry was helpful in constraining possible values for the mass ratio.

The "best" binary orbit for a symbiotic star may be that of T CrB as determined by Kraft (1958; also Sanford 1949) and modified by Paczyński (1965). Both M-type absorption features and strong emission lines are easily visible, and Kraft found the velocities of both components to be comparatively well-behaved. The orbital solution yields a mass ratio of roughly 1.4, with the giant component being the more massive. For an inclination of 68°, the masses of the hot and cool components in T CrB are 1.9 $M_\odot$ and 2.6 $M_\odot$, respectively. The 227 day period derived by Kraft was later verified by optical photometry (Bailey 1975), although the orbital inclination remains somewhat uncertain.

Orbital solutions have been obtained for a few other symbiotic stars (and candidate systems), and are perhaps less reliable (cf. Table A.5). Most of these studies have relied on radial velocities derived from the H I (and other) emission lines, although TiO absorption features have been measured in a few instances. Given the asymmetrical shape and variable emission line intensities found in most systems, it may not be a good idea to rely on these features to determine orbits.

A promising alternative to the traditional methods of measuring radial velocities in symbiotic stars involves high resolution echelle spectra centered at λ5200 and obtained with the intensified Reticon detector system discussed by Latham (1982). The radial velocity of the cool component may then be determined by cross-correlation with various velocity standards (Tonry and Davis 1979). This procedure has the advantage of reducing observer bias during the measurement of wavelengths, since all spectra are handled identically by a computer. Garcia (1983) has searched for radial velocity variations in a small sample of symbiotic stars, using this technique, with results as displayed in Table 3.8. Significant variations were observed in every system showing strong absorption lines at λ5200 (the cross-correlation technique was found to work poorly when the emission line spectrum dominated the absorption line spectrum), and most of these appear to follow photometric variations (Garcia 1983). The velocities represent the orbital motion of the cool star, and reliable periods for these systems should be available within a few years.

3.4   Summary

In spite of the immense, and sometimes contradictory, amount of observational data for quiescent symbiotic stars, a few important points need to be emphasized before moving on to discuss the outburst properties of these interesting variables.

1. While it is now commonly accepted that most, if not all, symbiotic systems are binary stars, the lack of reliable binary periods for a significant fraction of these objects must be considered a great failure (cf. Table 3.1). Broad-band UBVRI photometry remains the most straightforward method for identifying periodicities, although additional spectroscopic data are required to confirm orbital motion. High resolution spectroscopic observations combined with cross-correlation analysis and reliable photometric data may ultimately prove to be the most effective technique for measuring orbital motion in symbiotic systems - the program initiated by M. Garcia and collaborators at Harvard-Smithsonian is a beginning, and hopefully other investigators will take advantage of new high resolution detector packages in operation at major observatories.

2. One of the great successes in the field is the identification of S and D-type systems by Webster and Allen (1975). This separation remains

Table 3.8. *Radial velocity variations in symbiotic stars*[1]

Symbiotic	Δv (km s^{-1})
EG And	20
TX CVn	15
T CrB	35
AG Dra	15
RW Hya	10
BX Mon	5
RS Oph	20

[1] adapted from Garcia (1983)

important since a large number of observational properties (including radio emission, high excitation [Fe VII] and λ6830 emission, and featureless optical/ultraviolet continua) continue to correlate with the "D-ness" of a given symbiotic system. The persistent association of D-type systems with Mira variables is another triumph of the classification scheme, and leads to the natural separation of symbiotic objects into two distinct types of binaries: (i) short period S-type systems (P ∼ a few years) and (ii) long period D-type systems (P ∼ a few decades). Mira variables remain to be detected in some D-type symbiotics, and it is important to verify the nature of these systems through near-infrared (1-4 μm) monitoring projects such as that underway at the South African Astronomical Observatory (e.g., Whitelock, et al. 1983a,b).

In spite of these successes, very little progress has been made in understanding the mass transfer process in symbiotic binaries. Lobe-filling giants are essential in S-type systems containing accreting main sequence stars, and significant mass loss rates ($> 10^{-7}\ M_\odot\ \mathrm{yr}^{-1}$) may be required in other systems as well. Reliable diagnostic criteria for both the spectral type and luminosity class of the late-type giants should provide additional insight into mass transfer in these binaries.

Many properties of the D-type systems also remain ambiguous. It is not yet clear if these objects are dominated by circumstellar dust *emission* or dust *absorption* (or both). Patchy dust absorption/emission has been invoked to explain the behavior of R Aqr and HM Sge, but satisfactory support for this hypothesis is lacking. Simultaneous optical and high resolution infrared spectra may be needed to resolve this question on a case-by-case basis.

3. Another major advance was initiated by the launch of the *IUE* satellite. The ultraviolet spectroscopic observations demonstrate that the S-type systems display two distinct classes of continua, independent of interstellar reddening. Many objects resemble the central stars of planetary nebulae in the UV, and, indeed, detailed fits by numerous investigators confirmed that this basic picture applies to the majority of systems. A smaller subset of symbiotics is not easily explained by this interpretation; rather, these objects appear to be accreting main sequence stars, confirming an idea originally proposed by Kuiper in the early 1940's. It is comforting that some of these systems contain lobe-filling giants, which is essential if the accretion rates derived earlier are correct. The relative numbers of these systems are ∼ 7-8:1 based on a

classification scheme employing the λ6830 emission band, with accreting main sequence stars making up the minority of S-type objects.

4. The structure of the gaseous nebulae have so far eluded rational explanation, although good progress has been made in deriving basic physical quantities (n_e, T_e, and R_{neb}) in a number of systems. These have been sufficient to demonstrate that *symbiotic binaries possess ionized nebulae that are very different from those of planetary nebulae.* The densities in symbiotic systems are > 100 times larger than those found in a typical planetary nebula, while the electron temperatures seem marginally higher. The behavior of the density-sensitive intercombination lines also shows that the gaseous nebulae in symbiotic binaries are neither homogeneous nor spherically symmetric. It is unfortunate that high resolution spectroscopic observations (both optical and ultraviolet) have not been of more help in this regard; the detailed monitoring projects have served more to illustrate the complexity of these systems than to clarify our understanding.

Since symbiotic systems are now commonly accepted as binaries, it may be appropriate to abandon the traditional nebular analysis (which, after all, assumes spherically symmetric shells surrounding single stars), and apply results derived for systems that are more closely related to symbiotics. An obvious parallel with symbiotic stars is the well-studied cataclysmic variable systems, which display many similar features and undergo nearly identical outburst cycles. A less obvious parallel lies with active galaxies, such as Seyferts and QSO's. Spectra of the Seyfert galaxy III Zw 77 (Osterbrock 1981) quite closely resemble spectra of V1016 Cyg and HM Sge, and models for the spectra of active galaxies therefore may be applicable to models of symbiotic systems. A final, and perhaps less obvious, parallel may be found among Harbig-Haro objects and FU Orionis-type stars. Accretion-driven instabilities recently have been proposed as explanations for the outbursts of FU Ori and V1057 Cyg (Lin 1984), and wind-driven outflow from the disk is probably sufficient to produce the collimated outflow observed in some H-H objects (Mundt and Hartmann 1983). The accretion luminosities inferred for FU Ori and V1057 Cyg are comparable to those observed in CI Cyg and AX Per, implying that outflow from disks in symbiotic stars might be important. Such winds might provide an explanation for the peculiar behavior of some emission lines observed in these objects.

4

The Outburst Phase - Theory

4.1 Introduction

As we shall see in Chapter 5, the outbursts of symbiotic stars are very exciting events which last from a few weeks to many decades. The appearance of P Cygni-like emission line profiles in the early stages of these outbursts is naturally explained by mass ejection at a velocity of ≈ 100 km s^{-1} (e.g., Beals 1951), suggesting that some type of explosive event causes the 2-7 mag increases in brightness (cf. Figures 1.1 and 5.1-5.3). Spectroscopic observations generally show A or F-type supergiant absorption features at maximum visual light (e.g., Belyakina 1979), which diminish in intensity shortly thereafter. In this respect, symbiotics resemble classical novae, which also have A or F-type spectra near visual maximum (Payne-Gaposchkin 1957). A few systems tend to resemble planetary nebulae at maximum, with fairly prominent Wolf-Rayet emission features pointing to rapid mass ejection via a stellar wind (Thackeray 1977).

With the development of the binary model by Berman and Hogg in the 1930's, the outburst was presumed to be associated with the hot component, although the nature of the instability was left to the imagination. Once single-star models were developed for quiescent symbiotic stars, mechanisms were proposed in which such an object could undergo multiple outbursts. Bruce (1955, 1956) likened symbiotics to tremendous thunderstorms, with outbursts resulting from electrical discharges in the extended atmosphere of an evolved red giant. Gauzit (1955), a champion of the coronal model, postulated that scaled-up versions of solar flares would explain the behavior of eruptive symbiotics, an interpretation supported by Aller and Menzel. Expanding on another idea of Menzel's, Wdowiak (1977) hypothesized that a large magnetic field formed during a helium shell flash in an evolved red giant would rise up through the photosphere due to its intrinsic buoyancy and eject material at high velocities.

A more reasonable single-star scenario was proposed by Ahern, et al. (1977) and Kwok (1977) in connection with the recent outbursts of V1016 Cyg and HM Sge (Chapter 5). This model envisions a normal red giant ejecting mass in a stellar wind at velocities of 10-30 km s^{-1}. Eventually, most of the giant's convective envelope will be lost in this slow wind, exposing a hot, compact core. It is this unveiling of the hot core that results in· an optical outburst, and the large luminosity and high effective temperature of this component produce a fast wind ($v \sim 1000$ km s^{-1}), which overtakes material in the slow wind. Shocks and hard photons emitted by the compact remnant ionize the expanding winds to produce high excitation emission lines.

In spite of these innovative single-star scenarios, recent theoretical attempts to understand the eruptions have emphasized binary models. These efforts were encouraged by the association of classical novae with short period binary systems by Kraft (1964). Kraft found that a few old novae (V603 Aql, T Aur, DQ Her, and GK Per) were close binaries ($P < 1$ day) consisting of a lobe-filling red dwarf star and a hot, compact star. The standard model for classical novae was developed from these and subsequent observations, and identifies the hot "star" as an accretion disk surrounding an otherwise normal white dwarf, with thermonuclear runaways in the white dwarf's accreted envelope the most promising eruption mechanism (e.g., Gallagher and Starrfield 1978). Since the visual magnitudes achieved by classical novae ($M_V = -5$ to -10) are roughly comparable to those of outbursting symbiotic systems ($M_V = -3$ to -8), Tutukov and Yungel'son (1976) and Paczyński and Żytkow (1978) suggested such a mechanism might be applicable to symbiotics, as well.

The major alternative to nuclear burning models involves a parallel with the less-luminous relatives of classical novae, the dwarf novae. Dwarf nova systems are essentially identical in nature to classical novae, but are usually 1-3 mag fainter in quiescence. Every few months, a 2-3 mag eruption boosts the luminosity of a given dwarf nova for 1-2 weeks, after which it fades rapidly to its pre-outburst level (cf. Warner 1976). It is currently thought that (i) an increase in the mass loss rate from the lobe-filling red dwarf, or (ii) an instability in the accretion disk results in an overall increase in the brightness of the disk surrounding the white dwarf in a typical "cataclysmic" binary. Bath (1977; also Webbink 1975) and Bath and Pringle (1982) have suggested that the behavior of some symbiotic stars might be understood if these objects are simply scaled-up dwarf novae, with a lobe-filling red giant losing material to an accreting main sequence star. If so, the cool giants in these systems must fill their tidal lobes, a constraint which is satisfied for systems such as CI Cyg and AR Pav.

4.2 Thermonuclear Outbursts

4.2.1 The Theory of the Outburst

Thermonuclear runaways appear to be the most promising mechanism for classical nova outbursts, as noted above. In these objects, matter lost by the red dwarf flows into an extended disk, and viscous stresses within the disk transfer material onto the white dwarf companion star. Eventually, the increase in density at the base of the accreted material initiates hydrogen-burning reactions (either the proton-proton chain or the CNO cycle; e.g., Kutter and Sparks 1980), leading to a rapid release of energy over a short timescale. The resulting outburst is truly a spectacular event, as the bolometric luminosity increases by a factor of 10^6 in the space of a few days (or perhaps hours; Gallagher and Starrfield 1978). After this abrupt rise in brightness, a typical nova fades to insignificance in a matter of months, presumably to begin the outburst cycle once again.

The development of this thermonuclear runaway is a complicated process which depends on a variety of physical parameters, including (i) the mass accretion rate ($\dot{M}$ in $M_\odot$ yr^{-1}), (ii) the mass and luminosity of the white dwarf (M_{wd} in $M_\odot$ and L_{wd} in $L_\odot$, respectively), and (iii) the composition of the accreted material. Numerical calculations described by Paczyński and Żytkow (1978) first demonstrated that accreted material could burn stably on a white dwarf, in a manner analogous to that which occurs in normal red giant stars. The minimum accretion rate needed to achieve steady hydrogen burning has been investigated by Iben (1982), who finds:

$$\dot{M}_{steady} \approx 1.32 \times 10^{-7} M_{wd}^{3.57} M_\odot \text{ yr}^{-1}. \tag{4.1}$$

Naturally, the maximum accretion rate is fixed by the point at which gravity is balanced by radiation pressure:

$$\dot{M}_{max} \approx 1.2 \times 10^{-3} R_{wd} M_\odot \text{ yr}^{-1}, \tag{4.2}$$

where R_{wd} is the radius of the white dwarf in $R_\odot$. For $\dot{M}_{steady} < \dot{M} < \dot{M}_{max}$, incoming material accumulates above the burning shell and expands to giant dimensions. The structure of such an accreting white dwarf resembles that of a single red giant quite closely (although the effective temperature may not be as low as in a normal red giant), and can remain in this state as long as $\dot{M} > \dot{M}_{steady}$.

When the accretion rate falls below $\dot{M}_{steady}$, material again cannot be processed through the burning shell as rapidly as it is accreted. At these low accretion rates, the hydrogen-rich envelope does not expand to giant dimensions, but remains close to the white dwarf's surface. Detailed calculations have demonstrated that hydrogen burning is thermally unstable under such conditions (Fujimoto 1982a,b; Iben 1982; MacDonald 1983). Since the energy released from hydrogen burning (in erg g^{-1}) exceeds the binding energy of the accreted material, a period of mass ejection may follow the onset of the runaway.

A wide variety of numerical and semi-analytical calculations has been performed for accreting white dwarfs, and presents a fairly coherent picture of the evolution of these interesting objects (e.g., Fujimoto 1982a,b; MacDonald 1980, 1983; Nariai, Nomoto, and Sugimoto 1980; Starrfield, Sparks, and Truran 1984; Prialnik, et al. 1982). The most important physical parameter in these calculations is the ratio of the amount of energy produced via nuclear reactions before the envelope expands to red giant dimensions, E_{nuc}, to the binding energy of the accreted envelope, E_{bind}. Very violent runaways occur if $E_{nuc}/E_{bind} \gg 1$, and these eject material at velocities of 1,000-3,000 km s^{-1}. The shell flash is bound to be rather weak if $E_{nuc}/E_{bind} \leq 1$, and matter may not be ejected.

The behavior of hydrogen shell flashes as a function of $\dot{M}$ and M_{wd} is summarized in Figure 4.1 for a cold white dwarf ($L_{wd} \ll 1 L_\odot$). Strong flashes with rapid mass ejection are possible if the following conditions are satisfied:

(1) $M_{wd} \geq 1 M_\odot$ - increasing the mass of the white dwarf decreases the amount of material required for the onset of a runaway (thereby reducing E_{bind}), and increases the pressure at the base of the accreted envelope (thereby increasing E_{nuc});

(2) $L_{wd} \leq 1 L_\odot$ - decreasing the white dwarf's luminosity increases the degeneracy at the base of the accreted envelope: since the envelope cannot expand until the degeneracy is lifted, additional nuclear energy can be produced before the expansion of the envelope quenches the shell flash;

(3) $\dot{M} \leq 10^{-9} M_\odot$ yr^{-1} - the kinetic energy of the infalling matter cannot be radiated effectively at high accretion rates, decreasing the degeneracy of the accreted material and reducing the strength of the flash; and

(4) $Z_{CNO} > Z_{CNO, solar}$ - nearly all of the nuclear energy is produced via the CNO cycle: increasing the amount of CNO nuclei in the envelope thus increases E_{nuc}.

In general, weak flashes result if these conditions are not fulfilled; mass ejection may still occur during a weak flash, but high velocites are not expected.

The recurrence timescale, Δt, for a typical thermonuclear event may be estimated from the time spent at quiescence ($\Delta t_{off} \sim M_{env}/\dot{M}$) and the time spent in eruption ($\Delta t_{on} \sim E_{nuc}/L_{on}$). Characteristic values for a 1 $M_\odot$ white dwarf accreting at various rates are listed in Table 4.1 (cf. Iben 1982), and Δt ($= \Delta t_{on} + \Delta t_{off}$) ranges from ~ 60 to $\sim 10^5$ yr. As long as $\dot{M}$ does not greatly exceed a few $\times 10^{-8}$ $M_\odot$ yr^{-1}, the ratio $\Delta t_{off}/\Delta t_{on} \gg 1$, implying that symbiotic stars undergoing ther-

Figure 4.1 - Behavior of hydrogen shell flashes as a function of the mass of the white dwarf (M_{wd} in $M_\odot$) and the mass accretion rate ($\dot{M}$ in $M_\odot$ yr^{-1}; Fujimoto 1982a,b; MacDonald 1983). The accreting white dwarf is assumed to be non-luminous; luminous white dwarfs tend to produce weaker shell flashes as noted in the text. Strong shell flashes are those that produce rapid ejection of material ($v > 100$ km s^{-1}), while weak shell flashes produce little, if any, ejection of material by dynamical processes. Mass loss during a weak shell flash is likely to occur via a stellar wind.

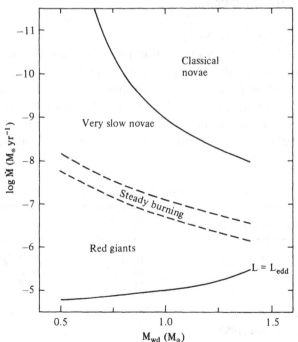

monuclear runaways should spend the bulk of their time in a quiescent state. Rapid recurrence timescales are possible only at very high accretion rates; symbiotic stars displaying fairly frequent outbursts (e.g., Z And - Figure 1.2) are therefore expected to be very intense UV sources at quiescence if their eruptions are a result of thermonuclear runaways (cf. Chapter 2).

An additional complication with accreting white dwarfs is the binary separation, A. In a classical nova system, $A \approx 1 \; R_\odot$; thus during the phases near visual maximum, the red dwarf companion is actually inside the photosphere of the outbursting nova!! It is expected that frictional forces induced by the red dwarf's motion through the expanding envelope will produce an extra energy source, and cause a more rapid ejection of the envelope. MacDonald's (1980) calculations suggest that relatively rapid mass ejection might occur during a weak shell flash if this gravitational stirring mechanism operates as expected. Indeed, the various numerical calculations imply that gravitational stirring is essential if slow classical novae (those having ejection velocities of $\approx 100 - 700$ km s^{-1}; Sparks, Starrfield, and Truran 1978) are to decline to minimum in a period of months, rather than years or decades. This mechanism should not be important in a symbiotic system, since the photosphere of the outbursting star is significantly smaller than the binary separation.

4.2.2   Application to Symbiotic Stars

Tutukov and Yungel'son (1976) first proposed that accreting white dwarfs might explain the blue continuum and high ionization emission lines in symbiotic stars. They envisioned a symbiotic star containing a white dwarf with a steady hydrogen shell source fed continuously by matter lost via a stellar wind from its red giant companion star. Small variations in the mass inflow rate would result in

Table 4.1. *Timescales for accreting white dwarf stars*[1]

Accretion rate ($M_\odot$ yr^{-1})	Δt_{on}	Δt_{off}
10^{-10}	110 yr	3.3×10^5 yr
10^{-9}	49	1.5×10^4
10^{-8}	21	630
10^{-7}	18	38

[1] adapted from Iben (1982)

abrupt changes in the character of the shell source, and could give rise to outbursts.

Later, Tutukov and Yungel'son (1982) investigated this model in more detail, and noted that the appearance of a potential symbiotic system depends on the mass loss rate of the giant, $\dot{M}_g$, and the binary separation, A. The value of $\dot{M}_g$ is limited by (i) the point at which substantial amounts of dust are formed in the expanding wind of the giant (as in OH/IR sources; Zuckerman 1980), and (ii) the point below which the rate of accretion onto the hot component is insufficient to produce high ionization emission lines (cf. Chapter 2). Thus, $\dot{M}_g$ is approximately restricted to $10^{-5} M_\odot \text{ yr}^{-1} < \dot{M}_g < 10^{-7} M_\odot \text{ yr}^{-1}$. Such large mass loss rates are expected in evolved red giants and supergiants. The limits on $\dot{M}_g$ force additional constraints on the binary separation, since some of the material lost by the giant is assumed to power the hot component; thus $2 \text{ AU} < A < 20 \text{ AU}$. Assuming a total binary mass of 3 $M_\odot$, this latter condition implies binary periods of 1-50 years, which is in reasonable agreement with observations summarized in Chapter 3. Larger separations, and hence longer binary periods, are probably appropriate for those symbiotics with substantial dust envelopes (e.g., the D-type systems described in Chapter 3).

Applications of the various numerical calculations to actual outbursts of symbiotic stars have been discussed by Paczyński and Rudak (1980; also Paczyński and Żytkow 1978) and Kenyon and Truran (1983). Paczyński and Rudak proposed that the outbursts of symbiotic stars naturally divide into two categories (type I - Z And, CI Cyg, AG Dra and type II - V1016 Cyg, V1329 Cyg, HM Sge), which parallels the concepts of steady burning and hydrogen shell flashes in accreting white dwarf stars. They suggested that type I symbiotics possess accreting white dwarfs with stable hydrogen burning shells; outbursts then result from variations in the accretion rate (as described by Tutukov and Yungel'son). Type II symbiotic systems, on the other hand, could be white dwarfs accreting at rates below $\dot{M}_{steady}$, in which case their outbursts would be caused by hydrogen shell flashes (cf. Kenyon and Truran 1983). The light curves of these objects should decline on the nuclear timescale:

$$\tau_{nuc} = 400 \; [(M_{env}/10^{-4} M_\odot)/(L_{on}/2 \times 10^4 L_\odot)] \text{ yr}. \qquad (4.3)$$

Rapid declines in some symbiotics necessitate low M_{env} and high L_{on}; Paczyński and Rudak suggest a 1.2 $M_\odot$ white dwarf accreting matter at $4 \times 10^{-7} M_\odot$ yr^{-1} could explain the eruptive behavior of Z And (Figure 1.2).

Kenyon and Truran pointed out that MacDonald's interpretation of slow classical novae argued against rapid thermonuclear outbursts in the longer period symbiotic binaries. The much larger dimensions of a symbiotic binary guarantee that, even when it has expanded to supergiant dimensions, the outbursting component never engulfs its red companion as in classical nova outbursts. The evolution following maximum thus proceeds as if the nova-like component were an isolated (single) star, and possible effects attributable to its binary nature are unimportant. If gravitational stirring is indeed necessary for the decline of slow classical novae, then only very slow nova-like outbursts would be possible in symbiotic binaries. Such outbursts would decline on the nuclear timescale given in equation (4.3), modified to include the effects of wind-driven mass loss as in O-B stars.

4.2.3   *Evolution of a Thermonuclear Outburst in a Symbiotic Binary*

Consider a symbiotic star containing a red giant and a white dwarf, as illustrated in Figure 4.2. The white dwarf accretes matter

Figure 4.2 - Schematic representation of a symbiotic binary containing a red giant star and a white dwarf accreting matter from the giant's stellar wind. Mass is lost spherically symmetrically by the cool giant, and some of this material is accreted by the white dwarf. High energy photons emitted by the hot white dwarf ionize a portion of the wind (H$^+$ region), although some of the circumbinary nebula remains neutral. Note that the particle trajectories do *not* account for binary rotation.

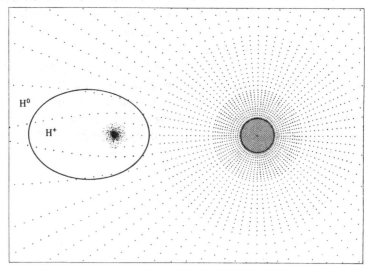

from the stellar wind of its binary companion, and emits radiation that ionizes a fraction of the circumbinary gas. This binary appears as a symbiotic star only if the luminosity generated by accretion or thermonuclear burning exceeds the criteria developed in Chapter 2 (e.g., $L_h \geq 10\ L_\odot$ for $T_h > 30,000$ K), and is likely to be a strong radio source as well.

Once the white dwarf accretes a sufficient amount of hydrogen, a thermonuclear runaway begins. The rise to visual maximum during this event is characterized by two phases: (i) a rapid increase in bolometric luminosity at nearly constant radius, followed by (ii) a slow expansion at constant bolometric luminosity (Figure 4.3). During

Figure 4.3 - Behavior of a white dwarf undergoing a thermonuclear runaway. The top panel shows the evolution in the H-R diagram, while the lower panel illustrates the evolution of the visual magnitude as a function of time (Paczyński and Żytkow 1978; Iben 1982; Kenyon and Truran 1983).

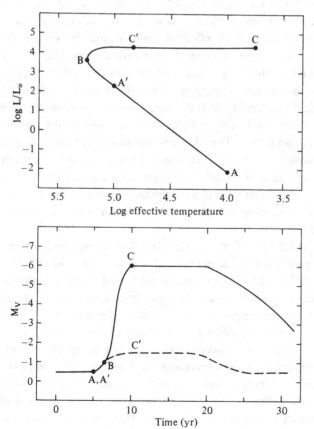

phase (i) the visual brightness of the outbursting component remains fairly constant, although ionization of material expelled by the giant should produce optical emission lines (see below). Once the eruptive star enters phase (ii), its visual brightness increases significantly and may overwhelm that of its giant companion. The visual luminosity at maximum can be estimated from a core mass-luminosity relation for such objects ($L_{on} \approx 59250 \, [M - 0.522 \, M_\odot] \, L_\odot$) to be $\approx 4000 \, L_\odot$ for a typical ($0.6 \, M_\odot$) white dwarf, and can be compared to the $100 - 200 \, L_\odot$ expected of a normal giant star. Note that the evolutionary sequence outlined in Figure 4.3 and the subsequent discussion assumes a strong flash in a predominantly degenerate accreted envelope. Weaker flashes result in smaller excursions in the H-R diagram (e.g., A' to C' in Figure 4.3), with a correspondingly smaller increase in visual brightness (Iben 1982; Kenyon and Truran 1983).

The spectroscopic evolution of this symbiotic binary can be outlined using the synthetic spectra discussed in Chapter 2. The nebular spectrum should grow dramatically in strength and excitation during phase (i), as the hot star increases in effective temperature. A comparison of successive frames increasing in effective temperature in Figure 2.1 serves to illustrate this phase. During the expansion phase (ii), the hot component evolves from a hot white dwarf ($R_h = 0.1 \, R_\odot$, $T_h \approx 250,000$ K) to an A-F supergiant ($R = 100 \, R_\odot$, $T_h = 6000$ K). Most of the optical brightening of the system occurs in this phase, with the nebular spectrum weakening substantially relative to the rapidly rising visible continuum. The higher-excitation nebular species disappear successively from the spectrum. Lower-excitation features decrease as well in equivalent width, although their absolute fluxes may at first increase. The spectroscopic development in late phase (ii) may be visualized by comparing successive frames decreasing in effective temperature in Figure 4.4.

The physical evolution of the underlying hot component in this model closely parallels that of slow classical novae (Paczyński and Żytkow 1978; Iben 1982), but with one major difference from an observational standpoint: in classical novae, the vast increase in ultraviolet flux during phase (i) escapes unprocessed to infinity (except for potential thermal heating of the companion star) for lack of any significant circumsystem gas. Thus these objects display a nebular spectrum only after visual maximum, as their expanding photospheres turn optically thin. In symbiotic binaries, on the other hand, a fairly dense ($n_e \approx 10^8$ cm^{-3}) circumstellar envelope generated by the stellar wind of the giant companion is already in place, and there is every reason to suppose that the rise to maximum should be characterized by the appearance first of a nebular spectrum. In fact, however, phase (i) and phase (ii) may be so brief in duration (of the order of weeks or months) as to stand little chance of optical detection before the system reaches late phase (ii).

In the decline from optical maximum, the hot component retraces its evolution in the Hertzsprung-Russell diagram, as do the outbursting components in classical novae. During this phase novae typically possess dense outflowing winds, a circumstance which we may anticipate pertains to the hot components of these symbiotic stars as well. In novae, outflow velocities during decline are generally 1000 km s^{-1}, i.e., much higher than wind velocities in normal late-type giants. If this holds true for thermonuclear symbiotic outbursts as well, the higher-excitation lines of the nebular spectrum as they reappear during optical decline should be substantially broader than at minimum or during the rise to maximum.

Figure 4.4 - Synthetic spectra of a symbiotic binary undergoing a thermonuclear runaway (Kenyon and Webbink 1984). These panels combine an M4 II giant with Kurucz model atmospheres at the indicated effective temperatures and a constant bolometric luminosity of $10^4 \, L_\odot$.

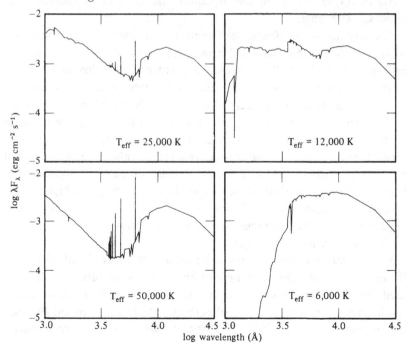

4.3   Accretion Driven Outbursts

4.3.1   *Basic Theory*

Theories for accretion-driven outbursts are derived from Kuiper's (1940) original proposal that β Lyr and other peculiar systems might contain a lobe-filling giant and a low mass, main sequence companion. Matter lost by the giant through the inner Lagrangian point would form a ring or disk around its companion, giving rise to a variable blue continuum and various emission lines. Sudden variations in the mass flow from the giant could result in optical outbursts, in which matter might be expelled through the outer Lagrangian points. Indeed, Paczyński (1965) demonstrated that a convective star (such as an evolved red giant) tends to expand as it loses mass, and thus mass loss might be maintained for long periods of time once such a star encountered its tidal surface.

Current models for the structure of very evolved red giants imply that these objects are composed of a compact, degenerate carbon-oxygen core surrounded by an extended atmosphere of roughly solar composition (Becker 1979). Hydrogen burning at the base of this atmosphere deposits helium into a thin shell atop the inert core, while helium burning reactions convert newly formed helium into carbon and oxygen. Since hydrogen burning reactions proceed more rapidly than those involving helium in this situation, the mass of the helium shell increases with time.

The physics of this *double-shell phase* of red giant evolution was first studied by Schwarzschild and Harm (1965), who showed that the thin helium shell is thermally unstable. As material is added to the shell, increased energy production via the triple alpha process results in a slow increase in the local thermal energy. This would normally result in an expansion of the burning region, leading to a steady-state configuration consisting of two distinct burning shells above an inert core. However, the massive convective envelope above the helium shell prevents this expansion, and the local temperature rises significantly. Since the energy generation rate from helium burning is highly temperature sensitive ($\varepsilon_{nuc} \propto T^{40}$ for $T \sim 10^8$ K; Clayton 1968), small increases in the local temperature result in a rapid release of energy known as the *helium shell flash*.

Recent detailed evolutionary calculations of evolved red giants by Becker (1979 and references therein) illustrate the evolution of the thin shell instability quite dramatically, as powerful helium shell flashes result once a sufficient amount of material has been processed through the hydrogen burning shell. The helium-burning luminosity increases significantly during the flash, resulting in a rapid expansion through the convective atmosphere that extinguishes the hydrogen burning shell. The atmosphere effectively absorbs most of the momentum of the expanding helium shell, and thus the observable parameters of the star are not significantly affected. Eventually, the reduction in density and the conversion of helium to carbon and oxygen during the expansion phase quench the flash, and the small, remnant helium shell contracts to leave a rekindled hydrogen burning shell in its wake. Since hydrogen burning commences as the flash powers down, one expects a series of helium shell flashes that gradually add material to the degenerate carbon-oxygen core (Becker 1979). This phenomenon is expected to be periodic, with P (in years) given by:

$$\log P = 3.05 - 4.5 \, (M_{core} - 1.0 \, M_{\odot}). \tag{4.4}$$

While periodic shell flashes may not have observational significance in single red giants, interesting consequences may result in a symbiotic binary. The small variations in the photospheric radius predicted by model calculations may be enhanced by the binary potential, and induce epochs of relatively rapid mass transfer (as in the Bath mechanism discussed below). R Aqr appears to experience phases of rapid mass loss every 44 years, which have been recently explained as an effect of periastron passage in an eccentric binary (Willson, Garnavich, and Mattei 1981). The 44 year period, if due to recurrent shell flashes, implies a core mass of 1.3 $M_{\odot}$ or $M_{bol} \sim -7$, using an appropriate core mass-luminosity relation $(L = 59250 \, [M_{core} - 0.522 \, M_{\odot}] \, L_{\odot})$. This is tolerably close to the observed bolometric magnitude of R Aqr (-5.5), given the uncertainties in equation (4.4) and the distance estimate.

An alternative mechanism for inducing mass transfer in a binary system has been discussed by Bath (1975, 1977, and references therein) in connection with the outbursts of dwarf novae and symbiotic stars. Bath noted that material flowing out of the envelope of a mass-losing star must pass through a series of ionization zones, and gain energy from recombination processes. He envisions that a small amount of mass outflow from a lobe-filling giant will tend to give rise to a dynamical instability which taps the available recombination energy $(\sim 10^{13}$ erg g^{-1} for a hydrogen ion), and rapidly ejects the star's outer atmosphere $(\sim 10^{-4} \, M_{\odot})$. This material flows into an accretion disk surrounding the secondary star, resulting in an optical outburst. The instability is expected to recur once the giant's atmosphere has recovered to once again fill its tidal lobe ($\sim$ a few years). Bath's hydrodynamic calculations of lobe-filling radiative stars lent some support to this idea, as these models rapidly developed dynamical instabilities recurring on timescales of weeks to years (depending on the nature of the lobe-filling star). However, calculations including the effects of convective energy transport demonstrated that lobe-filling red giants lose mass continuously, rather than in periodic bursts (Wood 1977). This argues against mass transfer instabilities in symbiotic binaries with circular orbits. If symbiotic systems should have eccentric orbits, the additional mass transfer expected near periastron passage might induce an outburst. The recent behavior experienced by CI Cyg supports this possibility, as outbursts appear to be associated with a particular orbital phase (Belyakina 1979).

A final possibility for repetitive accretion-driven outbursts is the disk instability model (e.g., Faulkner, Lin, and Papaloizou 1983, and references therein), which was also developed to explain dwarf nova outbursts. While the physics of disks is not understood in great detail (Pringle 1981), it is clear that the evolution of the disk is largely determined by the dependence of the effective temperature $(T_d \propto L_d^{1/4} \propto \dot{M}_d^{1/4}$; which measures the ability of the disk to radiate energy $- \dot{M}_d$ is the rate matter flows through the disk) on the surface density $(\Sigma$; which measures the amount of energy generated by viscosity). If the effective temperature increases as mass is added to the disk (i.e., $T_d \propto \Sigma^\beta$ with $\beta > 0$), radiative losses can compensate for increases in the viscous dissipation. However, situations may arise in which an increase in viscous dissipation leads to lower disk temperatures (i.e., $T_d \propto \Sigma^\beta$ where $\beta < 0$; this occurs in a hydrogen recombination zone for example). The disk is then unable to adjust quasi-statically to small changes in Σ, and must search for a new state in which β exceeds zero.

The various disk instability mechanisms are based on the "limit cycle" shown in Figure 4.5. The disk is initially assumed to be in a quasi-static steady-state, in which radiative losses compensate for viscous dissipation $(T_d \propto \Sigma^\beta$ with $\beta > 0)$. The subsequent evolution of this disk (which will lie on the lower curve if $\dot{M}_d < \dot{M}_B$) is determined by the rate at which mass is supplied by the lobe-filling star, $\dot{M}_{in}$. As the disk searches for an equilibrium state in which $\dot{M}_d$ equals $\dot{M}_{in}$, it must move up (or down) the locus shown in Figure 4.5. Exciting activity is anticipated if $\dot{M}_{in}$ exceeds the maximum accretion rate allowed on the lower curve, $\dot{M}_B$. The disk then encounters a series of quasi-static steady-states as it moves up the curve, until it reaches point B. Small increases in the surface density, Σ, beyond point B result in a drop in the temperature (and radiative losses), but a rise in the viscous dissipation. The disk must then jump to point D on the upper curve, when the radiative losses can once again compensate for viscous dissipation. If $\dot{M}_{in}$ still exceeds $\dot{M}_d$, the disk moves up the upper curve in its search for equilibrium. It is more likely that $\dot{M}_d$ exceeds $\dot{M}_{in}$, and the disk evolves down the curve towards point A. Radiative losses overcompensate for viscous dissipation should the disk reach point A, and the disk must drop to the cold state at point C to maintain equilibrium. Note that if $\dot{M}_B < \dot{M}_{in} < \dot{M}_A$, the disk can never reach an equilibrium state, and must repeat this instability cycle until some physical process causes $\dot{M}_{in}$ to change. Models for dwarf novae suggest that $\dot{M}_B \sim$ a few $\times 10^{-9} M_\odot \text{ yr}^{-1}$, a value which might

Figure 4.5 - Limit cycle behavior of the disk temperature, T_d, and the mass flow rate through the disk, $\dot{M}_d$, as a function of its surface density, Σ, for a disk instability model. Each solid line is a locus of points in which the local heating rate due to viscous stresses is balanced by radiative losses. On either side of these lines, the heating and cooling rates are unequal, as indicated. When the disk reaches point B, local heating exceeds local cooling, and the disk must undergo a "phase transition" to find a place where radiation can compensate for viscous heating (point D). The reverse is true at point A, as radiation cools the disk more rapidly than viscosity can generate energy. The disk must then drop to the lower curve (point C), where radiation is less efficient and once again balances viscous heating.

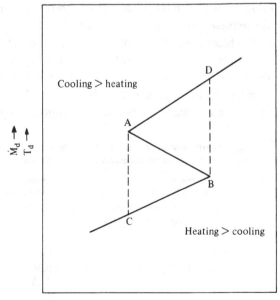

increase substantially for the larger disks present in symbiotic stars (cf. Faulkner, Lin, and Papaloizou 1983).

4.3.2   Application to Symbiotic Stars

The observed amplitude of erupting symbiotics places serious constraints on theoretical models, since $M_V < -3$ must be achieved at visual maximum. Thermonuclear runaways provide such luminosities naturally (e.g., $L \sim 1,000 - 10,000 \ L_{\odot}$) and can emit the bulk of this radiation in the optical if the object expands to giant dimensions. Accretion-driven outbursts are somewhat more problematical, since the disk is not able to expand in size unless the accretion rate exceeds the Eddington limit. Since the evolution of a super-Eddington disk resembles a thermonuclear runaway (see below), accretion-driven outbursts can be identified only if $M < 10^{-5} \ M_{\odot} \ \mathrm{yr}^{-1}$ ($10^{-3} \ M_{\odot} \ \mathrm{yr}^{-1}$) when the accreting star is a white dwarf (main sequence star).

The expected amplitude of a given accretion event can be estimated from the synthetic spectra presented in Chapter 2. The predicted magnitude at $\lambda 5820$, m_{5820}, is listed as a function of accretion rate for each type of accreting star in Table 4.2; a normal M2 III star with $M_V = -0.5$ has $m_{5820} = 0.17$ on this scale. An accreting white dwarf is obviously incapable of producing a substantial outburst in a symbiotic system unless the accretion rate exceeds the Eddington limit (Bath and Pringle 1982), at which point the object evolves in a fashion similar to that of the thermonuclear runaway models described above (Bath 1977). An accreting main sequence star, on the other hand, can easily overwhelm the optical brightness of its cool companion as M approaches a few $\times 10^{-4} \ M_{\odot} \ \mathrm{yr}^{-1}$. Given the rates derived for quiescent systems in Chapter 3, this represents approximately a tenfold increase in $\dot{M}$ over its quiescent value. Calculations discussed by Webbink (1979) demonstrate that mass loss rates approaching

Table 4.2. *Optical magnitudes for accretion disk models*

Accretion rate ($M_{\odot} \ \mathrm{yr}^{-1}$	m_{5820}, white dwarf	m_{5820}, main sequence
10^{-8}	4.05	
10^{-7}	2.60	
10^{-6}	1.11	2.17
10^{-5}	-0.62	-0.36
10^{-4}		-2.45
10^{-3}		-4.31

$10^{-3} M_\odot$ yr^{-1} are inevitable for lobe-filling red giants, and thus accretion events involving main sequence stars may not be uncommon.

Both the Bath mechanism and disk instabilities are capable of producing the desired amplitude of an erupting symbiotic star, providing the accreting star is a main sequence star as noted above. Various results suggest that $\dot{M}_d$ increases by factors of 5-20 during a transition from a cold state to a hot state in an unstable accretion disk. Exact values of $\dot{M}_d$ during such transitions await detailed calculations of disks surrounding main sequence stars, but preliminary calculations for dwarf novae suggest that symbiotic stars may provide a new laboratory in which to test the disk instability theory. The amount of material required in the Bath mechanism can be estimated from the inferred change in the accretion rate (10^{-5} to a few $\times 10^{-4} M_\odot$ yr^{-1}) and the duration of a typical outburst (2-3 yr), yielding $\sim 5 \times 10^{-4} M_\odot$ if the outburst is approximated by an instantaneous rise followed by a linear decline. This is comparable to the amount of material contained within the ionization zones of a typical red giant, so this mechanism may also operate in symbiotic systems. The high ionization emission lines provide an opportunity to distinguish between the two models: if these lines are a result of hard photons emitted by the boundary layer, the blue continuum should begin to rise before the lines can react if material is introduced into the outer regions of the disk (as in the Bath mechanism). If a disk instability begins in the inner regions of the disk, these emission lines may display dramatic activity before the blue continuum rises substantially.

4.3.3 Evolution of an Accretion-powered Outburst

A symbiotic star experiencing accretion-driven outbursts is expected to have the structure portrayed in Figure 4.6; a lobe-filling giant is essential to achieve the high $\dot{M}$ required for an accreting main sequence star, and such giants have been confirmed in CI Cyg and AR Pav. Since most of the matter is transferred to the secondary component, any ionized region is expected to be very small, perhaps confined to the tidal lobe of the accreting star. Such systems are not expected to be strong radio sources as noted in Chapter 2.

An accretion-powered outburst is distinguished observationally from a thermonuclear event by an increase in the emitted flux at all wavelengths, rather than the flux redistribution which characterizes the rise to optical maximum (phase ii) and early optical decline in the thermonuclear case. During decline the systems are expected qualitatively to retrace this evolution. The details of outburst development, however, depend critically upon the nature of the accreting object.

The outburst development for a white dwarf accretor can be visualized by following successively higher mass accretion rates in Figure 2.3, and reversing the sequence in the return to minimum. The most outstanding feature of this sequence is that, provided the accretion rate remains below the Eddington limit ($\dot{M} \approx 10^{-5} \, M_\odot \, \mathrm{yr}^{-1}$), the accretion disk is incapable of producing a substantial increase (more than a factor of 2) in the continuum redward of the Balmer jump, nor does it ever develop the strong A-F continuum typical of symbiotic stars at maximum (Table 4.2). These models are incapable of producing outburst amplitudes greater than 1 magnitude at sub-Eddington accretion rates onto white dwarfs (as noted earlier), and therefore *cannot* explain the 2-3 mag outbursts of a typical symbiotic system. Should the accretion rate exceed the Eddington limit, however, radiation pressure will disrupt the disk and lead to a rapid degradation of the emitted flux to the optical region, in a manner closely paralleling phase (ii) in the development of the thermonuclear outburst discussed above, and resulting in an A-F supergiant optical spectrum (Bath 1977).

Outburst development for a main sequence star accretor can be traced in Figure 2.2, again following sequences of increasing mass accretion rates. The flux distributions in this case are much softer than

Figure 4.6 - Schematic representation of a symbiotic binary consisting of a lobe-filling red giant star and an accreting main sequence star. Matter lost by the giant flows through the L_1 point, and forms a disk surrounding the main sequence star. Radiation emitted by the disk gives rise to a blue continuous spectrum and various high ionization emission lines.

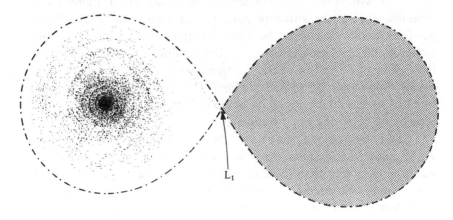

those of white dwarf accretors of equal bolometric luminosity. As a result, the disk continuum is capable of completely overwhelming that of the late-type giant, presenting the appearance of an F-type supergiant as $\dot{M}$ nears $10^{-3} \, M_{\odot} \, yr^{-1}$, well before radiation pressure (the Eddington limit) disrupts the disk (Table 4.2). The simple blackbody models of the boundary layer predict as well that the nebular emission increases both in excitation and in the equivalent widths of individual lines. The predicted Hβ equivalent width at maximum in this type is roughly 60 Å, comparable to that observed in the 1973 outburst of CI Cygni (58.5 Å; Belyakina 1975). It must be stressed, however, that the behavior of the higher excitation lines depends crucially on the assumptions that (i) the nebula is photoionized and (ii) the inner edge of the disk and the boundary layer do not expand in size (and therefore decrease in temperature) during an outburst. If mechanical heating is the dominant excitation mechanism near minimum, as discussed by Kenyon and Webbink (1984), it is possible that increased accretion rates actually lead to less efficient nebular excitation, since the typical Mach number in the boundary layer flow decreases with increasing mass flux (the Keplerian velocity in the disk is unchanged - or perhaps even reduced if pressure support of the disk becomes important -- whereas the sound speed increases with the increased boundary layer temperature). The weakening or disappearance near maximum of the highest excitation features seen in CI Cygni (Chapter 5) lends some support to this idea.

4.4 Testing the Theories

The basic theories for the outbursts of symbiotic stars predict essentially identical behavior at or near visual maximum, including (i) 2-5 mag outbursts, (ii) F-type spectra, and (iii) low ejection velocities. However, the predictions of the various theories could be tested if observations could be obtained over an entire outburst cycle.

1. The spectroscopic evolution of a thermonuclear runaway is characterized by a redistribution of the emitted energy at roughly constant luminosity (phase [ii]), while an increase of the observed flux at all wavelengths is the signature of an accretion-driven outburst.

2. The line profiles during an accretion event are expected to maintain a constant shape, although they may increase in overall intensity due to the increased flow of matter through the disk. A large variation in line widths is expected during a thermonuclear event: as the photosphere contracts towards the

white dwarf during the declining phases, lines are expected to become broader.

3. The absolute intensity of hydrogen lines should *decrease* during phase (ii) of a thermonuclear event, as the number of H-ionizing photons decreases with the declining effective temperature. Hydrogen line intensities (but not equivalent widths) should *increase* in an accretion event.

4. The cool giants in eclipsing systems provide a means whereby the luminosity profile of the eclipsed hot star can be determined. Accretion disks should have a very centrally condensed surface brightness profile, which is significantly different from that expected from a star.

As we shall see in the following Chapter, complete outburst cycles have been observed only at optical wavelengths. Hopefully, future outbursts can be followed in many wavelength regimes, so that the models presented here can be tested adequately.

5

The Outburst Phase - Observations

5.1   Introduction

The outbursts of symbiotic stars have generally attracted much attention, and each eruption is a new adventure in complexity. Since the hot component is the source of the instability, observations acquired throughout the event reveal little of the nature of the cool giant star. Indeed, infrared photometry of CH Cyg and CI Cyg suggests the giant maintains a roughly constant luminosity during a complete outburst cycle (Swings and Allen 1972; Szkody 1977; Kenyon and Gallagher 1983; Luud 1980, and references therein). Thus, the giant appears not to play a direct role in the evolution of an eruption, although it certainly has a good view!!

A typical symbiotic eruption commences with a 2-7 mag rise in visual brightness, which is accompanied by remarkable spectroscopic changes (Belyakina 1979). The B-V color decreases from 1.0-1.5 to 0.0-0.6, suggesting that a relatively hot source dominates the light at visual maximum. While the evolution of B-V as a function of V may be similar for nearly all outbursting symbiotic stars, the duration of and the spectroscopic changes observed in a given outburst naturally divide these systems into three distinct groups:

   (i)   recurrent novae (e.g., T CrB),
   (ii)   classical symbiotic stars (e.g., CI Cyg), and
   (iii) symbiotic novae (e.g., AG Peg; also known as very slow novae).

Representative light curves for each group are shown in Figures 5.1-5.3; the outbursts of the slow nova-like systems require many decades, while those of CI Cyg and other classical symbiotic systems are much more rapid. The eruptions of recurrent novae are extremely rapid, lasting less than a few months.

Recurrent outbursts are characteristic of a number of symbiotic stars, and systems such as Z And, CI Cyg, and AX Per undergo eruptions every 10-15 years (cf. Figures 1.2 and 5.2). However, only 20% of those objects having symbiotic optical spectra have ever been observed in a high state; this could be somewhat of an underestimate, as historical records generally have not been consulted for outbursts of newly discovered objects in the southern hemisphere. Rapid eruptions of the type displayed by T CrB and RS Oph could also have been missed in some systems. Indeed, nearly all ($\approx 75\%$) of the well-studied northern symbiotics have been observed to undergo at least one eruption (the exceptions being EG And, UV Aur, YY Her, V443 Her, RW Hya, and BX Mon). The light curves displayed in Figures 5.1-5.3 therefore may be considered as typical, with the caveat that some systems appear content to remain in low states for long periods of time. These latter objects may be candidates for symbiotic nova eruptions, as the results of Chapter 4 suggest such systems spend the bulk of their time in a low state.

5.2   The Recurrent Novae

Of the eight known recurrent novae, only T CrB, RS Oph, and V1017 Sgr appear to contain late-type giant stars (Webbink 1978). In quiescence each of these systems appears to be a normal, low excita-

Figure 5.1 - Optical light curve for the 1946 outburst of T Coronae Borealis (adapted from Larsson-Leander 1947). The rise to primary maximum is nearly instantaneous, and is followed by a rapid decline. The evolution throughout the secondary maximum is slow, and resembles the primary outbursts of systems like CI Cyg.

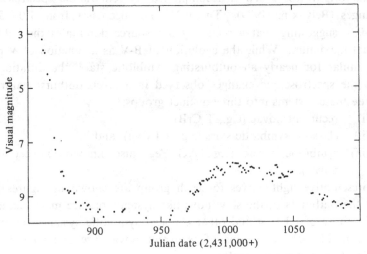

Figure 5.2 - Optical light curve for CI Cygni following its 1975
eruption (adapted from Belyakina 1979, 1984). Arrows indicate times
of minima from the ephemeris listed in Table A.4.

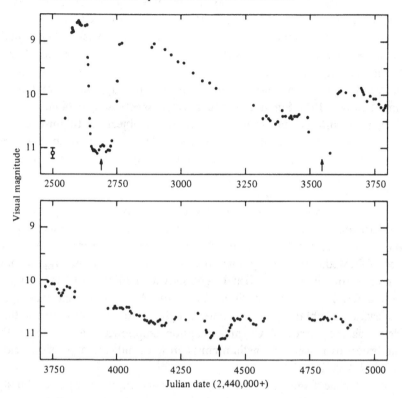

Figure 5.3 - Optical light curve for the single outburst of AG Pegasi
(Lundmark 1921; Rigollet 1947; Belyakina 1970; AAVSO). Optical
spectra were not obtained during the rise to maximum; data secured
since 1900 suggest the system has maintained a constant bolometric
luminosity.

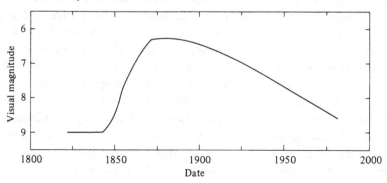

tion symbiotic star, with late-type absorption features and H I and He I emission lines. Higher ionization lines, such as [O III] and He II, are sometimes observed, but they are generally quite weak. T CrB is one of a few symbiotic stars with a well-determined binary orbit (P = 227 days), as Sanford (1949; cf. Kraft 1958; Paczynski 1965) found cyclic velocity variations in both the emission and the absorption lines. Radial velocity variations with a comparable period recently have been observed in RS Oph (Garcia 1983), and it is perhaps only a matter of time before V1017 Sgr is also a confirmed spectroscopic binary. It is during an outburst, however, that these three objects distinguish themselves from the majority of symbiotic stars. Their outbursts are sufficiently similar that only those of T CrB will be described in detail.

5.2.1   The Evolution of the Outburst

Two major T CrB extravaganzas have been reported by modern observers, and published accounts of the two outbursts suggest that they were nearly identical. While the rise to maximum has never been followed, negative sightings before the 1866 eruption suggest that the enormous increase in visual light shown in Figure 5.1 may require only a few hours (see the discussion in the Appendix). Spectroscopic observations obtained at maximum light tend to substantiate this claim, as the narrow P Cygni absorption components associated with the broad hydrogen and helium emission lines indicate expansion velocities of 4500 km s^{-1}.

The decline from maximum has been very rapid in each outburst, and approaches 0.5 mag day^{-1} during the early stages. The strong, wide lines of H I and He I fade dramatically within a few hours of light maximum, to be replaced by somewhat narrower emission lines from He II and N III (Morgan and Deutsch 1947). Intense coronal lines ([Fe X] and [Fe XIV]) appear 2-3 days later, when He II and N III have already begun to vanish. These extremely high excitation lines disappear rapidly once V declines below 7th magnitude, although they rivaled Hβ in intensity early in the outburst. Within three weeks of the initial rise, both TiO absorption bands and [Fe II] emission lines appear as weak features. These strengthen markedly as V falls to 9th magnitude, and they regain their pre-outburst intensities ≈ 50-60 days after visual maximum. Bright [O III] and [Ne III] lines also appear during this time; the small [O III] λ4363/λ5007 intensity ratio indicates a relatively low density forbidden line region (Bloch and Tcheng 1953).

The rise to a secondary maximum that began ≈ 100 days after the first outburst in 1866 was quite unexpected by observers, since previous novae had displayed only single maxima. This unusual behavior was repeated during the 1946 outburst, and was characterized by a rapid quenching of the He II and N III emission lines and the slow development of a strong blue continuous spectrum. Weak lines from H I, He I, Fe II, and [Fe II] remained visible during this rise, while the [O III] and [Ne III] lines nearly vanished (McLaughlin 1947; Bloch and Tcheng 1953). This behavior is in sharp contrast with classical novae (e.g., DQ Her), whose spectra are dominated by intense emission lines superimposed on a weak continuous spectrum during their secondary maxima (Payne-Gaposchkin 1957). Small fluctuations in the light curve are very pronounced during this maximum, and they continued to be prominent during the subsequent decline. This phase of the outburst resembles the primary maxima exhibited by CI Cyg and related systems (discussed below), and may be produced by a similar physical mechanism.

The decline from secondary maximum is a comparatively leisurely affair: nearly two months pass before V drops by 1 mag, and a few years pass before the system returns to a normal quiescent state. High ionization lines (including He II, N III, and [Fe X]) return as the blue continuum vanishes. These lines do not remain strong, however; they begin to fade once the visual magnitude drops below $V = 9$. As these lines disappear, the absorption spectrum becomes more prominent. The high ionization lines are rarely visible once the normal minimum level is reached, and the spectrum is dominated by late-type absorption features and low ionization emission lines as in Figure A.7.

5.2.2 The Cause of the Outburst

The most popular explanations for the outbursts of symbiotic stars currently involve (i) thermonuclear events in the accreted envelopes of white dwarf stars, or (ii) sudden increases in the accretion rate onto a main sequence star or a white dwarf star (Chapter 4). The 7 magnitude increase in the visual luminosity observed in T CrB and RS Oph implies $L_V \approx 65,000 \, L_\odot$ if the giant component has a visual magnitude, $M_V \approx -0.5$. This luminosity corresponds to the Eddington limit for a $1.7 \, M_\odot$ star, which is tolerably close to the Chandrasekhar limit. Either model can apparently produce the required amplitude, and may also be capable of explaining the spectroscopic development as well. Although detailed calculations have not been made, shocks accompanying either (i) the rapid ejection of matter from the surface of a white dwarf or (ii) the formation of an accretion ring surrounding a main sequence star may be sufficiently strong to ionize iron to

[Fe XIV]. At present though, the observed mass of the secondary ($1.9\ M_\odot$), the evolution of the spectrum during secondary maximum, and the lack of an intense quiescent UV source tend to support the accretion model proposed by Webbink, as described below.

5.2.2.1 The Thermonuclear Model. As discussed in Chapter 4, it is now commonly believed that a classical nova is the result of a thermonuclear runaway on the surface of an accreting white dwarf in a short period binary. While recurrence timescales of 10^4 to 10^6 yr are suggested by most investigations of classical nova outbursts (e.g., Mac-Donald 1983), it was suspected that recurrence timescales as short as 30 yr could not be produced by thermonuclear events on the surface of any cold white dwarf. Recent hydrodynamic calculations presented by Starrfield and collaborators (1984) suggest that short recurrence times are possible on very massive white dwarfs ($1.4\ M_\odot$) accreting material at moderate rates ($10^{-8}\ M_\odot\ yr^{-1}$; resulting in an accretion luminosity of $\approx 100\ L_\odot$, which has yet to be observed in a quiescent T CrB; Cassatella, et al. 1982). The material ejected by the outbursts in these calculations has velocities of $\sim 2{,}000\text{-}3{,}000$ km s^{-1} , which approaches the inferred velocity of material in T CrB. The calculated light curves are also fairly similar to the primary maxima experienced by T CrB and RS Oph, but do not show a secondary maximum.

The secondary maximum experienced by T CrB remains a problem for these models, as a thermonuclear runaway does not naturally undergo a secondary outburst. It may be possible, however, for the material ejected at primary maximum to collide with a pre-existing nebula and produce an additional radiation source (Gratton 1952 and references therein). This material could be the remnant of a previous outburst of the hot component, or matter lost by the cool giant in a stellar wind. There is some evidence for the existence of a cloud of material surrounding T CrB (Williams 1977), and the observed interval between primary and secondary maxima (100 days) implies a nebular radius of ≈ 250 AU for an ejection velocity of 4500 km s^{-1}. If this nebula moves slowly (≈ 10 km s^{-1}), the kinetic energy available in the ejecta allows one to estimate the amplitude of the secondary maximum to be $\Delta m \approx 1$ mag, which is comparable to the observed amplitude in T CrB. However, most of this energy should be detected as strong emission lines and nebular continuum radiation, rather than as an A-type continuum source with very weak emission lines, as is actually observed.

5.2.2.2 The Accretion Model. A somewhat more natural explanation for the behavior of T CrB involves an accretion-powered outburst similar to that described in Chapter 4 (Webbink 1976). This event begins with the dynamical ejection of a blob of gas by the giant component. The matter takes up an initially highly eccentric orbit around the hot secondary; the orbit is then circularized by collisions within the orbiting gas. It is this circularization process that results in the primary maximum of T CrB (and also the rapid ejection of a small fraction of the material from the newly-formed ring); energetic arguments require $5 \times 10^{-4}\, M_\odot$ to be injected into a ring of material surrounding the hot star. Once the ring is formed, viscous dissipation broadens it into a disk which slowly approaches the surface of the central star. Since $L_{\text{disk}} \propto 1/R$, the disk luminosity increases as its edge moves inward, and it reaches a maximum when this edge reaches the stellar surface. The predicted height of the secondary maximum depends severely on the surface potential of the accreting star; $V = 7.77$ is predicted for a main sequence star, and is very close to the observed height (Figure 5.1).

The time interval (Δt) between the formation of the ring and secondary maximum is a measure of the viscous timescale of the disk, τ_v ($\tau_v = R_d^2/6v$; where R_d is the radius of the disk and v is the viscosity). In a constant-viscosity time-dependent accretion disk model (Lynden-Bell and Pringle 1974), Δt is proportional to τ_v, or:

$$\Delta t \approx 0.4\, (R_d/R_\odot)^2\, (M_1 R_1)^{-1/2}\, \text{days}, \qquad (5.1)$$

where M_1 and R_1 are the mass and radius of the central star (in solar units). Webbink derived $R_d \approx 20\, R_\odot$ for his disk model, and adopting $M_1 = 1.9\, M_\odot$ and $R_1 = 1.8\, R_\odot$ results in $\Delta t \sim 90$ days. This is in excellent agreement with the observed value of ≈ 110 days (Figure 5.1).

Finally, the viscous timescale (and therefore Δt) should also measure the rate at which the disk is emptied of matter during the decline from secondary maximum. The outer radius of the disk is limited by the radius of the Roche lobe ($\approx 100\, R_\odot$), implying Δt is ~ 2000 days. One would therefore expect the system to approach minimum within a few years, and observations described by McLaughlin (1947, 1953) show that the system assumed a normal quiescent state in 1950-51. These observations are therefore consistent with the decay of a massive disk following secondary maximum.

5.3   Classical Symbiotic Stars

A typical classical symbiotic star (i.e., one fulfilling Merrill's original classification criteria) is CI Cyg (others are Z And, BF Cyg, V443 Her, RW Hya, SY Mus, AR Pav, and AX Per, as discussed in the Appendix). This system usually can be found to display an M4 absorption spectrum and prominent H I and He II emission lines. Every 855.25 days, the V magnitude drops from $V \approx 10.8$ to $V \approx 11.1$, while the color index increases from $B - V \approx 1.1$ to $B - V \approx 1.5$. The intensities of the strong optical and ultraviolet emission lines and the Balmer continuum also decrease in step with V, while the level of the late-type continuum remains roughly constant. Belyakina (1979) suggested these phenomena were the result of an eclipse of the hot component in CI Cyg by its cool giant companion. This interpretation is now well-established, and the 855.25 day period is accepted as the binary period of CI Cyg.

5.3.1   The Evolution of the Outburst

The spectroscopic changes triggered by CI Cyg's occasional 2-3 mag outbursts are quite interesting, and are remarkably similar from one event to the next (Mammano, Rosino, and Yildizdogdu 1975; Belyakina 1979; Yildizdogdu 1981). The rise to visual maximum begins with an increase in the intensity of the blue continuum coupled with a decrease in the *relative* intensity of the strong emission lines (cf. Belyakina 1979 and references therein). The high ionization lines (and the TiO absorption bands) generally vanish sometime during this rise, but it is not obvious if this is a result of (i) the overall increase in the brightness of the blue continuum, or (ii) an actual decrease in the absolute line intensities. From photographic spectra available in the literature (for other systems as well as CI Cyg), it seems that both (i) and (ii) contribute to the observed behavior of lines such as He II λ4686 (cf. Aller 1954). Detailed spectrophotometry with digital detectors is needed to confirm this opinion. However, it is apparent that the absolute intensity of Hβ does increase in outburst: observations of CI Cyg suggest the Hβ equivalent width decreases by a factor of 2-3 during the tenfold increase in the blue continuum (implying an increase in the total Hβ flux of a factor of ≈ 3; cf. Belyakina 1975, 1979; Kenyon 1983).

Once CI Cyg reaches visual maximum, its optical spectrum is dominated by numerous strong absorption lines (e.g., Na I, Ca II, Ti II) and a few prominent emission lines (Hα and Hβ), as in the accompanying spectrum of the related system CH Cyg (Figure 5.4). The intensities of Ca II H and K and the G band at λ4305 are consistent with those expected from an F5 star (as is the color index, $B - V \approx$ 0.6), while the additional absorption lines from Ba II and La II suggest an F supergiant rather than an F giant. Recent outbursts of CI Cyg have been interrupted by an eclipse (Figure 5.2), during which time the F-type spectrum is replaced by an M absorption spectrum. The depths of these eclipses demonstrate that the M star is essentially unchanged by the dramatic evolution of its binary companion (this is also confirmed by infrared photometry obtained throughout the eruption cycle; Swings and Allen 1972; Szkody 1977; Kenyon and Gallagher 1983). As the hot star emerges from eclipse, its F-type absorption spectrum returns to dominate the optical spectrum (Belyakina 1979).

The 2-3 mag decline from outburst in systems such as CI Cyg parallels the 2 mag decline from secondary maximum in T CrB, and is substantially different from the evolution of the symbiotic novae described below. All the available evidence points to a gradual diminishing of the "blue" continuum, which is accompanied by the return of the pre-outburst spectrum (the late-type absorption features and the high ionization emission lines). *IUE* spectrophotometry suggests that the inten-

Figure 5.4 - Optical spectrum of CH Cygni, which is in the midst of a major outburst. Strong TiO bands visible for λ > 5500 are consistent with a late M spectral type, while H I and Ca II absorption in the blue are indicative of the outbursting component. Optical spectra of CI Cygni near visual maximum show a much more dominant F component with little evidence for TiO absorption features as in CH Cygni.

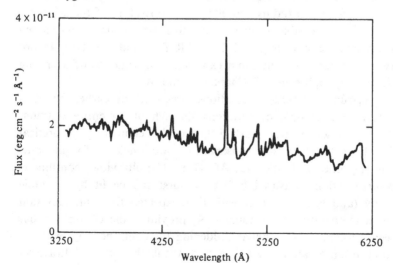

sities of the emission lines actually *increase* during the late decline of CI Cyg and AX Per, but simultaneous optical/ultraviolet data are not available for a complete outburst (cf. Stencel, et al. 1982). Thus, it is not possible to say for certain if the bolometric luminosity of the hot source *decreases* following maximum visual light.

5.3.2   *The Cause of the Outburst*

Thermonuclear runaways and accretion events have been proposed as outburst mechanisms for the classical symbiotics, and both appear to be capable of reproducing qualitatively the amplitude and the spectroscopic evolution observed in a given system. Bath (1977) first noted that the behavior of Z And was consistent with a low mass main sequence star accreting matter at rates approaching the Eddington limit, with outbursts resulting from small increases in the accretion rate. Bath and Pringle (1982) later successfully reproduced the visual light curve of CI Cyg through three outbursts using a time-dependent disk evolution code. They found that CI Cyg's outbursts could only be understood if the accreting star is a solar-type main sequence star (which has been subsequently confirmed by the UV continuum diagnostics discussed in Chapters 2 and 3).

A different view is held by Paczynski and Rudak (1980), who suggested that these systems (type I symbiotics in their terminology) could be explained in terms of the model originally proposed by Tutukov and Yungel'son (1976). They concluded that steady hydrogen burning could provide the background luminosity of a classical symbiotic star, while variations in the accretion rate were responsible for occasional large excursions in the radius and effective temperature of the accreting white dwarf. A possible problem with this hypothesis is that many classical symbiotic stars (e.g., CI Cyg, AR Pav, and AX Per) do not display the intense UV continuum expected of a white dwarf accreting material at very high rates (cf. Chapters 2 and 3).

The spectrum synthesis calculations presented in earlier chapters imply that spectroscopic observations throughout an entire outburst would allow one to distinguish between the competing theoretical models. Unfortunately, such observations are available for the outburst of only one symbiotic star, AG Dra. The ultraviolet continuous energy distribution measured before outburst can be fit by a simple blackbody (and is inconsistent with that expected from an accretion disk) at a temperature of $\approx 130,000$ K, providing the continuum has not been affected by interstellar reddening (as confirmed by independent optical observations; Kenyon and Webbink 1984). Ultraviolet observations secured during the decline from this maximum show that the UV continuum of AG Dra was ≈ 2.5 mag brighter with little change in overall shape (Viotti, et al. 1982, 1983). Most of the high ionization lines increased in intensity as well, although their *relative* intensities showed little evidence for a change. When this behavior is combined with the evolution of the system in the optical (which was somewhat different from that described above for CI Cyg as noted in the Appendix), it suggests that the hot component increased dramatically in luminosity while undergoing only a modest increase in its effective temperature. This type of evolution is consistent with that expected from an accreting white dwarf undergoing a weak shell flash, as described in Chapter 4 (cf. Figure 4.3).

Although detailed spectroscopic observations of a representative sample of outbursts are not yet available, some progress can be made from an examination of the rate of decline from visual maximum and the time interval between major eruptions. The decay time for a thermonuclear outburst is independent of the binary period (or, equivalently, the size of the accretion disk), depending, rather, on the mass of the accreted envelope, M_{env}, and the luminosity of the burning shell, L_{on} (p. 77). Given reasonable values for the envelope mass and the luminosity $(M_{env} \approx 10^{-5}\, M_\odot,\; L_{on} \approx 5 \times 10^3\, L_\odot),\; \Delta t_{on} \approx 150$ yr. Shorter timescales are possible only if M_{env} is reduced sharply; $\Delta t_{on} \approx 20$ yr requires $M_{env} \approx 10^{-7}\, M_\odot$, implying a very high mass, luminous white dwarf (cf. Table 4.1).

The approximate timescales, τ_d, required for various symbiotic stars to decline 1 mag from maximum visual light, as listed in Table 5.1, are not characteristic of a typical thermonuclear runaway. The values are consistent with weak shell flashes and accretion-driven outbursts, as noted below. According to the theoretical results, hydrogen shell flashes with recurrence timescales < 10 yr are possible only if very massive white dwarfs $(M > 1.2\, M_\odot)$ accrete matter at very high rates $(\dot{M} \sim 10^{-7}\, M_\odot\, \mathrm{yr}^{-1})$. Such systems spend only a small fraction of time $(\leq 25\%)$ in a high state, which is consistent with the recent eruptive activity observed in AG Dra.

The timescale for the decline in an accretion-powered outburst is given in equation (5.1); for a $1\, M_\odot$ main sequence star in a binary with $P \approx 2 - 4$ yr, $\Delta t \approx$ a few years. The values listed in Table 5.1 are all consistent with this possibility, and ultraviolet spectroscopic observations have identified some of these systems as accreting main sequence stars (Chapter 3). The recurrence timescales for accretion-driven outbursts are not as limited as are thermonuclear runaways, but do depend on the nature of the mass-losing star and the eruption mechanism. Systems with an unstable cool component must wait for this star to recover ($\sim$ a few years; the thermal timescale), while objects containing an unstable disk must allow time for matter to accumulate (also a few years; the viscous timescale). If mass loss from the cool giant is enhanced by periastron passage, outbursts might occur at particular orbital phases. This seems to be a reasonable interpretation for the historical behavior of CI Cyg and AX Per, as discussed in the Appendix.

5.4   The Symbiotic Novae

While multiple outbursts are quite common among symbiotic stars, a few systems have undergone a single, protracted outburst. Seven of these *very slow* or *symbiotic novae* were identified by Allen (1980a), and their basic properties are listed in Table 5.2. It may soon be possible to add PU Vul to this exclusive company, as its recent development parallels the early outburst phases experienced by RT Ser and RR Tel.

The evolution of these objects is explained most naturally by thermonuclear runaways on the surfaces of accreting white dwarf stars,

Table 5.1. *Outburst decline timescale for classical symbiotic stars*

Symbiotic star	τ_d (days)
Z And	200
BF Cyg	1000
CH Cyg	400
CI Cyg	500
AG Dra	1000
AR Pav	500
AX Per	500

Table 5.2. *Properties of symbiotic novae*

Symbiotic nova	Year	Δt_{on} (yr)	Spectral type at maximum
AG Peg	1850:	> 130	A or later
RT Ser	1909	~ 40	F supergiant
AS 239	1940::	10?	?
RR Tel	1944	> 40	F supergiant
V1329 Cyg	1956	> 30	planetary nebula
V1016 Cyg	1964	> 20	planetary nebula
HM Sge	1975	> 10	planetary nebula

although various single-star hypotheses have been proposed (Chapter 4). The very slow decay of M_V observed in each of these systems is comparable to the nuclear timescale, τ_{nuc}, for material of solar composition burning on the surface of a white dwarf star as derived in Chapter 4. Also, estimated values of the mass of ionized material and the luminosity of the hot component are consistent with those expected from a slow classical nova outburst (e.g., $\sim 10^{-4} M_{\odot}$ and $1,000 - 10,000 L_{\odot}$, respectively). It seems likely that both weak shell flashes (V1016 Cyg, V1329 Cyg, and HM Sge) and strong shell flashes (AG Peg, RT Ser, and RR Tel) can occur in a symbiotic nova, and that this is relatively independent of the nature of the cool component (V1016 Cyg, HM Sge, and RR Tel contain Miras, while V1329 Cyg, AG Peg, and RT Ser appear to contain normal giants) and the binary period (P = 827 days for AG Peg; P = 950 days for V1329 Cyg).

All known symbiotic novae, with the exception of RR Tel, have been discovered in outburst, and thus detailed pre-outburst observations are not available in most instances. The picture emerging from a few scattered records is that pre-outburst symbiotic novae had spectra of M giants, with perhaps some indication of strong emission lines. These systems therefore may have appeared as symbiotic stars *before* their eruptions - this is almost certainly true for V1016 Cyg as discussed in the Appendix. Mira-like variability was noted in RR Tel before its 1944 eruption, and V1016 Cyg was classified as a long-period variable by Nassau and Cameron (1954); similar behavior has not been reported in other symbiotic novae preceding their increases in visual brightness. Besides V1329 Cyg (950 days) and AG Peg (827 days), binary periods are also unavailable for symbiotic novae, although infrared observations have demonstrated that a number of these objects (V1016 Cyg, HM Sge, and RR Tel) contain Mira variables having pulsational periods of 350-500 days. The binary periods of these latter systems are probably fairly long (perhaps decades) if the Mira is to be contained within its tidal lobe, and upper limits on radial velocity variations in V1016 Cyg and HM Sge (Wallerstein, et al. 1984) are consistent with this possibility.

5.4.1 The Evolution of the Outburst

5.4.1.1 Optical Observations. The evolution of a symbiotic nova parallels that of a classical nova, but on an extended timescale (Figure 5.3). The rise to maximum is somewhat rapid, as the color index decreases and the blue continuum strengthens over a period of weeks to months. There is some evidence that B-V reaches a minimum value *before* visual maximum, and then increases slightly. Both RT Ser and RR Tel were found to have B-V $\sim$ 0.2 well before visual maximum; B-V subsequently increased to $\sim$0.6 once visual maximum had been achieved. This phenomenon appears not to have occurred in V1016 Cyg and HM Sge, as these objects have always been very blue.

The spectroscopic appearance of a symbiotic nova at maximum visual light varies dramatically from system to system, but can be generalized into two broad categories:

1) systems resembling A-F supergiants (AG Peg, RT Ser, and RR Tel), and

2) systems resembling planetary nebulae (V1016 Cyg, V1329 Cyg, and HM Sge).

The spectra of AG Peg, RT Ser, and RR Tel closely resemble spectra of classical novae in outburst, with strong H I and Ca II absorption lines and many weak lines from various singly ionized metals. The development of the emission line spectrum is very illuminating: the hydrogen lines appear slowly, beginning with Hα and evolving up the Balmer series through Hβ, Hγ, and higher members (Thackeray 1977; Merrill 1959; Joy 1931; and references therein). The color temperature of the continuum increases in pace with the emission lines; once the

Figure 5.5 - Optical/ultraviolet spectrum of V1016 Cygni. Spectra of this system (also HM Sge and RR Tel) closely resemble those of planetary nebulae, as the strong emission lines dominate a weak continuum. However, the strong CO and H$_2$O absorption bands commonly observed in Mira variables are present on near-infrared spectra.

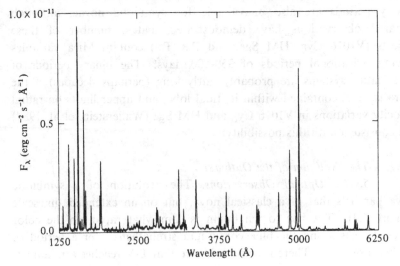

Balmer lines are prominent emission features, He I lines appear in absorption, and as He I develop emission components, He II and Si IV absorption lines emerge.

The early stages of the decline of AG Peg and related systems are similar to the appearance of V1016 Cyg-type objects at maximum, and closely resemble spectra of planetary nebulae (Figure 5.5; Andrillat, Ciatti, and Swings 1982). However, the nebulae of symbiotic novae are considerably more dense ($n_e \sim 10^6 \, \mathrm{cm}^{-3}$) and somewhat hotter ($T_e > 10^4$ K) than a typical planetary nebula, and there is overwhelming evidence for H_2O, CO, TiO, and VO absorption bands on near-infrared and optical spectra of each system (Allen 1980a; Andrillat 1982; also the Appendix). The H I and [O III] emission lines are extremely intense features during this time, while He I and He II are also very prominent. P Cygni absorption components sometimes are associated with the stronger H I and He I lines, indicating expansion velocities of a few hundred km s^{-1}. Multiple absorption features have been reported in AG Peg (Merrill 1951) and, more recently, in PU Vul (Iijima and Ortolani 1984), and suggest the ejection of several shells of material. Similar behavior is observed in most classical novae (which also exhibit multiple emission line systems; Payne-Gaposchkin 1957), strengthening the association between these objects and the symbiotic novae.

Multiple components in various emission lines have been noted primarily in V1016 Cyg, HM Sge, and RR Tel (Figure 5.6). The majority of these lines are double-peaked with velocity separations of 50-70 km s^{-1}, although more complicated profiles may be present in the higher ionization lines such as [A V] and [Fe VII]. The profiles of the low ionization lines (such as [N II] and [S III]) resemble line profiles of

Figure 5.6 - Profiles for various optical emission lines ([O I] λ6300, [S III] λ6312, and the λ6830 band) in V1016 Cyg. The velocity separation of the twin peaks is ~ 50-70 km s^{-1}, which is in overall agreement with the velocity deduced from twin emission peaks observed in the radio (Stauffer 1984).

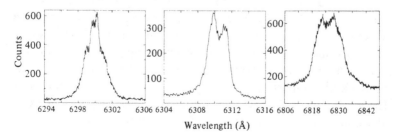

planetary nebula, but at somewhat higher velocity. However, the trend of increasing velocity separation as a function of ionization potential is *opposite* to the trend experienced by planetary nebulae, as noted by Thackeray (1977).

Models for the complex line profiles in symbiotic novae have been suggested by Solf (1983), Stauffer (1984), and Wallerstein, et al. (1984). Since the line profiles of the high ionization lines are very similar to those observed in classical novae, Hutchings' (1972) model (in which the line emission is assumed to be emitted from two polar blobs and an equatorial ring) may also be applicable. Both Solf and Stauffer propose that the lines are formed in a biconical region perpendicular to the plane of the binary orbit, and are excited by radiation from the hot star. The alternative discussed by Wallerstein, et al. places the line-emitting region in a curved sheath of material that partially surrounds the cool component, which is produced by the interaction of low velocity matter ejected by the cool star with high velocity material lost during the eruption of the hot star. While these models can generally explain the shape of the line profiles and the overall intensity of X-ray and radio emission, they may be severely tested when attempting to explain high resolution radio maps, as discussed below.

It now appears that all symbiotic novae pass through a stage during which they resemble the more luminous Wolf-Rayet stars. AG Peg could currently be classified as a Wolf-Rayet star (WN type), of not for its conspicuous TiO absorption bands (Boyarchuk 1967). A profile of the He II λ1640 band in AG Peg appears in Figure 5.7. Broad emission lines of He II and N III have also been observed in optical spectra of V1016 Cyg, V1329 Cyg, HM Sge, and RR Tel (Andrillat and Houziaux 1976; FitzGerald and Pilavaki 1974; Thackeray 1977; Wallerstein 1978), but the W-R phases of these systems ended before *IUE* spectra could be secured. It is not clear if spectra of RT Ser and AS 239 ever contained such wide bands, but the coverage of these objects during their declines is so incomplete that a short-lived Wolf-Rayet phase, as in HM Sge, might easily have been missed. It is interesting that the rapid rate of mass loss currently observed in AG Peg ($\sim 10^{-6} M_\odot$ yr^{-1}), and inferred to have occurred in the other systems, may have some bearing on wind theories for hot stars. The fact that the observed mass loss rates among symbiotic novae are a factor of ten less than those of W-R stars which are an order of magnitude more luminous lends support to wind models based on radiative processes.

The final stage of the decline is heralded by the disappearance of the W-R features and the development of strong high ionization lines from

[Ne V] and [Fe VII]. The He II λ4686/Hβ intensity ratio increases dramatically, suggesting that a rapid increase in the effective temperature of the outbursting star results in the quenching of the W-R lines. In fact, Gallagher, et al. (1979) concluded that the historical evolution of AG Peg was consistent with a slow increase in effective temperature at roughly constant *bolometric luminosity*. Recent observations of HM Sge and V1016 Cyg, as well as Thackeray's detailed observations of the decline of RR Tel, lend further support to the statement that *all* symbiotic novae evolve towards higher effective temperatures at constant luminosity following visual maximum.

Besides [Ne V] and [Fe VII], two other, as yet unidentified, emission bands at λ6830 and λ7088 appear during the late decline of a symbiotic nova (Allen 1980b). These lines tend to be very broad ($v \sim 1000$-2000 km s^{-1}) and are composed of multiple components as in classical novae. Various identifications have been proposed for these bands, including P V, Fe VI, Fe VII, Fe VIII, Ne VIII, Na IX, and Ca X; permitted transitions of Fe VII seem most promising, but the highly excited levels of Fe VII are not known with sufficient accuracy to associate unambiguously the two symbiotic bands with a given pair of transitions (Allen 1980a).

Of all the symbiotic novae, only RT Ser has returned to what might be described as quiescence, although AG Peg is slowly approaching a similar state. The former system appears to be a normal symbiotic

Figure 5.7 - Line profile for He II λ1640 in AG Pegasi. The width of this line indicates a wind velocity of ~ 800 km s^{-1} (cf. Keyes and Plavec 1980a,b).

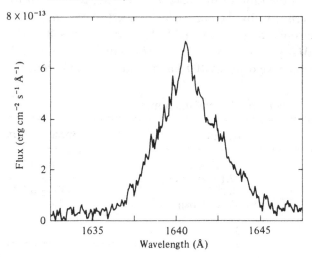

star, with strong TiO absorption bands and prominent H I and He II emission lines. High ionization lines ([Fe VII], the λ6830 feature) are also very intense, and thus the hot component in this system is still fairly luminous. It remains to be seen if the other symbiotic novae will return to a quiescent state and maintain their symbiotic appearance.

5.4.1.2 Radio Observations. Radio data have been quite helpful in understanding the outbursts of symbiotic novae, as all but AS 239 have been detected. The spectral data are consistent with a power law, $S_v \propto v^\alpha$; observed values of α are listed in Table 5.3. Only AG Peg and RR Tel have α sufficiently close to 0.6 to allow derivation of the mass loss rate from radio data alone; the observed flux density yields $\dot{M} \sim 10^{-6}$ (d/0.6 kpc)$^{2/3}$ $M_\odot$ yr^{-1} for AG Peg and $\sim 4 \times 10^{-6}$ (d/1 kpc)$^{2/3}$ $M_\odot$ yr^{-1} for RR Tel. A nearly complete set of observations has been obtained for HM Sge, the most recent confirmed symbiotic nova. Data for other systems are necessarily incomplete, as these objects began their eruptions long before their radio emission could be detected.

The radio evolution of HM Sge is summarized in Figure 5.8, and shows that the source has slowly increased in intensity ($\sim$a factor of 2 since 1977) while maintaining a nearly constant spectral index ($\alpha \sim 1$; Purton, Kwok, and Feldman 1983). The large spectral index cannot be understood straightforwardly with spherically symmetric mass loss as in AG Peg, but can be explained by Taylor and Seaquist's (1984) radio model (cf. Chapter 3), which parameterizes the radio emission in terms of the binary separation, A, the photon luminosity of the ionization source, N_{ph}, and the ratio of the mass loss rate, $\dot{M}$, to the velocity of the wind, v. A crude fit by the author to the spectral data and various optical spectrophotometry of HM Sge yields $A \sim 40$ AU, $\dot{M}/v \sim 1.5 \times 10^{-7}$ $M_\odot$ yr^{-1} per km s^{-1}, and $N_{ph} \sim 10^{47-48}$ photon s^{-1}. Adopting an expansion velocity of 100 km s^{-1} for the wind from the

Table 5.3. *Radio observations of symbiotic novae*

Symbiotic nova	$I_{6\,cm}$ (mJy)	Spectral index
V1016 Cyg	46	0.8 ± 0.1
V1329 Cyg	2.1 (variable)	-
AG Peg	8.2	0.6 ± 0.1
HM Sge	33 (variable)	1.0 ± 0.1
RT Ser	1.3	0.9 ± 0.1
RR Tel	30	0.6 ± 0.1

cool star implies a mass loss rate of $\sim 10^{-5}\ M_\odot\ \mathrm{yr}^{-1}$, which is not unreasonable for a Mira-like star. The derived binary separation of 40 AU results in a orbital period of 100-200 yr, so it is not surprising that significant radial velocity variations have yet to be observed (the expected radial velocity amplitude is $\approx 20\ \mathrm{km\ s}^{-1}$ assuming equal mass components).

The factor of two increase in the total radio flux of HM Sge has been very dramatic, and resembles the optically thick expansion phase of classical novae. Hjellming, et al. (1979) suggested that the radio evolution of a typical classical nova could be understood in terms of an ejection of a shell of matter during the optical outburst, and their derived velocities tend to agree rather well with those inferred from optical spectra. The data in Figure 5.8 can be used to estimate an expansion velocity of $\sim 100\ \mathrm{km\ s}^{-1}$, which agrees very well with other estimates (see below). This lends further support to the idea that symbiotic novae are very slow versions of classical novae.

High resolution *VLA* maps have been secured for two systems, V1016 Cyg and HM Sge. The 1.3 cm map of HM Sge (Figure 5.9; Kwok, Bignell, and Purton 1984) shows a great deal of structure, and

Figure 5.8 - Evolution of the radio spectrum of HM Sge. The data have been obtained with various instruments, and demonstrate a slow rise in the overall flux density at roughly constant spectral index (Purton, Kwok, and Feldman 1983). Filled circles indicate data obtained in 1977, while observations acquired in 1978 and 1980 are denoted by open circles and squares, respectively.

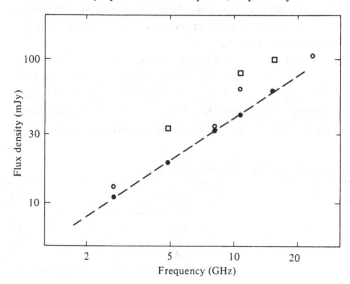

similar structure is present in maps of V1016 Cyg (Hjellming and Bignell 1982). Both systems appear to contain two condensations of gas
embedded in an elliptical shell. The two condensations are separated
by 0″.08 in HM Sge or $\sim$150 AU at an adopted distance of 1.8 kpc
(Kenyon 1983). Assuming these features were ejected in 1975 implies a
velocity of 75 km s^{-1} providing the binary inclination is 90°, nearly
identical to the expansion velocity derived from high resolution optical
data ($v \sim$ 50 km s^{-1}). These observations are therefore consistent with
an explosive event in 1975 that led to the optical brightening.

5.4.1.3 X-ray Observations. Many symbiotics were observed with the
Einstein satellite, and three symbiotic novae (V1016 Cyg, HM Sge and
RR Tel) were among the half-dozen positive detections. The actual
fluxes of these objects have been the subject of some discussion; fluxes
based on recent reprocessing of the data are listed in Table 5.4. Accurate X-ray luminosities for these objects are difficult to determine, since
each contains a Mira variable embedded in an extensive dust cloud
(which complicates distance estimates). Assuming the cool components
in each system are somewhat similar, a color index, X-K, may be used
to represent the ratio of the X-ray flux to the average K magnitude
($X = -2.5 \log F_x$-21.1). This is listed in the final column of Table 5.4.
As first noted by Allen (1981), the X-ray flux is a decreasing function
of the time interval since outburst, and X-K declines linearly with time.
Assuming the eruptions were fairly similar, this result suggests the X-
ray flux will decline by a factor of 2 in $\sim$10 yr. Indeed, the observed
flux of HM Sge declined by 30% over 1979-81 - continued monitoring
of this system as it declines should provide more information concerning the source of X-ray emission.

The X-ray spectra of these objects are consistent with thermal
bremsstrahlung at $T \sim 10^7$ K, and cannot be blackbody emission unless
the source is a neutron star (which is ruled out by a lack of hard X-
rays). The preferred mechanism for X-ray production is therefore a
shock, presumably the result of a collision between high velocity

Table 5.4. *X-ray observations of symbiotic novae*

Symbiotic nova	$F_x(10^{-12}$ erg cm^{-2} s$^{-1})$	X-K
V1016 Cyg	0.82	3.7
HM Sge	5.0	3.2
RR Tel	0.67	5.1

matter ejected by the hot star with low velocity material lost by its cool giant companion (e.g., Kwok and Leahy 1984; Willson, et al. 1984). The temperature of the emitting gas requires a shock velocity ≥ 700 km s^{-1} (Raymond 1979), implying a mass flow rate through the shocked region of $\leq 10^{-8}$ $M_\odot$ yr^{-1} to supply the observed X-ray luminosity (~ 0.5 $L_\odot$ for $d = 1.8$ kpc). This mass loss rate is comparable to that expected from a declining classical nova (cf. Chapter 4), and is well below that determined to be lost by the hot component in AG Peg (which begs the question why this latter system, having a wind velocity of ~ 1000 km s^{-1}, is not also an X-ray source).

5.5 Summary

Since complete spectroscopic coverage of symbiotic outburst cycles is available only for AG Dra and a few symbiotic novae, it has been difficult to assign the eruptions of any given system to a particular outburst mechanism with any degree of certainty. The published

Figure 5.9 - Radio map of HM Sge at 1.3 cm obtained with the VLA. The two peaks of emission inside the elliptical ring are separated by $\sim 0''.08$ (Kwok, Bignell, and Purton 1984).

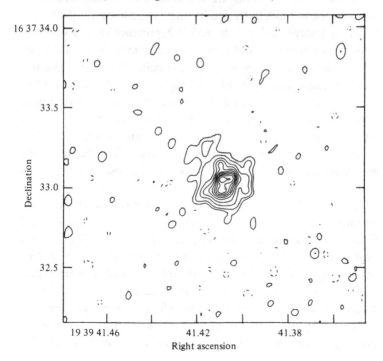

data do suggest some tendencies which serve as a useful framework until more detailed data have been secured.

1) Symbiotic nova eruptions are due to thermonuclear runaways on the surface of an accreting white dwarf as in classical novae. This is envisioned as a four-step process.

   A. The erupting white dwarf ejects a shell of matter ($v \sim 100$ km s^{-1}) at maximum visual light which interacts with material lost by the cool giant companion ($v \sim 10$ km s^{-1}).

   B. The photosphere of the erupting star contracts and the effective temperature increases to 30,000-40,000 K, at which point a fast wind (~ 1000 km s^{-1}) interacts with local material to produce X-ray emission.

   C. The photosphere contracts further, ending the W-R phase (and the fast wind) - at this time, the material ejected during the W-R phase has swept up additional matter from the wind of the cool star, and has therefore decreased in velocity.

   D. The eruption ends, and the system returns to a normal quiescent state.

2) Recurrent nova eruptions are the result of accretion events, as thermonuclear models have difficulties explaining the lack of a bright quiescent UV source and the presence of a secondary optical maximum in outburst. These events begin with the ejection of a blob of material by the cool giant star. The initially eccentric orbits of this material are circularized by collisional processes, resulting in a bright primary maximum. Viscous dissipation broadens the ring of matter into a disk, which leads to a secondary maximum when the disk encounters the central main sequence star. The decline from secondary maximum is rather slow, as the disk is emptied of material.

3) Classical symbiotic stars have outbursts resulting from either accretion events onto main sequence stars or weak shell flashes, depending on the system. The weak shell flashes evolve in a fashion similar to that described above for symbiotic novae, although the outbursts are significantly more rapid. The accretion events are identical in nature to those experienced by recurrent novae, although a steady-state disk is already in place when the outburst begins. Thus, there is no primary maximum as in T CrB. Rather, the eruption closely

resembles the phase of secondary maximum in recurrent novae, as material ejected by the giant reaches the accreting star.

The next few decades will certainly see a number of new eruptions, which must necessarily be observed in many wavelength domains. The major questions these observations must address are:

1) the shape of the continuous energy distribution as a function of time;

2) the evolution of the emission line intensities and profiles as a function of time; and

3) the variations in (a) the integrated X-ray and radio flux densities and (b) the shape of the X-ray and radio spectra produced by an outburst.

Hopefully, the various types of symbiotic stars will oblige us by erupting when a wide range of satellite observatories is ready for them!!

6

The Formation and Evolution of Symbiotic Stars

6.1   Introduction

In the preceding Chapters, we have seen that symbiotic stars are neither an entirely homogeneous group of variables, nor a random collection of stellar misfits. Although they possess an exciting diversity of hot white dwarfs, hot subdwarfs, and accreting main sequence stars, the unifying factor among symbiotic systems is the evolved red giant star losing mass via stellar wind or tidal overflow. This giant, whether it be a Mira, a semi-regular, or a non-variable red star, severely limits the length of the symbiotic relationship, as the evolutionary timescale for this star is $< 10^6$ yr (Becker 1979; Iben 1967). This is a very small fraction of the lifetime of any binary system, and it is therefore important to understand (i) how a given binary manages to become symbiotic, and, once it finds itself in this unusual state of affairs, (ii) how the binary extricates itself from a period of symbiosis.

The formation of binary stars is a process that is not understood in detail, although considerable progress has been made in recent years. The currently most popular scenario for the formation of binary stars begins with the collapse of a rotating interstellar cloud threaded by a magnetic field. The loss of angular momentum by magnetic braking allows the cloud to contract, and instabilities encountered during the early phases of the collapse result in the formation of dense fragments (cf. Mouschovias 1978). Although detailed calculations for the subsequent evolution of the cloud are not yet possible, it is believed that the rapidly rotating fragments undergo fission into smaller components which then contract into normal stars. This mechanism predicts a single maximum in the distribution of binaries as a function of binary period; observational results by Abt and Levy (1976, 1978) tend to confirm such a distribution for binaries with $P < 100$ yr.

Previous attempts to understand the formation and evolution of symbiotic systems have concentrated on models involving a red giant star and its white dwarf companion (e.g., Rudak 1982; Tutukov and Yungel'son 1982), although Plavec, Ulrich, and Polidan (1973) and Webbink (1979b) considered main sequence companions in evolutionary scenarios for T CrB and AX Mon. It is clear from earlier chapters that multiple formation processes may be required for these objects, as both main sequence and white dwarf stars are plausible companions to the red giant. In the following discussion, symbiotics are assumed to consist of a red giant star and a hot companion, which might be an evolved compact star (e.g., a neutron star, white dwarf, or hot subdwarf) or an unevolved solar-type main sequence star. Massive main sequence star companions to the giant are unlikely for a variety of reasons, including (i) the hot components in symbiotic binaries are not as luminous as massive main sequence stars, (ii) massive stars do not remain on the main sequence for a long period of time (necessary if the primary is already a giant), and (iii) the space distribution of symbiotic stars favors a total binary mass $\leq 2 - 3 \ M_\odot$. A final assumption is that symbiotic stars can be explained within the context of normal stellar evolution, and are not the result of a peculiar evolutionary event. This, of course, can be evaluated *a posteriori*.

6.2   Formation and Evolution of Symbiotic Systems

6.2.1   Basic Principles

The evolution of a given binary star depends on the masses of the two components (M_1, the primary, and M_2, the secondary; $M_1 > M_2$) and the binary period, P. If the binary period is sufficiently long ($P \geq$ 5-10 days), the two components can initially evolve as if they were single stars, with main sequence lifetimes given by (Iben 1967):

$$t_{ms} \approx 10^{10} \ (M/M_\odot)^{-3.3} \ \text{yr.} \tag{6.1}$$

The main sequence lifetime is such a strong function of the stellar mass that stars with $M < 1 \ M_\odot$ are unlikely sources of red giants. If the masses of symbiotic primaries are restricted to $M_1 \sim 1 - 3 \ M_\odot$, the number of binaries that might become symbiotic systems can be estimated from the local birthrate of single white dwarfs, ν_{wd}. If f_{bin} represents the fraction of all stars in binaries, the binary formation rate is:

$$\beta_{bin} \sim \pi \ R_o^2 \ h \ [f_{bin}/(1 - f_{bin})] \ \nu_{wd} \ \text{yr}^{-1}, \tag{6.2}$$

where R_0 is the radius of the galaxy and h is the thickness of the disk. Adopting $R_0 = 9$ kpc, $h = 500$ pc, and $f_{bin} \sim 0.5$ (Popova, Tutukov, and Yungel'son 1982), the observed $v_{wd} \sim 10^{-12}$ pc^{-3} yr^{-1} (Weidemann 1979) implies $\beta_{bin} \sim 0.14$ yr^{-1}.

The binary period, P, places other restrictions on systems that seek to become symbiotic, as shown in Figure 6.1 (Webbink 1979b). Primaries in systems with $P \leq 50\text{-}100$ days encounter their Roche lobes near the base of the giant branch; these stars cannot become bright red giants, and must be content to spend a few years in an Algol binary. Binaries with $P \geq 50\text{-}100$ days can avoid becoming Algols, although the primary still must fill its Roche lobe somewhere on the giant branch if $P < 5,000\text{-}10,000$ days. Various calculations suggest that the mass loss

Figure 6.1 - Limiting binary periods for the onset of tidal instability at various stages of evolution of a low mass star (Webbink 1979b). The curves indicate the critical periods during which a star of a given mass will encounter its Roche lobe at the base of the red giant branch (dashed curve) and the tip of the red giant branch (dot-dashed curve). Very evolved red giants will eject a planetary nebula or undergo carbon ignition in their cores when they reach periods plotted in the solid line.

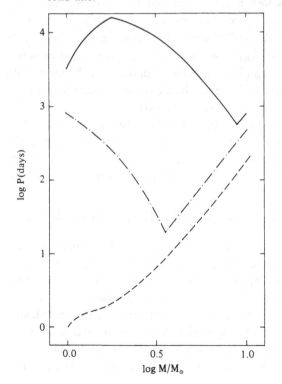

rate of an evolved lobe-filling giant exceeds $10^{-6} \, M_\odot \, \mathrm{yr}^{-1}$ (Webbink 1979b), and, thus binaries with $P \approx 100\text{-}10,000$ days eventually are identified as symbiotic stars. It is expected that many of these systems evolve into common envelope binaries, during which P is drastically reduced (see below).

If a binary with $P < 10,000$ days can survive the common envelope phase as a long period binary, it can enter another symbiotic phase once the secondary reaches red giant dimensions. This system will be "activated" once the newly-formed white dwarf can accrete sufficient material from the cool giant's stellar wind ($\sim 10^{-9} \, M_\odot \, \mathrm{yr}^{-1}$), and will then be identified as a symbiotic star (Chapter 2). Eventually, the red giant secondary encounters its tidal lobe, and loses mass at a rate of $\sim 10^{-6} \, M_\odot \, \mathrm{yr}^{-1}$. The white dwarf then expands to red giant dimensions, which leads to the formation of a contact binary and perhaps a second phase of common envelope evolution.

For $P > 5,000\text{-}10,000$ days, the primary never fills its tidal lobe, and can lose mass only via a stellar wind. A small fraction ($\sim 1\%$) of this material can be captured by the companion star, and may be sufficient to power the hot component of a symbiotic system. Current estimates suggest red giant stars lose mass at rates of 10^{-5} to $10^{-7} \, M_\odot \, \mathrm{yr}^{-1}$ (Zuckerman 1980), so the hot component can potentially accrete 10^{-7} to $10^{-9} \, M_\odot \, \mathrm{yr}^{-1}$ from the stellar wind of its companion. Since rates in excess of $10^{-5} \, M_\odot \, \mathrm{yr}^{-1}$ are needed if an *accreting main sequence star* is to produce a symbiotic system (Chapter 2), binaries with $P > 5,000\text{-}10,000$ days can only be symbiotic if the secondary is a *compact, degenerate star* (Kwok 1981; Willson 1981).

Mass loss via a stellar wind or tidal overflow therefore appears inevitable once the primary or secondary reaches the giant branch, and the loss of mass and angular momentum can have significant effects on the binary orbit. The response of the binary period to mass loss can be written as:

$$\dot{P}/P = -3(1-2\mu)/(1-\mu) + \alpha\mu[(2+\mu)/(1-\mu) - 3\beta]\dot{M}_1/M_1 \quad (6.3)$$

where $\mu = M_1/(M_1 + M_2)$, $\alpha = \dot{M}/\dot{M}_1$ (representing the amount of mass lost by the binary system), and $\beta = \mathrm{d}\ln J/\mathrm{d}\ln M$ (representing the amount of angular momentum lost by the binary system). For mass lost solely in a stellar wind, $\dot{P}/P > 0$, providing the wind carries off the same specific angular momentum as that of the orbit of the mass-losing star. Thus, the period and the separation always *increase* if a red giant loses mass in a stellar wind. If mass and angular momentum are not lost in a binary with a lobe-filling giant, $\alpha = \beta = 0$, and the evolution of the binary depends only on the mass ratio, M_1/M_2. The period always *decreases* as the giant loses mass if $\mu > 1/2$, and can decrease quite rapidly if the mass loss rate is large. Rapid decreases in the period tend to destabilize the binary, and can result in even larger mass transfer rates.

6.2.2 Symbiotics with Accreting Main Sequence Stars

The simplified discussion of the previous section suggests that binaries with $P = 100\text{-}10,000$ days can become symbiotic stars when the primary component ascends the giant branch and fills its Roche lobe. Theoretical calculations indicate that an evolved red giant can remain in this state for $\sim 10^4$ yr, when its mass loss rate increases rapidly to $> 10^{-3} M_\odot \text{ yr}^{-1}$ (Webbink 1979b). This upper limit to the lifetime of the lobe-filling component can be used to estimate the number of such systems that might be found in the galaxy. Recalling that the formation rate for binaries, β_{bin}, is $\sim 0.14 \text{ yr}^{-1}$, the number of symbiotic stars in the galaxy is:

$$N_{\text{ss}} = \beta_{\text{bin}} \, f_P \, \tau_{\text{ss}}, \tag{6.4}$$

where f_P is the fraction of binaries with $P = 100\text{-}10,000$ days and τ_{ss} is the lifetime of a symbiotic system ($\sim 5 \times 10^3$ yr). Adopting $f_P \sim 0.2$ as a reasonable estimate implies that a new symbiotic star is formed every 35 years, leading to a total number of ~ 150 such systems in the galaxy. This may be compared with the observed number of ~ 125.

Evolution Beyond Symbiosis - Part I

The short phase of symbiosis described above is characterized by an ever-increasing rate of mass loss from the lobe-filling giant, as noted by Webbink (1979b) and Plavec, Ulrich, and Polidan (1973). The main sequence star expands in response to the accelerating rate of mass transfer, eventually reaching its tidal lobe and forming a contact binary. Although detailed calculations of the evolution of giant contact systems have not been performed, the loss of mass and angular momentum through the outer Lagrangian points (particularly L_2) is expected to shrink the binary orbit. The secondary eventually is engulfed by the expanding atmosphere of its companion, producing a common envelope binary with an essentially unevolved main sequence star and a hot white dwarf orbiting inside a dense nebula (Paczynski 1976, and references therein).

The nature of the binary that emerges from the common envelope depends on the mass of the envelope, M_{env}, and the binary separation,

A (Livio, Salzman, and Shaviv 1979, and references therein). The effects of the interaction during this phase can be estimated following the work of Paczynski (1976). The energy dissipated by frictional forces within the common envelope is roughly $L_D \sim GMA^2\rho/P$, where L_D is the drag luminosity and ρ is the mass density of nebular material. The total energy dissipated is simply $L_D\tau$, where τ is the timescale over which the ejected material escapes the binary (e.g., $\tau \sim A/v$, where v is the velocity of the binary components). Letting $\rho \sim M_{env}/A^3$, the total energy transferred into the envelope is:

$$E_D \sim GMM_{env}/Pv. \tag{6.5}$$

The energy imparted to the envelope is reflected in a loss of the binary's total binding energy, E. The fractional change in orbital energy can be written as:

$$\Delta E_{orb}/E_{orb} = 2MM_{env}A/M_1M_2Pv. \tag{6.6}$$

Substituting typical values for M (3 $M_\odot$), M_1 (2 $M_\odot$), and M_2 (1 $M_\odot$) results in $\Delta E_{orb}/E_{orb} \sim 0.5\, M_{env}/M_\odot$. Large changes in the orbital energy (and the binary period) are possible only if a massive common envelope is formed during the contact phase.

The mass of the common envelope is apt to be small for a long-period system ($P \sim 10$ years), as the giant is able to process most of its atmosphere through a hydrogen-burning shell before encountering its tidal lobe. Wind-driven mass loss during the ascent of the giant branch also acts to stabilize the system by reducing M_{env} and inflating the binary orbit. Adopting $M_{env} \sim 0.2\, M_\odot$ as typical for such objects leads to $\Delta E_{orb}/E_{orb} \sim 0.1$, and significant changes in the orbital parameters are not expected. Once the common envelope is ejected, these long-period systems will appear as normal planetary nebulae with perhaps some bipolar structure. However, the relatively unevolved main sequence star may eventually become a red giant, and could initiate a second symbiotic phase.

A much more exciting conclusion to the common envelope phase is expected in shorter period objects ($P \sim 1$-3 years). The giant components of these systems are unable to reduce M_{env} substantially by nuclear processing or a stellar wind before encountering their tidal lobes, and thus M_{env} tends to be fairly large. For $M_{env} \sim 1.0\, M_\odot$, $\Delta E_{orb}/E_{orb} \sim 0.5$; a substantial decrease in the orbital period is therefore needed to eject the common envelope. This may lead to additional mass loss from the giant as the orbit (and the radius of the

Roche lobe) decays. This scenario is considered to be the most promising mechanism for the production of the short period cataclysmic binaries (Paczynski 1976; Ritter 1976; Webbink 1977), although many details remain to be worked out. The formation rate for symbiotic binaries containing accreting main sequence stars (~ 1 every 35 years) is comparable to the formation rate for cataclysmic binaries recently derived by Patterson (1984; ~ 1-5 every 100 years), and lends some support to this notion.

6.2.3 Symbiotics Containing Accreting White Dwarfs

It is anticipated that a given binary can avoid the symbiotic phase described above if its period is initially greater than 5,000-10,000 days. The primary component can then live out its life as a normal red giant, losing mass in the form of a stellar wind and eventually ejecting its outer atmosphere as a planetary nebula. Since the main sequence companion can never accrete as much as $10^{-6} M_\odot$ yr^{-1}, the system would not be identified as a symbiotic star. However, satellite ultraviolet observations might reveal some of these objects as binaries, and the supergiant TV Gem (Michalitsianos, Hobbs, and Kafatos 1980) might be an example of such a system.

The period changes experienced by these long period binaries are never very dramatic, as mass is never lost by tidal interaction. If the giant's stellar wind carries away the same specific angular momentum as the orbit of the mass-losing star (i.e., d ln J/d ln $M = J_1/M_1$), then $\dot{P}/P = -2\dot{M}/M$, and the period always *increases* in response to the wind ($\dot{M} < 0$; cf. equation [6.3]). However, some reduction of the period may occur during a planetary nebula ejection. Recalling equation (6.6), the relative change in the orbital energy resulting from the loss of a 0.2-0.3 $M_\odot$ nebula is rather small, $\Delta E/E \sim 0.1$. This is comparable to the decrease in orbital binding energy resulting from a stellar wind, and thus only a negligible variation in orbital period is expected.

The shorter period systems ($P < 5000$ days) cannot escape an initial symbiotic phase, but some are likely to escape the catastrophic common envelope evolution, as noted above. These systems must suffer some decrease in orbital period (perhaps as much as a factor of 2), although the end result is essentially identical to the longer period counterparts, namely a hot white dwarf and a main sequence companion orbiting inside a planetary nebula.

Once the red giant atmosphere escapes the immediate vicinity of the binary, the remaining components are a main sequence star and a hot, very luminous white dwarf. Plavec (1982) has suggested that a "natural" symbiotic star might result if the secondary component could be ascending the giant branch before the newly-formed white dwarf can cool off (which will occur in $\sim 10^4$ yr; Paczynski 1971). This can only happen if the initial mass ratio is sufficiently close to unity. The *maximum* difference in main sequence lifetimes allowed in this scenario occurs if the primary ejects a planetary nebula just as the secondary begins its ascent of the giant branch, or $\Delta t_{ms} \lesssim 10^7$ yr. This implies a mass difference, $\Delta M = M_1 - M_2 \sim 0.01$ $M_\odot$ for $M_1 \sim 1 - 3M_\odot$. The relative fraction of secondaries with $M_2 \gtrsim M_1 - \Delta M$ is also ~ 0.01, given the distribution function for secondary masses derived by Abt and Levy (1976, 1978). Thus, a substantial fraction of newly-formed white dwarfs in long-period binaries must cool off before the secondary ascends the giant branch, with perhaps 1% involved in a natural symbiotic star. Based on observations discussed by Lutz (1984), it appears that the suspected symbiotic star HD 330036 is currently in this interesting evolutionary phase.

Since natural symbiotics are extremely unlikely, the binary must wait awhile for the secondary star to ascend the giant branch and begin to lose mass at a rate that approaches 10^{-7} to $10^{-6} M_\odot$ yr^{-1}. Recalling the results of Chapter 2, only $\sim 10^{-9} M_\odot$ yr^{-1} is required by the cold white dwarf, which is easily accomplished if P does not exceed 100-200 years (Livio and Warner 1984; Tutukov and Yungel'son 1982). The relative numbers of such systems depends on the fraction of binaries, f_M, in which both components exceed 1 $M_\odot$, and thus equation (6.4) can be written as:

$$N_{SS} \approx \beta_{bin} f_P f_M \tau_{SS} \qquad (6.7)$$

The distribution of secondary masses in long period binaries suggests $f_M \sim 0.2$ (Abt and Levy 1976, 1978), while f_P will be set to 0.2 as before. This results in the formation of a symbiotic system containing a red giant and a white dwarf every 175 years, which is a factor of 5 below the birthrate of systems containing accreting main sequence stars. If $\tau \sim 5 \times 10^5$ yr is a reasonable lifetime for this symbiotic phase, then ~ 3000 symbiotic binaries should exist in our galaxy.

The identification of these binaries as symbiotic stars is complicated by (i) dust formation in the outer atmosphere of the cool giant, and (ii) the binary separation. Merrill and Stein (1976) noted significant dust excesses in very late-type giants ($> M6$), Mira variables, and M-type supergiants, and this dust might act to enshroud the binary system. The radii of these very evolved red giants (> 1 AU) require binary separations exceeding 10 AU if the system is to avoid tidal interaction. Thus symbiotic stars with extensive circumbinary dust shells (e.g., the *D-type* systems; Chapter 3; Willson 1981; Willson, et al. 1984) should have long binary periods ($P > 20$ yr).

The binary separation also plays a role in the identification of symbiotic stars, as it determines the amount of material that can be accreted by the hot companion. Long periods ($P > 10$ yr) are not expected if the cool component is a normal giant ($R_g \sim 50 - 100 R_\odot$; $M_g \sim 10^{-7} M_\odot$ yr^{-1}), since only a small fraction of the wind ($\sim 0.1\%$; Tutukov and Yungel'son 1982) reaches the white dwarf primary. Thus, short period symbiotics with accreting white dwarfs should contain normal giants; these systems do not form dust and would be detected as *S-type* objects (cf. Chapter 3).

The observed binary period distribution (Abt and Levy 1976, 1978) implies that the natural division into short period S-type systems ($P \sim 2$-10 yr) and long period D-type systems ($P \sim 20$-100 yr) produce comparable numbers of symbiotic stars. The figure of ~ 1500 S-type systems is comparable to that of ~ 1000 derived by Tutukov and Yungel'son (1982) in an independent analysis, and these are supplemented by ~ 1500 additional D-type objects. The predicted abundance of S-type systems should be close to actual observations of symbiotic stars, since these objects are fairly easy to distinguish from normal red giant stars (cf. Figure 3.9). It may be more difficult to identify D-type systems, as dust may enshroud the entire binary system.

Evolution Beyond Symbiosis - Part II

(a) Short Period Systems ($P \sim 1000$ days)

The termination of this final symbiotic phase occurs as the giant approaches its tidal lobe and loses mass at an accelerated pace. Unlike a main sequence star, an accreting white dwarf cannot accept this increased flow of material: it must expand into a giant configuration once $\dot{M}$ exceeds $10^{-6} M_\odot$ yr^{-1}. A contact system is formed as $\dot{M}$ approaches $10^{-5} M_\odot$ yr^{-1} (the Eddington limit), producing a common envelope binary as noted earlier. In this case, the core of the common envelope is a white dwarf pair, rather than a white dwarf with a main sequence companion.

Once again, the mass of the common envelope and the binary separation decide the fate of the binary core. We can expect that most objects with short periods eject a massive envelope, and therefore suffer drastic reductions in orbital period. The end result is a close binary consisting of two white dwarfs surrounded by an expanding planetary nebula. If the two white dwarfs are sufficiently close, gravitational radiation may act to contract the binary orbit and form a semi-detached system in which one white dwarf transfers material to its more massive companion.

(b) Long Period Systems (P ~ 10,000 days)

Long-period symbiotic binaries never enter a phase of tidal interaction, and therefore do not evolve into contact systems. Thus, the red giant component eventually ejects a planetary nebula, which results in a small increase in the binary period. This planetary nebula will be excited by a binary nucleus composed of two hot white dwarfs separated by >10 AU. Such a system would be difficult to identify as a long-period binary; the formation rate derived above suggests <4% of all planetary nebulae might contain widely separated binary nuclei.

6.2.4   Comparison with Observations

The basic conclusions obtained above may be briefly summarized as:

1) systems with accreting main sequence stars and accreting white dwarfs are both likely to be detected as symbiotic stars, with white dwarf systems having larger orbital periods on the average, and

2) white dwarf symbiotics are expected to be more prevalent than main sequence systems by a factor of ~10.

A direct comparison of these results with the observations is complicated by the various selection criteria that have been used to classify symbiotic stars (cf. Chapter 1). In general, systems containing main sequence accretors should be straightforward to detect, as they (i) are very bright ($L \sim 300\ L_\odot$ for the hot component), and (ii) undergo frequent outbursts due to instabilities associated with the lobe-filling giant. Conversely, the predicted number of ~3,000 white dwarf symbiotics might somewhat overestimate their detection probability, since (i) some systems might be enshrouded in a dust shell, and (ii) the outbursts are rather infrequent.

Given these uncertainties, the estimates are in surprising (and hopefully not fortuitous) accord with the observations. Systems identified as main sequence accretors all have periods of ~200-1000 days, and several systems with hot white dwarfs are also known to have relatively short orbital periods (as discussed in the Appendix). These latter objects appear to be ~7 times more frequent than their main sequence counterparts (Chapter 3), which is in reasonable accord with the predicted value of ~10. The total number of known symbiotic stars is

~125, and various estimates yield ~1000 such systems in the entire galaxy (e.g., Boyarchuk 1975). This is somewhat less than the ~3000 derived above, which may reflect (i) an overestimate in the types of binaries that evolve into symbiotic stars (or their lifetimes in the symbiotic phase), and/or (ii) the lack of a comprehensive search for symbiotic systems.

Roughly 20% of all known symbiotic stars are D-type systems with Mira-like cool stars, and are identified as the long period objects predicted above. The period estimates for V1016 Cyg and HM Sge (> 10 years, based on a lack of radial velocity variations) and H1-36 (~100 years, from *VLA* observations) favor this interpretation, and it is important to obtain period estimates for other D-type systems to provide additional information on this interesting class of systems.

6.3   Symbiotic Stars and Type I Supernovae

Evolutionary scenarios leading to type I supernova explosions have become a popular topic in recent years, and generally involve accreting white dwarf stars (cf. Iben and Tutukov 1983). If the mass of this white dwarf increases as a result of accretion (which is not altogether obvious since hydrogen shell flashes may eject the accreted material, as noted in Chapter 4), it may eventually reach a mass exceeding the Chandrasekhar limit and explode rather violently. Since some symbiotic systems appear to contain accreting white dwarfs, it is appropriate to comment on the possibility of a spectacular explosive event occurring in one of these long period binaries.

The formation rates discussed earlier imply that a binary containing a red giant and a white dwarf is formed every 175 years, which is tolerably close to the observed frequency of type I supernovae (~1 every 100 years; Tammann 1982). The maximum amount of material such a white dwarf can accrete can be estimated to be ~0.1-0.2 $M_\odot$, based on (i) the adopted lifetime of a symbiotic star (~ 10^6 yr) and (ii) the maximum allowed accretion rate ($\sim 1 - 2 \times 10^{-7} M_\odot$ yr^{-1}, if $\dot{M}$ exceeds this rate, the white dwarf expands to giant dimensions and can no longer be classified as symbiotic, although such an object may still become a type I supernova). This mass is comparable to that required for a helium detonation in a low mass white dwarf ($M_{wd} < 1.1 M_\odot$), and hydrodynamic calculations suggest such an explosion might result in a type I supernova (Nomoto 1982). More massive white dwarfs could undergo a carbon deflagration supernova (Mazurek and Wheeler 1980), although symbiotic stars with 1.2-1.3 $M_\odot$ white dwarfs are probably rare.

It is very difficult to estimate the fraction of symbiotics that might contribute to supernova events, since the evolution of accreting white dwarfs is not a well-understood process (cf. Iben and Tutukov 1983). Assuming that most of the material accreted by the white dwarf is not subsequently ejected in a hydrogen shell flash (Chapter 4) implies that most, if not all, symbiotic stars eventually explode as type I supernovae. This results in ~ 5 supernova events every 1000 years, which is comparable to the frequency of ~ 4 per 1000 years derived independently by Iben and Tutukov (1983). This is slightly below the observed frequency of 10 every 1000 years, and Iben and Tutukov (1983) suggested that multiple scenarios might be needed to produce this observed rate. However, it is important to note that V2116 Oph and the related object HD 154791 (Garcia, et al. 1983) both consist of evolved red giants with neutron star companions, indicating that some type of supernova explosion occurred in each binary.

6.4   Summary

This Chapter has developed a simple scenario for the formation and evolution of symbiotic binaries. The major conclusion of this analysis is that symbiotic stars are a natural consequence of normal stellar evolution in a binary with $P > 100$ days. These objects are not a homogeneous group of binaries, and therefore require multiple formation mechanisms.

1. Most symbiotic stars (1 every 35 yr) are formed in a 100-10,000 day binary when the primary ascends the giant branch and fills its tidal lobe. The symbiotic optical spectrum is formed as a result of accretion onto a low mass main sequence star (Chapter 2). This phase lasts $< 10^4$ yr, as the giant rapidly causes the system to evolve into a common envelope binary. Thus, only a small fraction ($\sim 10\%$) of these systems constitute the set of known symbiotic stars. It appears that many of these objects evolve into short-period cataclysmic variables during the ejection of a common envelope.

2. Some ($\sim 10\%$) of the symbiotic stars described above escape the dramatic common envelope evolution, and begin a second symbiotic phase when the secondary climbs the giant branch and loses mass rapidly in a stellar wind. The white dwarf companion accretes some of this material, and provides a source of high energy photons which ionize the wind. Although these systems are formed less frequently than "first generation" symbiotic stars, they remain symbiotic for

$\sim 10^6$ yr and make up $\sim 50\%$ of the entire symbiotic population.

3. Long period binaries ($P > 10{,}000$ days) become symbiotic when the secondary evolves into a red giant and transfers matter to its white dwarf companion via a stellar wind. The large mass loss rates needed to energize the white dwarf favor very evolved red giants that form dust in their outer atmospheres. These D-type systems are likely to be as numerous as their shorter period counterparts, but may be more difficult to detect.

It is difficult to compare observations directly with theory, since complete searches for symbiotic stars are not available. Estimates based on ultraviolet and optical data suggest hot stellar sources outnumber accreting main sequence stars in the short period systems by $\sim 7{:}1$, which is close to the prediction of $\sim 10{:}1$. Roughly 20% of all symbiotics are D-type objects, which is a factor of ~ 2 below the prediction. Given the historical difficulty in classifying D-type systems as planetary nebulae (cf. Appendix), the lack of D-type systems is not discouraging. Finally, the number of symbiotic stars presumed to exist in the galaxy (~ 1000; Boyarchuk 1975) is in good agreement with the predicted value of ~ 3000.

7

Epilogue

The past two decades have seen a renaissance in the study of symbiotic stars, and modern results have established these objects as belonging to an independent class of binary system. The next decade will see a new generation of satellite observatories, and many cherished theories may crumble under the onslaught of information from Space Telescope, the Hopkins Ultraviolet Telescope, the Extreme Ultraviolet Explorer, and the Advanced X-ray Astronomy Facility (among others). However, the basic properties of symbiotic stars are now well-established, and it is important to understand these characteristics when planning future observational and theoretical projects.

1. Symbiotic stars are binaries with periods ranging from ~ 200 days to > 10 years. They are members of the old disk population, and therefore have masses of 1-3 $M_\odot$. Spectroscopic orbits are available for a few systems, and these generally confirm kinematic mass estimates.

2. One component of a symbiotic binary is a red giant. In long-period D-type systems ($P > 5$-10 yr), the giant is a Mira-like variable losing mass in a powerful stellar wind. Most of the giants in the shorter period S-type systems ($P < 5$-10 yr) tend not to exhibit Mira-like behavior, and lose mass at a considerably lower rate. A few short-period systems contain lobe-filling giants which transfer material tidally into an accretion disk surrounding a low mass companion.

3. The hot component in a majority of symbiotic systems is a hot white dwarf or subdwarf ($R_h \sim 0.01 - 0.1$ $R_\odot$, $T_h \sim 30,000$-150,000 K) accreting material from the stellar wind of its red giant companion. The total accretion luminosity is a negligible fraction of the total energy radiated by the hot component,

and thus this object resembles the central star of a planetary nebula. Ultraviolet photons emitted by the hot star ionize a portion of the surrounding nebula, giving rise to intense emission lines (H I, He I, and He II) and radio emission. A small percentage ($< 10\%$) of symbiotic stars contain a low mass main sequence star surrounded by an extensive accretion disk. These objects have rather small ionized nebulae, and have not been detected as radio sources.

4. Symbiotic stars display a bewildering variety of eruptive behavior. Many outbursts are a result of thermonuclear runaways in the accreted hydrogen-rich envelope of a hot white dwarf or subdwarf. These eruptions tend to last many decades, and resemble outbursts of classical novae. In some cases, the interaction between the high velocity wind ejected by the eruptive star and the low velocity wind lost by the cool giant forms a shock, which cools via X-ray and high ionization line radiation. Other eruptions are caused by accretion events in disks surrounding main sequence stars. These tend to be short eruptions (< 1-3 yr), although they can be quite spectacular (e.g., T CrB).

5. The current population of symbiotic stars can be understood in terms of normal stellar evolution within a low mass binary system. In most cases, the red giant is the system's original secondary component; only in objects containing accreting main sequence stars is the red giant the more massive binary component originally.

There are many aspects of symbiotic behavior that remain mysterious and confusing, and I have tried to emphasize some of these areas throughout this book. Three areas of research have generally been neglected by recent experimental efforts (and in this monograph), and might allow a new outlook on the symbiotic phenomenon.

1. Observations with better angular resolution would teach us much concerning the physical nature of symbiotic stars. The binary components of a typical long-period system such as V1016 Cyg have an angular separation exceeding 0".01 at a distance of 1 kpc, and thus these objects could be resolved with a long baseline interferometer. Shorter period objects are more challenging, as the binary components are typically separated by 0".001. Advanced speckle interferometric techniques using a narrow-band interference filter centered on a strong emission line might reveal important information concerning the distribution of gas in a symbiotic binary.

2. High resolution spectroscopic projects in the infrared could resolve the controversy regarding the nature of the late-type giant, and yield additional information from the H I Brackett series and other emission lines. Contemporaneous observations of a wide variety of optical emission lines (Hα, He II λ4686, [Ne V] λ3426, and [Fe VII] λ6087) would be most helpful in probing the dynamics of mass transfer within the binary.

3. Although symbiotics are not prolific X-ray emitters, it is important to (i) understand their lack of emission at wavelengths below 50-100 Å, and (ii) probe emission in the extreme ultraviolet (λλ100-1000). All models for the hot component predict substantial emission below 1000 Å, and such information is vital for theoretical analyses of the emission-line spectra.

There are, of course, many symbiotic phenomena that remain mysterious and confused, and new observations will continue to add new information regarding the symbiotic puzzle. I hope the next book on this topic can speak of the resolution of many of these problems, and that this treatise played some small part in increasing our awareness of these interesting binary systems.

References to Chapters 1-7

Chapter 1

Allen, D.A. 1979. in
IAU Colloquium No. 46, Changing Trends in Variable Star Research,
    ed. F.Bateson, J.Smak, and I.Urch (Hamilton, N.Z.: U. Waikato Press), p. 125.

Allen, D.A. 1980. *Mon. Not. Roy. Astr. Soc.,* **192**, 521.

Allen, D.A. 1983. *Astrophys. Space Sci.,* **99**, 101.

Blanco, B.M., Blanco, V.M., and McCarthy, M.F. 1978. *Nature,* **271**, 638.

Boyarchuk, A.A. 1969. in
Non-Periodic Phenomena in Variable Stars,
    ed. L. Detre (Budapest: Academic Press), p. 395.

Boyarchuk, A.A. 1975. in
IAU Symposium No. 67, Variable Stars and Stellar Evolution,
    ed. V.E. Sherwood and L. Plaut (Dordrecht: Reidel), p. 377.

Duerbeck, H.W. 1984. *Astrophys. Space Sci.,* **99**, 363.

Fleming, W.P. 1912. *Ann. Harv. Coll. Obs.,* **56**, 165.

Greenstein, N.K. 1937. *Bull. Harv. Coll. Obs.,* No. 906.

Hogg, F.S. 1932. *Pub. Astr. Soc. Pac.,* **44**, 328.

Kenyon, S.J. and Webbink, R.F. 1984. *Astrophys. J.,* **279**, 252.

Lindsay, E.M. 1932. *Bull. Harv. Coll. Obs.,* No. 888.

Merrill, P.W. 1919. *Pub. Astr. Soc. Pac.,* **31**, 305.

Merrill, P.W. 1936. *Astrophys. J.,* **83**, 272.

Merrill, P.W. 1958. *Etoiles a raies d'emission*
    (Cointe-Sclessin: Institut d'Astrophysique), p. 436.

Merrill, P.W. 1959. *Astrophys. J.,* **129**, 44.

Merrill, P.W. and Humason, M.L. 1932.
    Pub. Astr. Soc. Pac., **44**, 56.

Mihalas, D. and Binney, J. 1981. *Galactic Astronomy*
    (San Francisco: Freeman).

Nussbaumer, H. 1982. in
IAU Colloquium No. 70, The Nature of Symbiotic Stars,
    ed. M. Friedjung and R. Viotti (Dordrecht: Reidel), p. 85.

Payne-Gaposchkin, C. 1957. *The Galactic Novae*
    (Amsterdam: North Holland).

Plaskett, H.H. 1928. *Pub. DAO Victoria*, **4**, 119.
Wallerstein, G. 1981. *Obs.*, **101**, 172.
Yamamoto, I. 1924. *Bull. Harv. Coll. Obs.*, No. 810.

Chapter 2

Allen, C.W. 1973. *Astrophysical Quantities* (London: Athlone).
Aller, L.H. 1954. in
   Astrophysics - Nuclear Transformations, Stellar Interiors, and Nebulae
   (New York: Ronald), p. 180.
Berman, L. 1932. *Pub. Astr. Soc. Pac.*, **44**, 318.
Boyarchuk, A.A. 1969. in
   Non-Periodic Phenomena in Variable Stars,
   ed. L. Detre (Budapest: Academic Press), p. 395.
Boyarchuk, A.A. 1975. in
   IAU Symposium No. 67, Variable Stars and Stellar Evolution,
   ed. V.E. Sherwood and L. Plaut (Dordrecht: Reidel), p. 377.
Bruce, C.E.R. 1955. *Obs.*, **75**, 82.
Bruce, C.E.R. 1956. *Comptes Rendus*, **242**, 2101.
Cowley, A.P. 1964. *Astrophys. J.*, **139**, 817.
Cox, D.P. and Daltabuit, E. 1971. *Astrophys. J.*, **167**, 257.
Ferland, G.J. 1979. *Astrophys. J.*, **231**, 781.
Gallagher, J.S., Holm, A.V., Anderson, C.M., and Webbink, R.F. 1979.
   Astrophys. J., **229**, 994.
Gauzit, J. 1955a. *Comptes Rendus*, **241**, 741.
Gauzit, J. 1955b. *Comptes Rendus*, **241**, 793.
Hack, M. and Selvelli, P.L. 1982. *Astr. Astrophys.*, **107**, 200.
Hartmann, L., Dupree, A.K., and Raymond, J.C. 1981.
   Astrophys. J., **246**, 193.
Hartmann, L., Dupree, A.K., and Raymond, J.C. 1982.
   Astrophys. J., **252**, 214.
Hogg, F.S. 1934. *Pub. Am. Astr. Soc.*, **8**, 14.
Iben, I. Jr. 1982. *Astrophys. J.*, **259**, 244.
Iijima, T. 1981. in *Photometric and Spectroscopic Binary Systems*,
   ed. E.B. Carling and Z. Kopal (Dordrecht: Reidel), p. 517.
Johnson, H.L. 1966. *Ann. Rev. Astr. Astrophys.*, **4**, 193.
Kaler, J.B. 1978. *Astrophys. J.*, **220**, 887.
Kenyon, S.J. and Webbink, R.F. 1984. *Astrophys. J.*, **279**, 252.
Kuiper, G.P. 1940. *Pub. Am. Astr. Soc.*, **10**, 57.
Livio, M. and Warner, B. 1984. *Obs.*, **104**, 152.
Menzel, D.H. 1946. *Physica*, **12**, 768.
Merrill, P.W. 1940. *Spectra of Long Period Variable Stars*
   (Chicago: University of Chicago).
Meyer, D.M. and Savage, B.D. 1981. *Astrophys. J.*, **248**, 545.
O'Connell, R.W. 1973. *Astr. J.*, **78**, 1074.
Oliversen, N.G., Anderson, C.M., Cassinelli, J.P., Sanders, W.T., and
   Kaler, J.B. 1980. *Bull. Amer. Astr. Soc.*, **12**, 819.
Osterbrock, D.E. 1974. *Astrophysics of Gaseous Nebulae*
   (San Francisco: Freeman).

Paczynski, B. 1971. *Acta Astr.*, **21**, 417.

Savage, B.D. and Mathis, J.S. 1979.
Ann. Rev. Astr. Astrophys., **17**, 73.

Sobolev, V.V. 1960. *Moving Envelopes of Stars*
(Cambridge: Harvard Univ.), p. 82.

Stencel, R.E. and Ionson, J.A. 1979. *Pub. Astr. Soc. Pac.*, **91**, 452.

Stencel, R.E., Michalitsianos, A.M., Kafatos, M., and
Boyarchuk, A.A. 1982. *Astrophys. J. (Letters)*, **253**, L77.

Stencel, R.E. and Sahade, J. 1980. *Astrophys. J.*, **238**, 929.

Thompson, G.T., Nandy, K., Jamar, C., Monfils, A., Houziaux, L. Carnochan, D.J.,
and Wilson, R. 1978. *Catalogue of Stellar Ultraviolet Fluxes*
(Science Research Council).

Tutukov, A.V. and Yungel'son, L.R. 1982. in
IAU Colloquium No. 70, The Nature of Symbiotic Stars,
ed. M. Friedjung and R. Viotti (Dordrecht: Reidel), p. 283.

Watanabe, T. and Kodaira, K. 1978. *Pub. Astr. Soc. Japan*, **30**, 21.

Webbink, R.F. 1979. in
IAU Colloquium No. 46, Changing Trends in Variable Star Research,
ed. F.M Bateson, J. Smak, and I. Urch (Hamilton, N.Z.: U. Waikato Press), p. 102.

Wood, P.R. 1973. *Pub. Astr. Soc. Aust.*, **2**, 198.

Wu, C.-C., Faber, S.M., Gallagher, J.S., Peck, M., and Tinsley, B.M. 1980.
Astrophys. J., **237**, 290.

Zuckerman, B. 1980. *Ann. Rev. Astr. Astrophys.*, **18**, 263.

Chapter 3

Allen, D.A. 1980a. *Mon. Not. Roy. Astr. Soc.*, **190**, 75.

Allen, D.A. 1980b. *Mon. Not. Roy. Astr. Soc.*, **192**, 521.

Allen, D.A. 1982. in
IAU Colloquium No. 70, The Nature of Symbiotic Stars,
ed. M. Friedjung and R. Viotti (Dordrecht: Reidel), p. 27.

Allen, D.A. 1983. *Mon. Not. Roy. Astr. Soc.*, **204**, 113.

Allen, D.A. 1984. *Proc. Astr. Soc. Austr.*, **5**, 369.

Altenhoff, W.J., Braes, L.L.E., Olnon, F.M., and Wendker, H.J. 1976.
Astr. Astrophys., **46**, 11.

Altenhoff, W.J. and Wendker, H.J. 1973. *Nature*, **241**, 37.

Anderson, C.M., Cassinelli, J.P., and Sanders, W.T. 1981.
Astrophys. J. (Letters), **247**, L127.

Anderson, C.M., Cassinelli, J.P., Oliversen, N.G., Myers, R.V., and Sanders, W.T.
1982. in *IAU Colloquium No. 70, The Nature of Symbiotic Stars,*
ed. M. Friedjung and R. Viotti (Dordrecht: Reidel), p. 117.

Anderson, C.M., Oliversen, N.G., and Nordsieck, K.H. 1980.
Astrophys. J., **242**, 188.

Andrews, P.J. 1974. *Mon. Not. Roy. Astr. Soc.*, **167**, 635.

Andrillat, Y. 1982. in
IAU Colloquium No. 70, The Nature of Symbiotic Stars,
ed. M. Friedjung and R. Viotti (Dordrecht: Reidel), p. 47.

Andrillat, Y. and Houziaux, L. 1982. in
    IAU Colloquium No. 70, The Nature of Symbiotic Stars,
    ed. M. Friedjung and R. Viotti (Dordrecht: Reidel), p. 57.
Bailey, J. 1975. *J. Brit. Astr. Assoc.*, **85**, 217.
Baldwin, J.R., Frogel, J.A., and Persson, S.E. 1973.
    Astrophys. J., **184**, 427.
Bath, G.T., Evans, W.D., Papaloizou, J., and Pringle, J.E.
    1974. *Mon. Not. Roy. Astr. Soc.*, **169**, 447.
Bath, G.T. and Wallerstein, G. 1976. *Pub. Astr. Soc. Pac*, **88**, 759.
Becker, S.A. 1979. Ph. D thesis, Univ. Illinois.
Belyakina, T.S. 1968. *Astr. Zh.*, **45**, 139.
Belyakina, T.S. 1969. *Izv. Krym. Astrofiz. Obs.*, **40**, 39.
Belyakina, T.S. 1970. *Astrofizica*, **6**, 49.
Belyakina, T.S. 1979. *Izv. Krym. Astrofiz. Obs.*, **59**, 133.
Belyakina, T.S. 1984. *Izv. Krym. Astrofiz. Obs.*, **68**, 108.
Berman, L. 1932. *Pub. Astr. Soc. Pac*, **44**, 318.
Blair, W.P., Stencel, R.E., Feibelman, W., and Michalitsianos, A.G. 1983.
    Astrophys. J. Suppl., **53**, 573.
Boyarchuk, A.A. 1967. *Astr. Zh.*, **44**, 12.
Boyarchuk, A.A. 1969.
    in *Non-Periodic Phenomena in Variable Stars,*
    ed. L. Detre (Budapest: Academic Press), p. 395.
Boyarchuk, A.A. 1975. in
    IAU Symposium No. 67, Variable Stars and Stellar Evolution,
    ed. V.E. Sherwood and L. Plaut (Dordrecht: Reidel), p. 377.
Brocka, B. 1979. *Pub. Astr. Soc. Pac*, **91**, 519.
Burstein, D. and Heiles, C. 1982. *Astr. J.*, **87**, 1165.
Cahn, J.H. and Elitzur, M. 1979. *Astrophys. J.*, **231**, 124.
Cassatella, A., Patriarchi, P., Selvelli, P.L., Bianchi, L.,
    Cacciari, C., Heck, A., Perryman, M., and Wamsteker, W.
    1982. in *Third European IUE Conference*, ed. E. Rolfe, A. Heck
    and B. Battrick (ESA SP-176), p. 229.
Cester, B. 1968. *Inf. Bull. Var. Stars*, No. 291.
Cohen, N.L. and Ghigo, F.D. 1980. *Astr. J.*, **85**, 451.
Cordova, F. and Mason, K. 1984. *Mon. Not. Roy. Astr. Soc.*, **206**, 879.
Cowley, A. and Stencel, R. 1973. *Astrophys. J.*, **184**, 687.
Doty, J.A., Hoffman, J.A., and Lewin, W.H.G. 1981.
    Astrophys. J., **243**, 257.
Drake, S.A. and Ulrich, R.K. 1980. *Astrophys. J. Suppl.*, **42**, 351.
Engels, D. 1979. *Astr. Astrophys. Suppl*, **36**, 337.
Faraggiana, R. and Hack, M. 1980. *Inf. Bull. Var. Stars*, No. 1861.
Feast, M.W., Robertson, B.S.C., and Catchpole, R.M. 1977.
    Mon. Not. Roy. Astr. Soc., **179**, 499.
Feast, M.W., Whitelock, P.A., Catchpole, R.M., Roberts, G., Carter, B.S. and
    Roberts, G. 1983a. *Mon. Not. Roy. Astr. Soc.*, **202**, 951.
Feast, M.W., Catchpole, R.M., Whitelock, P.A., Carter, B.S., and
    Roberts, G. 1983b. *Mon. Not. Roy. Astr. Soc.*, **203**, 373.

Friedjung, M., Stencel, R.E., and Viotti, R. 1983.
Astr. Astrophys., **126**, 407.

Frogel, J.A., Persson, S.E., Aaronson, M., and Matthews, K. 1978.
Astrophys. J., **220**, 75.

Gallagher, J.S., Holm, A.V., Anderson, C.M., and Webbink, R.F. 1979.
Astrophys. J., **229**, 994.

Garcia, M.R. 1983. *Spectroscopic Orbits of Symbiotic Stars*,
Harvard College Observatory Research Exam.

Ghigo, F.D. and Cohen, N.L. 1981. *Astrophys.J.*, **245**, 988.

Gillett, F.C., Merrill, K.M. and Stein, W.A. 1971.
Astrophys. J., **164**, 87.

Greenstein, N.K. 1937. *Bull. Harv. Coll. Obs.*, No. 906.

Hack, M. and Selvelli, P.L. 1982. *Astr. Astrophys.*, **107**, 200.

Hjellming, R.M. and Wade, C.M. 1971. *Astrophys. J.*, **168**, L115.

Hoffleit, D. 1968. *Irish Astr. J.*, **8**, 149.

Hoffleit, D. 1970. *Inf. Bull. Var. Stars*, No. 469.

Hogg, F.S. 1934. *Pub. Am. Astr. Soc.*, **8**, 14.

Huang, C.-C. 1982. in *IAU Colloquium No. 70,*
The Nature of Symbiotic Stars,
ed. M. Friedjung and R. Viotti (Dordrecht: Reidel), p. 185.

Hutchings, J.B., Cowley, A.P., and Redman, R.O. 1975.
Astrophys. J., **201**, 404.

Hutchings, J.B., Cowley, A.P., Ake, T.B., and Imhoff, C.L. 1983.
Astrophys. J., **275**, 271.

Hutchings, J.B. and Redman, R.O. 1972. *Pub. Astr. Soc. Pac.*, **84**, 107.

Iben, I. Jr. 1982. *Astrophys. J.*, **259**, 244.

Iijima, T. 1981. in
Photometric and Spectroscopic Binary Syetems,
ed. E.B. Carling and Z. Kopal (Dordrecht: Reidel) p. 517.

Jewell, P.R., Eltizur, M., Webber, J.C., and Snyder, L.E. 1979.
Astrophys. J. Suppl., **41**, 191.

Johnson, H.M. 1980. *Astrophys. J.*, **237**, 840.

Kafatos, M., Michalitsianos, A.G., and Hobbs, R.W. 1980.
Astrophys. J., **240**, 114.

Kaler, J.B. 1981.
Astrophys. J., **245**, 568.

Kaler, J.B. and Hickey, J.P. 1983.
Pub. Astr. Soc. Pac., **95**, 759.

Kaler, J.B., Kenyon, S.J., and Hickey, J.P. 1983.
Pub. Astr. Soc. Pac., **95**, 1006.

Kenyon, S.J. 1982. *Pub. Astr. Soc. Pac.*, **94**, 165.

Kenyon, S.J. 1983. Ph. D. thesis, Univ. Illinois.

Kenyon, S.J. and Bateson, F. M. 1984. *Pub. Astr. Soc. Pac.*, **96**, 321.

Kenyon, S.J. and Gallagher, J.S. 1983. *Astr. J.*, **88**, 666.

Kenyon, S.J., Michalitsianos, A.G., Lutz, J.H., and Kafatos, M. 1985.
Pub. Astr. Soc. Pac., in press.

Kenyon, S.J. and Webbink, R.F. 1984. *Astrophys. J.*, **279**, 252.

Kenyon, S.J., Webbink, R.F., Gallagher, J.S., and Truran, J.W. 1982.
Astr. Astrophys., **106**, 109.

Keyes, C.D. and Plavec, M.J. 1980a. in
The Universe at Ultraviolet Wavelengths,
ed. R.D. Chapman (NASA CP-2171), p. 443.

Keyes, C.D. and Plavec, M.J. 1980b. in *IAU Symposium No. 88*,
Close Binary Stars: Observations and Interpretation,
ed. M.J. Plavec, D.M. Popper and R.K. Ulrich (Dordrecht: Reidel), p. 365.

Kraft, R.E. 1958. *Astrophys. J.*, **127**, 620.

Kwok, S. 1982. in
IAU Colloquium No. 70, The Nature of Symbiotic Stars,
ed. M. Friedjung and R. Viotti (Dordrecht: Reidel), p. 17.

Latham, D. 1982. in
*IAU Colloquium No. 67, Instruments for Astronomy with Large Optical
Telescopes*, ed. C.M. Humphries (Dordrecht: Reidel), p. 259.

Lepine, J.R.D., LeSqueren, A.M., and Scalise, E. 1978.
Astrophys. J., **225**, 869.

Lin, D. 1984. preprint.

Lynden-Bell, D. and Pringle, J.E. 1974.
Mon. Not. Roy. Astr. Soc., **168**, 603.

Mattei, J.A. and Allen, J. 1979. *J. Roy. Astr. Soc. Can.*, **73**, 173.

Mayall, M.W. 1937. *Ann. Harv. Coll. Obs.*, **105**, 491.

Meinunger, L. 1979. *Inf. Bull. Var. Stars*, No. 1611.

Meinunger, L. 1981. *Inf. Bull. Var. Stars*, No. 2016.

Menzies, J.W., Coulson, I.M., Caldwell, J.A.R., and Corben, P.M. 1982.
Mon. Not. Roy. Astr. Soc., **200**, 463.

Merrill, K.M. and Ridgway, S.T. 1979.
Ann. Rev. Astr. Astrophys., **17**, 9.

Merrill, P.W. 1933. *Astrophys. J.*, **77**, 44.

Merrill, P.W. 1959. *Astrophys. J.*, **129**, 44.

Michalitsianos, A.G. and Kafatos, M. 1982. in
IAU Colloquium No. 70, The Nature of Symbiotic Stars,
ed. M. Friedjung and R. Viotti (Dordrecht: Reidel), p. 191.

Michalitsianos, A.G. and Kafatos, M. 1984.
Mon. Not. Roy. Astr. Soc., **207**, 575.

Michalitsianos, A.G., Kafatos, M., and Hobbs, R.W. 1980.
Astrophys. J., **237**, 506.

Mirzoyan, L.V. and Bartaya, R.A. 1960.
Izv. Abastumani Astrofiz. Obs., No. 25, 121.

Mjalkovskij, M.I. 1977. *Perem. Zvezdy Prilozhenie*, **3**, 71.

Mundt, R. and Hartmann, L. 1983. *Astrophys. J.*, **268**, 766.

Nussbaumer, H. 1982. in
IAU Colloquium No. 70, The Nature of Symbiotic Stars,
ed. M. Friedjung and R. Viotti (Dordrecht: Reidel), p. 85.

O'Connell, R.W. 1973. *Astr. J.*, **78**, 1074.

Oliversen, N.G. and Anderson, C.M. 1982. in
IAU Colloquium No. 70, The Nature of Symbiotic Stars,
ed. M. Friedjung and R. Viotti (Dordrecht: Reidel), p. 71.

Oliversen, N.G. and Anderson, C.M. 1983. *Astrophys. J.*, **268**, 250.

Osterbrock, D.E. 1974. *Astrophysics of Gaseous Nebulae*
(San Francisco: W.H. Freeman and Company)

Osterbrock, D.E. 1981. *Astrophys. J.*, **246**, 696.

Paczynski, B. 1965. *Acta Astr.*, **15**, 197.

Panagia, N. and Felli, M. 1975. *Astr. Astrophys.*, **39**, 1.

Patterson, J. 1984. *Astrophys. J. Suppl.*, **54**, 443.

Payne-Gaposchkin, C. 1946. *Astrophys. J.*, **104**, 362.

Pettit, E. and Nicholson, S.B. 1933. *Astrophys. J.*, **78**, 320.

Pucinskas, A. 1970. *Bull. Vilniaus Univ. Astr. Obs.*, No. 27, 24.

Purton, C.R., Feldman, P.A., and Marsh, K.A. 1973.
Nature Phys. Sci., **245**, 5.

Ramsay, L.W. 1981. *Astr. J.*, **86**, 557.

Robinson, E.L. 1976. *Ann. Rev. Astr. Astrophys.*, **14**, 119.

Roman, N.G. 1953. *Astrophys. J.*, **117**, 467.

Rybski, P.M. 1973. *Pub. Astr. Soc. Pac.*, **85**, 653.

Sahade, J., Brandi, E., and Fontenla, J.M. 1984.
Astr. Astrophys., **56**, 17.

Sanford, 1947. *Pub. Astr. Soc. Pac.*, **59**, 333.

Sanford, R.F. 1949. *Astrophys. J.*, **109**, 81.

Seaquist, E.R. 1977. *Astrophys. J.*, **211**, 547.

Seaquist, E.R., Taylor A.R., and Button, S. 1984.

Shakura, N.I. and Sunyaev, R.A. 1973.
Astr. Astrophys., **22**, 337.

Sharpless, S. 1956. *Astr. J.*, **124**, 342.

Slovak, M.H. 1982. Ph. D. thesis, U. Texas.

Snyder, L.E. 1980. in *IAU Symposium No. 87, Interstellar Molecules*,
ed. B.H. Andrew (Dordrecht: Reidel), p. 525.

Stein, W.A., Gaustad, J.E., Gillett, F.C., and Knacke, R.F. 1969.
Astrophys. J., **155**, L3.

Stencel, R.E. 1984. *Astrophys. J. (Letters)*, **281**, L75.

Stencel, R.E., Michalitsianos, A.M., Kafatos, M., and Boyarchuk, A.A. 1982.
Astrophys. J. (Letters), **253**, L77.

Stencel, R.E. and Mullan, D.J. 1980. *Astrophys J.*, **238**, 221.

Swings, J.P. and Allen, D.A. 1972. *Pub. Astr. Soc. Pac.*, **84**, 523.

Swings, P. 1970. in *Spectroscopic Astrophysics*,
ed. G. Herbig (Berkeley: U. California), p. 189.

Taranova, O.G. and Yudin, B.F. 1982. *Soviet Astr. J.*, **26**, 57.

Taranova, O.G. and Yudin, B.F. 1983. *Soviet Astr. Lett.*, **9**, 322.

Taylor, A.R. and Seaquist, E.R. 1984. *Astrophys. J.*, **286**, 263.

Thackeray, A.D. 1959. *Mon. Not. Roy. Astr. Soc.*, **119**, 629.

Thackeray, A.D. 1977. *Mem. Roy. Astr. Soc.* **83**, 1.

Thackeray, A.D. and Hutchings, J.B. 1974. *Mon. Not. Roy. Astr. Soc.*, **167**, 319.

Thompson, G.T., Nandy, K., Jamar, C., Monfils, A., Houziaux, L.,
Carnochan, D.J., and Wilson, R. 1978.
Catalogue of Stellar Ultraviolet Fluxes, (Science Research Council).

Thronson, H.A. and Harvey, P.M. 1981. *Astrophys. J.*, **248**, 584.

Tonry, J. and Davis, M. 1979. *Astr. J.*, **84**, 1511.

Townley, S.D., Cannon, A.J., and Campbell, L. 1928.
  Ann. Harv. Coll. Obs., **79**, 161.
Tylenda, R. 1977. *Acta Astr.*, **27**, 3.
Viotti, R. 1976. *Astrophys. J.*, **204**, 293.
Walker, A.R. 1977. *Mon. Not. Roy. Astr. Soc.*, **179**, 587.
Wallerstein, G. 1968. *Obs.*, **88**, 111.
Wallerstein, G. 1983. *Pub. Astr. Soc. Pac.*, **95**, 135.
Webbink, R.F. 1979. in
  IAU Colloquium No. 46, Changing Trends in Variable Star Research,
  ed. F.M Bateson, J. Smak, and I. Urch (Hamilton, N.Z.: U. Waikato Press), p. 102.
Webster, B.L. and Allen, D.A. 1975.
  Mon. Not. Roy. Astr. Soc., **171**, 171.
Whitelock, P.A., Feast, M.W., Catchpole, R.M., Carter, B.S., and
  Roberts, G. 1983a. *Mon. Not. Roy. Astr. Soc.*, **203**, 351.
Whitelock, P.A., Catchpole, R.M., Feast, M.W., Roberts, G., and
  Carter, B.S. 1983b. *Mon. Not. Roy. Astr. Soc.*, **203**, 363.
Willis, A.J. and Wilson, R. 1978. *Mon. Not. Roy. Astr. Soc.*, **182**, 589.
Wolff, R.S. and Carlson, E.R. 1982. *Astrophys. J.*, **257**, 161.
Wright, A.E. and Allen, D.A. 1978.
  Mon. Not. Roy. Astr. Soc., **184**, 893.
Wright, A.E. and Barlow, M.J. 1975.
  Mon. Not. Roy. Astr. Soc., **170**, 41.
Zahn, J.-P. 1966. *Ann d'Astrophys.*, **29**, 313.
Zuckerman, B. 1979. *Astrophys. J.*, **230**, 442.
Zuckerman, B. 1980. *Ann. Rev. Astr. Astrophys.*, **18**, 263.

Chapter 4
Ahern, F.J., FitzGerald, M.P., Marsh, K.A., and Purton, C.R. 1977.
  Astr. Astrophys., **58**, 35.
Bath, G.T. 1975. *Mon. Not. Roy. Astr. Soc.*, **171**, 311.
Bath, G.T. 1977. *Mon. Not. Roy. Astr. Soc.*, **178**, 203.
Bath, G.T. and Pringle, J. E. 1982.
  Mon. Not. Roy. Astr. Soc., **201**, 345.
Beals, C. 1951. *Pub. DAO Victoria*, **9**, 1.
Becker, S.A. 1979. Ph. D. thesis, Univ. Illinois.
Belyakina, T.S. 1975. in
  IAU Symposium No. 67, Variable Stars and Stellar Evolution,
  ed. V.E. Sherwood and L. Plaut (Dordrecht: Reidel), p. 397.
Belyakina, T.S. 1979. *Izv. Krym. Astrofiz. Obs.*, **59**, 133.
Bruce, C.E.R. 1955. *Obs.*, **75**, 82.
Bruce, C.E.R. 1956. *Comptes Rendus*, **242**, 2101.
Clayton, D.D. 1968. *Principles of Stellar Evolution and Nucleosynthesis*
  (New York: McGraw-Hill).
Faulker, J., Lin, D.N.C., and Papaloizou, J. 1983.
  Mon. Not. Roy. Astr. Soc., **205**, 539.
Fujimoto, M.Y. 1982a. *Astrophys. J.*, **257**, 752.
Fujimoto, M.Y. 1982b. *Astrophys. J.*, **257**, 767.

Gallagher, J.S. and Starrfield, S.G. 1978.
    Ann. Rev. Astr. Astrophys., **16**, 171.
Gauzit, J. 1955. *Comptes Rendus*, **241**, 793.
Iben, I. Jr. 1982. *Astrophys. J.*, **259**, 244.
Kenyon, S.J. and Truran, J.W. 1983. *Astrophys. J.*, **273**, 280.
Kenyon, S.J. and Webbink, R.F. 1984. *Astrophys. J.*, **279**, 252.
Kraft, R.E. 1964. *Astrophys. J.*, **139**, 457.
Kuiper, G.P. 1940. *Pub. Am. Astr. Soc.*, **10**, 57.
Kutter, G.S. and Sparks, W.M. 1980. *Astrophys. J.*, **239**, 988.
Kwok, S. 1977. *Astrophys. J.*, **214**, 437.
MacDonald, J. 1980. *Mon. Not. Roy. Astr. Soc.*, **191**, 933.
MacDonald, J. 1983. *Astrophys. J.*, **267**, 732.
Nariai, K., Nomoto, K., and Sugimoto, D. 1980.
    Pub. Astr. Soc. Japan, **32**, 472.
Paczynski, B. 1965. *Acta Astr.*, **15**, 89.
Paczynski, B. and Rudak, B. 1980. *Astr. Astrophys.*, **82**, 349.
Paczynski, B. and Zytkow, A. 1978. *Astrophys. J.*, **222**, 604.
Payne-Gaposchkin, C. 1957. *The Galactic Novae* (Amsterdam: North Holland).
Prialnik, D., Livio, M., Shaviv, G., and Kovetz, A. 1982.
    Astrophys. J., **257**, 312.
Pringle, J.E. 1981. *Ann Rev. Astr. Astrophys.*, **19**, 137.
Schwarzschild, M. and Harm, R. 1965. *Astrophys. J.*, **159**, 903.
Sparks, W.M., Starrfield, S., and Truran, J.W. 1978.
    Astrophys. J., **220**, 1063.
Starrfield, S., Sparks, W.M., and Truran, J.W. 1984,
    Astrophys. J., **291**, 136.
Thackeray, A.D. 1977. *Mem. Roy. Astr. Soc.* **83**, 1.
Tutukov, A.V. and Yungel'son, L.R. 1976. *Astrofizika*, **12**, n21.
Tutukov, A.V. and Yungel'son, L.R. 1982. in
    IAU Colloquium No. 70, The Nature of Symbiotic Stars,
    ed. M. Friedjung and R. Viotti (Dordrecht: Reidel), p. 283.
Warner, B. 1976. in
    IAU Symposium 73, Structure and Evolution of Close Binary Systems,
    ed. P. Eggleton, S. Mitton, and J.A.J. Whelan (Dordrecht: Reidel), p. 85.
Wdowiak, T.J. 1977. *Pub. Astr. Soc. Pac.*, **89**, 569.
Webbink, R.F. 1975. Ph. D. thesis, Cambridge Univ.
Webbink, R.F. 1979. in
    IAU Colloquium No. 46, Changing Trends in Variable Star Research,
    ed. F.M Bateson, J. Smak, and I. Urch (Hamilton, N.Z.: U. Waikato Press), p. 102.
Willson, L.A., Garnavich, P., and Mattei, J.A. 1981.
    Inf. Bull. Var. Stars, No. 1961.
Wood, P.R. 1977. *Astrophys. J.*, **217**, 530.
Zuckerman, B. 1980. *Ann. Rev. Astr. Astrophys.*, **18**, 263.

Chapter 5
Allen, D.A. 1980a. *Mon. Not. Roy. Astr. Soc.*, **190**, 75.
Allen, D.A. 1980b. *Mon. Not. Roy. Astr. Soc.*, **192**, 521.
Allen, D.A. 1981. *Mon. Not. Roy. Astr. Soc.*, **197**, 739.

Aller, L.H. 1954. *Pub. DAO Victoria*, **9**, 321.

Andrillat, Y. 1982. in
IAU Colloquium No. 70, The Nature of Symbiotic Stars,
ed. M. Friedjung and R. Viotti (Dordrecht: Reidel), p. 47.

Andrillat, Y., Ciatti, F., and Swings, J.P. 1982.
Astrophys. Space. Sci., **83**, 423.

Andrillat, Y. and Houziaux, L. 1976. *Astr. Astrophys.*, **52**, 119.

Bath, G.T. 1977. *Mon. Not. Roy. Astr. Soc.*, **178**, 203.

Bath, G.T. and Pringle, J. E. 1982.
Mon. Not. Roy. Astr. Soc., **201**, 345.

Belyakina, T.S. 1970. *Astrofizika*, **6**, 49.

Belyakina, T.S. 1975. in
IAU Symposium No. 67, Variable Stars and Stellar Evolution,
ed. V.E. Sherwood and L. Plaut (Dordrecht: Reidel), p. 397.

Belyakina, T.S. 1979. *Izv. Krym. Astrofiz. Obs.*, **59**, 133.

Belyakina, T.S. 1984. *Izv. Krym. Astrofiz. Obs.*, **68**, 108.

Bloch, M. and Tcheng, M.-L. 1953. *Ann. d'Astrophys.*, **16**, 73.

Boyarchuk, A.A. 1967. *Soviet Astr. J.*, **11**, 8.

Cassatella, A., Patriarchi, P., Selvelli, P.L., Bianchi, L.,
Cacciari, C., Heck, A., Perryman, M. and Wamsteker, W.
1982. in *Third European IUE Conference*, ed. E. Rolfe, A. Heck
and B. Battrick (ESA SP-176), p. 229.

FitzGerald, M.P. and Pilavaki, A. 1974. *Astrophys. J. Suppl.*, **28**, 147.

Gallagher, J.S., Holm, A.V., Anderson, C.M., and Webbink, R.F. 1979.
Astrophys. J., **229**, 994.

Garcia, M.R. 1983. *Spectroscopic Orbits of Symbiotic Stars*,
Harvard College Observatory Research Exam.

Gratton, L. 1952. *Trans. IAU*, **8**, 849.

Hjellming, R.M. and Bignell, N. 1982. *Science*, **216**, 1279.

Hjellming, R.M., Wade, C.M., Vandenberg, N.R., and Newell, R.T. 1979.
Astr. J., **84**, 1619.

Hutchings, J.B. 1972. *Mon. Not. Roy. Astr. Soc.*, **158**, 177.

Iijima, T. and Ortolani, S. 1984. *Astr. Astrophys.*, **136**, 1.

Joy, A.H. 1931. *Pub. Astr. Soc. Pac.*, **43**, 353.

Kenyon, S.J. 1983. Ph. D. thesis, Univ. Illinois.

Kenyon, S.J. and Gallagher, J.S. 1983. *Astr. J.*, **88**, 666.

Kenyon, S.J. and Webbink, R.F. 1984. *Astrophys. J.*, **279**, 252.

Keyes, C.D. and Plavec, M.J. 1980a. in
The Universe at Ultraviolet Wavelengths,
ed. R.D. Chapman (NASA CP-2171), p. 443.

Keyes, C.D. and Plavec, M.J. 1980b. in *IAU Symposium No. 88*,
Close Binary Stars: Observations and Interpretation,
ed. M.J. Plavec, D.M. Popper and R.K. Ulrich (Dordrecht: Reidel), p. 365.

Kraft, R.E. 1958. *Astrophys. J.*, **127**, 625.

Kwok, S., Bignell, N., and Purton, C.R. 1984. *Astrophys. J.*, **279**, 188.

Kwok, S. and Leahy, D.A. 1984. *Astrophys. J.*, **283**, 675.

Larsson-Leander, G. 1947. *Ark. f. Mat., Astr., Fys.*, **34**, 16.

Lundmark, K. 1921. *Astr. Nachr.*, **213**, 93.

Luud, L. 1980. *Astrofizica*, **16**, 443.

Lynden-Bell, D. and Pringle, J.E. 1974.
Mon. Not. Roy. Astr. Soc., **168**, 203.

MacDonald, J. 1983. *Astrophys. J.*, **267**, 732.

McLaughlin, D.B. 1947. *Pub. Astr. Soc. Pac.*, **63**, 8.

McLaughlin, D.B. 1953. *Astrophys. J.*, **117**, 279.

Mammano, A., Rosino, L., and Yildizdogdu, S. 1975. in
Non-periodic Phenomena in Variable Stars,
ed. L Detre (Dordrecht: Reidel), p. 415.

Merrill, P.W. 1951. *Astrophys. J.*, **113**, 605.

Merrill, P.W. 1959. *Astrophys. J.*, **129**, 44.

Morgan, W.W. and Deutsch, A.J. 1947. *Astrophys. J.*, **106**, 362.

Nassau, J.J. and Cameron, D.M. 1954. *Astrophys. J.*, **119**, 75.

Paczynski, B. 1965. *Acta Astr.*, **15**, 197.

Paczynski, B. and Rudak, B. 1980. *Astr. Astrophys.*, **82**, 349.

Payne-Gaposchkin, C. 1957. *The Galactic Novae* (Amsterdam: North Holland).

Purton, C.R., Kwok, S., and Feldman, P.A. 1983. *Astr. J.*, **88**, 1825.

Raymond, J.C. 1979. *Astrophys. J. Suppl.*, **39**, 1.

Rigollet, R. 1947. *l'Astr.*, **61**, 247.

Sanford, R.F. 1949. *Astrophys. J.*, **109**, 81.

Solf, J. 1983. *Astrophys. J. (Letters)*, **266**, L.113.

Starrfield, S., Sparks, W.M., and Truran, J.W. 1984,
Astrophys. J., **291**, 136.

Stauffer, J.R. 1984. *Astrophys. J.*, **280**, 695.

Stencel, R.E., Michalitsianos, A.M., Kafatos, M., and
Boyarchuk, A.A. 1982. *Astrophys. J. (Letters)*, **253**, L77.

Swings, J.P. and Allen, D.A. 1972. *Pub. Astr. Soc. Pac.*, **84**, 523.

Szkody, P. 1977. *Astrophys. J.*, **217**, 140.

Taylor, A.R. and Seaquist, E.R. 1984. *Astrophys. J.*, **286**, 263.

Thackeray, A.D. 1977. *Mem. Roy. Astr. Soc.*, **83**, 1.

Tutukov, A.V. and Yungel'son, L.R. 1976. *Astrofizika*, **12**, 521.

Viotti, R., Altamore, A., Baratta, G.B., Cassatella, A., Friedjung, M.,
Giangrande, A., Ponz, D., and Ricciardi, O. 1982.
in *Advances in Ultraviolet Astronomy,*
ed. Y. Kondo, J.M. Mead, and R.D. Chapman (NASA CP-2238), p. 446.

Viotti, R., Ricciardi, O., Ponz, D., Giangrande, A., Friedjung, M.,
Cassatella, A., Baratta, G.B., and Altamore, A. 1983.
Astr. Astrophys., **119**, 255.

Wallerstein, G. 1978. *Pub. Astr. Soc. Pac.*, **90**, 36.

Wallerstein, G., Willson, L.A., Salzer, J., and Brugel, E. 1984.
Astr. Astrophys., **133**, 137.

Webbink, R.F. 1976. *Nature*, **262**, 271.

Webbink, R.F. 1978. *Pub. Astr. Soc. Pac.*, **90**, 57.

Williams, R.E. 1977. in
*IAU Colloquium No. 42, Interaction of Variable Stars with their
Environment*, ed. R. Kippenhahn, J. Rahe, and W. Strohmeier,
Veroff. Bamberg Remeis-Sternwarte, **11**, 242.

144

Willson, L.A., Wallerstein, G., Brugel, E., and Stencel, R.E. 1984.
 Astr. Astrophys., 133, 154.
Yildizdogdu, F.S. 1981. Publ. Instanbul Univ. Obs., No. 117.

Chapter 6
Abt, H.A. and Levy, S.G. 1976. Astrophys. J. Suppl., 30, 273.
Abt, H.A. and Levy, S.G. 1978. Astrophys. J. Suppl., 36, 241.
Becker, S.A. 1979. Ph. D. thesis, Univ. Illinois.
Boyarchuk, A.A. 1975. in
 IAU Symposium No. 67, Variable Stars and Stellar Evolution,
 ed. V.E. Sherwood and L. Plaut (Dordrecht: Reidel), p. 377.
Garcia, M.R., Baliunas, S.L., Doxsey, R., Elvis, M., Fabbiano, G.,
 Koenigsberger, G., Patterson, J., Schwartz, D.A., Swank, J., and Watson, M.
 1983. Astrophys. J., 267, 291.
Iben, I. Jr. 1967. Ann. Rev. Astr. Astrophys., 5, 571.
Iben, I., Jr. and Tutukov, A.V. 1983. in
 Stellar Nucleosynthesis, ed. C. Chiosi and A. Renzini
 (Dordrecht: Reidel), p. 181.
Kwok, S. 1981. in Proceedings of the North American Workshop on
 Symbiotic Stars, ed. R.E. Stencel (Boulder: JILA), p. 24.
Livio, M., Salzman, J., and Shaviv, G. 1979.
 Mon. Not. Roy. Astr. Soc., 188, 1.
Livio, M. and Warner, B. 1984. Obs., 104, 152.
Lutz, J.H. 1984. Astrophys. J., 279, 714.
Mazurek, T.J. and Wheeler, J.C. 1980. Fund. Cosmic Phys., 5, 193.
Merrill, K.M. and Stein, W.A. 1976. Pub. Astr. Soc. Pac., 88, 285.
Michalitsianos, A.G., Hobbs, R.W., and Kafatos, M. 1980. in
 The Universe at Ultraviolet Wavelengths, ed. R.D. Chapman
 (NASA CP-2171), p. 367.
Mouschovias, T. Ch. 1978. in Protostars and Planets,
 ed. T. Gehrels (Tucson, Univ. of Arizona), p. 209.
Nomoto, K. 1982. Astrophys. J., 257, 780.
Paczynski, B. 1971. Acta Astr., 21, 417.
Paczynski, B. 1976. in
 IAU Symposium No. 73, Structure and Evolution of Close Binary Systems,
 ed. P. Eggleton, S. Mitton, and J. Whelan (Dordrecht: Reidel), p. 75.
Patterson, J. 1984. Astrophys. J. Suppl., 54, 443.
Plavec, M. 1982. in
 IAU Colloquium No. 70, The Nature of Symbiotic Stars,
 ed. M. Friedjung and R. Viotti (Dordrecht: Reidel), p. 231.
Plavec, M., Ulrich, R.K., and Polidan, R.S. 1973.
 Pub. Astr. Soc. Pac., 85, 769.
Popova, E.I., Tutukov, A.V., and Yungel'son, L.R. 1982.
 Astrophys. Space Sci., 88, 55.
Ritter, H. 1976. Mon. Not. Roy. Astr. Soc., 175, 279.
Rudak, B. 1982. in
 IAU Colloquium No. 70, The Nature of Symbiotic Stars,
 ed. M. Friedjung and R. Viotti (Dordrecht: Reidel), p. 275.

Tammann, G.A. 1982. in
  Supernovae: A Survey of Current Research,
  Proc. of NATO Adv. Study Inst., ed. M.J. Rees and R.J. Stoneham
  (Dordrecht: Reidel), p. 371.
Tutukov, A.V. and Yungel'son, L.R. 1982. in
  IAU Colloquium No. 70, The Nature of Symbiotic Stars,
  ed. M. Friedjung and R. Viotti (Dordrecht: Reidel), p. 283.
Webbink, R.F. 1977. *Astrophys. J.,* **215,** 851.
Webbink, R.F. 1979a. in
  IAU Colloquium No. 46, Changing Trends in Variable Star Research,
  ed. F.M Bateson, J. Smak, and I. Urch (Hamilton, N.Z.: U. Waikato Press), p. 102.
Webbink, R.F. 1979b. in
  IAU Colloquium No. 53, White Dwarfs and Variable Degenerate Stars,
  ed. H.M. Van Horn and V. Weidemann (Rochester: Univ. of Rochester Press) p. 426.
Weidemann, V. 1979. in
  IAU Colloquium No. 53, White Dwarfs and Variable Degenerate Stars,
  ed. H.M. Van Horn and V. Weidemann (Rochester: Univ. of Rochester Press) p. 206.
Willson, L.A. 1981. in *Proceedings of the North American Workshop on
  Symbiotic Stars,* ed. R.E. Stencel (Boulder: JILA), p. 23.
Willson, L.A., Wallerstein, G., Brugel, E., and Stencel, R.E. 1984.
  Astr. Astrophys., **133,** 154.
Zuckerman, B. 1980. *Ann. Rev. Astr. Astrophys.,* **18,** 263.

Appendix

A.1   Introduction

In the preceding pages, I have tried to present the observed properties of and to develop a theoretical understanding for the group of unusual variables known as symbiotic stars. While excellent reviews of this subject have appeared elsewhere (Allen 1979a; Boyarchuk 1964a, 1975; Feast 1982; Hack 1982; Ilovaisky and Wallerstein 1968; Sahade 1960, 1976), it has been many years since Payne-Gaposchkin (1957) and Swings (1970) discussed the peculiarities of *individual* symbiotic systems (lists of symbiotic stars also appeared in Bidelman 1954 and Wackerling 1970). Over one hundred stars with combination spectra have been identified since Swings' review; most of these discoveries resulted from detailed objective prism searches conducted by Merrill and Burwell (1933, 1943, 1950), Henize (1967, 1976), and Sanduleak and Stephenson (1973). A number of systems were originally classified as stellar planetary nebulae (e.g., Minkowski 1946, 1948; Haro 1952; Perek 1960, 1962, 1963; Perek and Kohoutek 1967; Frantsman 1962; Thé 1962, 1964; Wray 1966), and later were correctly identified as symbiotic stars by Allen (1979a,b, 1984, and references therein).

In addition to these exciting new discoveries, large amounts of new data have been gathered for the classical symbiotics known by Merrill, especially in the infrared (cf. Allen 1974b, 1980a; Allen and Glass 1974, 1975; Glass and Webster 1973; Feast and Glass 1980; Swings and Allen 1972; Feast, et al. 1983a). It seems appropriate, then, to publish a new summary of individual systems, before the anticipated deluge of material from Space Telescope and other satellite projects buries the recent discoveries made by *Einstein*, *IUE*, and various ground-based observatories.

The goal of this Appendix is to provide astronomers with a

comprehensive and relatively unbiased database and bibliography for objects that are either known or suspected to be symbiotic stars. This material is believed to be complete up to the end of 1985, and additional material from 1986 has been added when possible. There are three major sections: (i) tabular data (including equatorial and galactic coordinates, ephemerides and orbital elements), (ii) historical summaries of selected symbiotic stars (including plotted spectra and light curves), and (iii) the bibliography. I hope that this section will furnish scientists with a broader understanding of the symbiotic phenomenon, and act as an aid in planning new observational and theoretical projects.

A.2   Tabular Data

Since the discovery of CI Cyg, RW Hya and AX Per in 1932, roughly 130 objects have been identified unambiguously as stars with combination spectra. The systems appearing in Table A.1 are ordered by right ascension for epoch 1950.0 and epoch 2000.0. Galactic coordinates are also provided for symbiotics that inhabit the Milky Way Galaxy (cf. Allen 1984).

A short list of systems suspected of having combination spectra is given in Table A.2. Many of these were taken from a similar compilation by Allen (1984); additional systems were added based on descriptions in the General Catalog of Variable Stars. These systems are also listed in order of 1950.0 right ascension; V and K magnitudes are given when available. Aside from V748 Cen, HD 330036 and BL Tel, very little is known about these objects - further study is needed to determine if any should be transferred to Table A.1.

A brief summary of the observed properties of the known symbiotic stars appears in Table A.3. These systems are listed in order of right ascension as in Table A.1, and the preferred object identification is given in the first column. Alternative designations are listed in Table A.8 and in the individual summaries. The observed X-ray fluxes (in units of 10^{-12} erg cm^{-2} s^{-1}) are given in column (2), and most of the values derived to date are upper limits (Allen 1981a,b). These fluxes are somewhat model dependent, but it is apparent that symbiotic stars are not copious X-ray emitters regardless of the model used to analyze the available observations.

The *IUE* satellite has been very important in improving our general understanding of symbiotic stars, and spectra covering $\lambda\lambda 1000$-3000 have been acquired for most bright systems (Sahade and Brandi 1980; Sahade, Brandi, and Fontenla 1984; Baratta and Viotti 1983). A "y"

Table A.1 *The known symbiotic stars*

Star	RA(1950)	Dec(1950)	RA(2000)	Dec(2000)	l^{II}	b^{II}
	h m s	° ′ ″	h m s	° ′ ″	°	°
EG And	00 41 52.7	40 24 22	00 44 37.2	40 40 46	121.5	−22.1
SMC N60	00 55 30.2	−74 29 30	00 57 06.0	−74 13 18		
SMC Ln 358	00 57 42.3	−75 21 29	00 59 12.2	−75 05 19		
SMC N73	01 03 18.9	−76 04 28	01 04 39.1	−75 48 24		
AX Per	01 33 06.0	54 00 18	01 36 16.9	54 15 35	129.5	−8.0
V741 Per	01 55 32.9	52 39 15	01 58 49.7	52 53 48	133.1	−8.6
UV Aur	05 18 33.3	32 27 51	05 21 49.0	32 30 44	174.2	−2.3
LMC ---	05 46 02.7	−71 17 13	05 45 19.7	−71 16 10		
LMC S63	05 48 52.4	−67 37 02	05 48 44.2	−67 36 13		
BX Mon	07 22 54.0	−03 30 00	07 25 23.9	−03 35 59	220.0	5.8
Wra 157	08 04 32.2	−28 23 18	08 06 35.1	−28 31 59	246.6	1.9
RX Pup	08 12 28.2	−41 33 18	08 14 12.4	−41 42 28	258.5	−3.9
Hen 160	08 23 26.7	−51 18 46	08 24 52.8	−51 28 35	267.7	−7.8
AS 201	08 29 36.8	−27 35 20	08 31 42.8	−27 45 32	249.1	6.9
He2-38	09 53 03.7	−57 04 39	09 54 43.3	−57 18 53	290.8	−2.2
SS 29	11 06 27.3	−65 31 02	11 08 27.5	−65 47 17	292.6	−5.0
SY Mus	11 29 55.0	−65 08 36	11 32 10.5	−65 25 10	294.8	−3.8
BI Cru	12 20 40.3	−62 21 39	12 23 26.4	−62 38 16	299.7	0.0
TX CVn	12 42 17.9	37 02 14	12 44 42.2	36 45 49	115.1	80.2
He2-87	12 42 48.3	−62 44 09	12 45 47.1	−63 00 32	302.3	−0.1
Hen 828	12 48 01.6	−57 34 28	12 50 58.0	−57 50 46	302.9	5.0
SS 38	12 48 21.8	−64 43 40	12 51 26.3	−64 59 58	302.9	−2.1
Hen 863	13 04 49.0	−47 44 22	13 07 43.8	−48 00 22	305.7	14.7
St2-22	13 11 22.4	−58 36 01	13 14 30.6	−58 51 52	305.9	3.8
Hen 905	13 27 23.2	−57 42 50	13 30 37.2	−57 58 17	308.1	4.5
RW Hya	13 31 32.0	−25 07 29	13 34 18.2	−25 22 49	315.0	36.4
Hen 916	13 31 59.1	−64 30 25	13 35 28.9	−64 45 44	307.6	−2.2
V835 Cen	14 10 22.7	−63 11 45	14 14 09.0	−63 25 44	312.0	−2.0
BD-21°3873	14 13 45.8	−21 32 59	14 16 34.3	−21 46 51	327.9	36.9
He2-127	15 21 10.4	−51 39 15	15 24 49.8	−51 49 48	325.5	4.1
Hen 1092	15 42 29.9	−66 19 59	15 47 10.8	−66 29 16	319.2	−9.3
Hen 1103	15 45 00.3	−44 09 50	15 48 28.4	−44 19 00	333.2	7.9
He2-139	15 50 48.6	−55 20 48	15 54 44.7	−55 29 36	326.9	−1.4
T CrB	15 57 24.5	26 03 39	15 59 30.2	25 55 12	42.4	48.1
AG Dra	16 01 23.2	66 56 25	16 01 41.0	66 48 09	10.3	40.9
W16-202	16 03 15.0	−49 18 40	16 06 57.1	−49 26 41	332.8	1.9
He2-147	16 09 55.6	−56 51 56	16 14 00.8	−56 59 31	327.9	−4.3
UKS-Ce1	16 12 31.2	−22 04 49	16 15 29.3	−22 12 16	353.0	20.2
Wra 1470	16 20 16.1	−27 33 14	16 23 21.6	−27 40 10	350.1	15.2
He2-171	16 30 47.0	−34 59 10	16 34 04.2	−35 05 23	346.0	8.5
Hen 1213	16 31 23.0	−51 36 14	16 35 15.2	−51 42 24	333.9	−2.8
He2-173	16 32 59.0	−39 45 38	16 36 24.6	−39 51 42	342.8	5.0
He2-176	16 37 54.3	−45 07 22	16 41 31.2	−45 13 05	339.4	0.7
KX TrA	16 40 00.3	−62 31 40	16 44 35.5	−62 37 13	326.4	−10.9
AS 210	16 48 15.7	−25 55 25	16 51 20.5	−26 00 27	355.5	11.5
HK Sco	16 51 29.8	−30 18 16	16 54 40.8	−30 23 04	352.5	8.2

Table A.1 (*cont.*)

Star	RA(1950)	Dec(1950)	RA(2000)	Dec(2000)	l^{II}	b^{II}
	h m s	° ′ ″	h m s	° ′ ″	°	°
CL Sco	16 51 40.3	− 30 32 30	16 54 51.9	− 30 37 17	352.3	8.2
V455 Sco	17 04 04.1	− 34 01 19	17 07 21.8	− 34 05 14	351.2	3.8
Hen 1341	17 05 42.5	− 17 22 41	17 08 36.7	− 17 26 29	5.0	13.3
Hen 1342	17 05 53.2	− 23 19 50	17 08 55.1	− 23 23 37	0.1	9.9
AS 221	17 08 56.8	− 32 34 14	17 12 12.3	− 32 37 48	353.0	3.9
H2-5	17 12 04.7	− 31 30 42	17 15 18.6	− 31 34 03	354.2	4.0
Th3-7	17 17 51.9	− 29 19 55	17 21 02.6	− 29 22 51	356.7	4.2
Draco C-1	17 19 08.5	57 53 01	17 19 57.6	57 50 05		
Th3-17	17 24 21.4	− 29 00 28	17 27 31.8	− 29 03 56	357.8	3.2
Th3-18	17 25 17.1	− 28 36 09	17 28 26.9	− 28 38 33	358.2	3.3
Hen 1410	17 25 54.9	− 29 41 01	17 29 06.4	− 29 43 22	357.4	2.6
V2116 Oph	17 28 57.8	− 24 42 45	17 32 02.1	− 24 45 53	1.9	4.7
Th3-30	17 30 34.2	− 28 05 20	17 33 43.4	− 28 07 21	359.3	2.6
Th3-31	17 31 05.6	− 29 27 12	17 34 16.8	− 29 29 11	358.2	1.8
M1-21	17 31 20.5	− 19 07 23	17 34 17.3	− 19 09 21	7.0	7.3
Pt-1	17 35 46.5	− 23 52 24	17 38 49.7	− 23 54 03	3.5	3.9
RT Ser	17 37 04.1	− 11 55 03	17 39 51.9	− 11 56 37	13.9	9.9
AE Ara	17 37 19.7	− 47 01 50	17 41 04.9	− 47 03 20	344.0	− 8.6
SS 96	17 38 04.8	− 36 46 14	17 41 28.3	− 36 47 42	352.8	− 3.3
UU Ser	17 39 46.3	− 15 23 18	17 42 38.4	− 15 24 40	11.2	7.6
V2110 Oph	17 40 30.8	− 22 44 16	17 43 32.5	− 22 45 34	5.0	3.6
SSM 1	17 40 32.6	− 36 02 07	17 43 54.8	− 36 03 24	353.7	− 3.4
V917 Sco	17 44 41.5	− 36 07 19	17 48 03.9	− 36 08 18	354.1	− 4.1
H1-36	17 46 24.1	− 37 00 36	17 49 48.2	− 37 01 27	353.5	− 4.9
RS Oph	17 47 31.6	− 06 41 39	17 50 13.2	− 06 42 27	19.8	10.3
He2-294	17 48 28.5	− 32 54 10	17 51 45.4	− 32 54 53	357.3	− 3.1
AS 245	17 48 59.5	− 22 18 50	17 52 00.7	− 22 19 31	6.3	2.3
Bl3-14	17 49 14.1	− 29 45 19	17 52 26.0	− 29 45 59	0.1	− 1.7
Bl L	17 50 00.7	− 30 17 25	17 53 13.5	− 30 18 01	359.7	− 2.1
AS 255	17 53 47.8	− 35 15 15	17 57 08.8	− 35 15 34	355.8	− 5.3
V2416 Sgr	17 54 15.6	− 21 41 10	17 57 15.9	− 21 41 28	7.6	1.4
SS 117	17 59 07.6	− 31 59 14	18 02 23.1	− 31 59 10	359.2	− 4.6
Ap 1-8	18 01 19.8	− 28 21 49	18 04 29.6	− 28 21 36	2.6	− 3.2
SS 122	18 01 33.1	− 27 09 26	18 04 41.1	− 27 09 12	3.7	− 2.7
AS 270	18 02 35.2	− 20 20 52	18 05 33.7	− 20 20 34	9.7	0.4
H2-38	18 02 51.5	− 28 17 23	18 06 01.2	− 28 17 03	2.8	− 3.5
SS 129	18 03 54.0	− 29 36 50	18 07 05.7	− 29 36 25	1.8	− 4.3
V615 Sgr	18 04 17.5	− 36 06 47	18 07 40.0	− 36 06 20	356.1	− 7.6
Hen 1591	18 04 25.8	− 25 54 13	18 07 32.0	− 25 53 46	5.1	− 2.6
AS 276	18 05 37.2	− 41 13 58	18 09 09.5	− 41 13 25	351.6	− 10.2
Ap 1-9	18 07 19.5	− 28 08 20	18 10 28.9	− 28 07 41	3.4	− 4.3
AS 281	18 07 34.6	− 27 58 31	18 10 43.8	− 27 57 50	3.6	− 4.3
V2506 Sgr	18 07 51.6	− 28 33 22	18 11 01.7	− 28 32 40	3.1	− 4.6
SS 141	18 08 53.9	− 33 11 29	18 12 11.3	− 33 10 42	359.1	− 7.0
AS 289	18 09 34.7	− 11 40 55	18 12 22.2	− 11 40 07	18.1	3.1
Y CrA	18 10 47.3	− 42 51 26	18 14 22.9	− 42 50 30	350.6	11.8

Table A.1 (*cont.*)

Star	RA(1950)	Dec(1950)	RA(2000)	Dec(2000)	l^{II}	b^{II}
	h m s	° ′ ″	h m s	° ′ ″	°	°
V2756 Sgr	18 11 22.5	−29 50 20	18 14 34.5	−29 49 23	2.4	−5.9
HD 319167	18 12 11.5	−30 32 56	18 15 24.6	−30 31 55	1.8	−6.4
YY Her	18 12 25.9	20 58 20	18 14 34.1	20 59 18	48.1	17.2
He2-374	18 12 30.8	−21 36 26	18 15 31.0	−21 35 24	9.7	−2.2
AS 296	18 12 33.0	−00 19 53	18 15 06.4	−00 18 59	28.5	7.9
AS 295B	18 12 51.9	−30 52 16	18 16 05.5	−30 51 12	1.6	−6.6
AR Pav	18 15 24.6	−66 06 07	18 20 28.6	−66 04 48	328.5	21.6
Hen 1674	18 17 12.2	−26 24 06	18 20 19.0	−26 22 44	6.0	−5.4
He2-390	18 17 51.3	−26 49 52	18 20 58.7	−26 48 27	5.7	−5.7
V3804 Sgr	18 18 13.8	−31 33 31	18 21 28.4	−31 32 04	1.5	−8.0
V443 Her	18 20 02.8	23 25 48	18 22 07.7	23 27 20	51.2	16.6
V3811 Sgr	18 20 28.5	−21 54 45	18 23 29.0	−21 53 09	10.3	−3.9
CD-28°14567	18 22 16.7	−28 37 42	18 25 26.7	−28 35 57	4.5	−7.4
V1017 Sgr	18 28 53.0	−29 26 06	18 32 04.1	−29 23 53	4.5	−9.1
V2601 Sgr	18 35 00.7	−22 44 30	18 38 02.1	−22 41 50	11.2	−7.3
AS 316	18 39 33.4	−21 20 46	18 42 32.8	−21 17 47	12.9	−7.6
MWC 960	18 44 58.1	−20 09 13	18 47 55.8	−20 05 51	14.5	−8.2
AS 327	18 50 13.4	−24 26 42	18 53 16.7	−24 22 57	11.1	11.2
FN Sgr	18 50 58.2	−19 03 30	18 53 54.4	−18 59 42	16.2	−9.0
Pe 2-16	18 51 30.9	−04 42 41	18 54 10.0	−04 38 52	29.1	−2.7
V919 Sgr	19 00 51.6	−17 04 24	19 03 45.1	−16 59 54	19.0	10.3
CM Aql	19 00 57.9	−03 07 43	19 03 35.1	−03 03 14	31.6	−4.0
AS 338	19 01 32.0	16 21 49	19 03 46.8	16 26 19	49.0	4.7
Ap3-1	19 08 05.4	02 44 33	19 10 36.0	02 49 31	37.6	−2.9
BF Cyg	19 21 55.2	29 34 34	19 23 53.4	29 40 28	62.9	6.7
CH Cyg	19 23 14.2	50 08 31	19 24 33.1	50 14 29	81.9	15.5
Hen 1761	19 37 20.8	−68 14 45	19 42 25.9	−68 07 41	327.7	29.7
HM Sge	19 39 41.4	16 37 33	19 41 57.0	16 44 39	53.6	−3.1
AS 360	19 43 35.7	18 29 23	19 45 49.3	18 36 45	55.6	−3.0
CI Cyg	19 48 21.0	35 33 24	19 50 12.2	35 41 03	70.9	4.7
V1016 Cyg	19 55 19.8	39 41 30	19 57 05.0	39 49 36	75.2	5.6
RR Tel	20 00 20.1	−55 52 04	20 04 18.5	−55 43 34	342.2	32.2
PU Vul	20 19 01.1	21 24 43	20 21 13.3	21 34 18	62.6	−8.5
He2-467	20 33 42.6	20 01 07	20 35 57.3	20 11 33	63.4	12.1
He2-468	20 39 20.5	34 34 07	20 41 18.9	34 44 52	75.9	−4.4
V1329 Cyg	20 49 02.6	35 23 37	20 51 01.2	35 34 54	77.8	−5.4
CD-43°14304	20 56 48.6	−42 50 34	21 00 06.3	−42 38 50	358.6	41.1
V407 Cyg	21 00 26.0	45 34 36	21 02 11.7	45 46 28	87.0	−0.4
AG Peg	21 48 36.2	12 23 27	21 51 02.0	12 37 31	69.3	30.8
Z And	23 31 15.3	48 32 32	23 33 40.0	48 49 06	110.0	0.0
R Aqr	23 41 14.3	−15 33 42	23 43 49.5	−15 17 02	66.5	70.3

in column (3) indicates that at least one spectrum of the given symbiotic star has been obtained by *IUE*.

Optical and infrared data for symbiotic stars are fairly comprehensive; although the V magnitude is contaminated by emission lines, mean values are listed in column (4). Three columns have been devoted to the nature of the cool giant; its spectral type, K magnitude, and infrared continuum classification (e.g., S- or D-type) appear in columns (5-7), respectively (cf. Allen 1982a).

Numerous surveys have searched for radio emission from symbiotic stars (Lepine and Rien 1974; Altenhoff, et al. 1976; Bath and Wallerstein 1976; Brocka 1979; Johnson, Balick, and Thompson 1979; Milne 1979; Wright and Allen 1978; Seaquist, Taylor, and Button 1984). Most of the results listed in column (8) are upper limits at 6 cm; recent limits derived by the *VLA* tend to be significantly lower than those deduced from earlier measurements.

The reader should remember that all symbiotic stars are variable; significant fluctuations in the IR or radio fluxes are denoted by "V" in columns (6) and (8).

Table A.2 *Stars suspected of being symbiotic*

Star	RA(1950)	Dec(1950)	V	K
	h m s	° ′ ″		
LW Cas	02 53 06	60 29 09	16.9	9.1
GH Gem	07 00 48	12 06 48	14.6	
Hen 461	10 37 03.9	−51 08 33		3.9
RT Car	10 42 44.7	−59 06 46		1.9
Hen 653	11 23 16.5	−59 40 02		5.4
CD-36°8436	13 13 12.0	−36 44 18		5.7
V704 Cen	13 51 30.5	−58 12 42	13	8.4
He2-104	14 08 33.3	−51 12 18		6.9
V748 Cen	14 56 30	−33 13 42	9.9	8.1
HD 330036	15 47 38.0	−48 35 54		7.6
Sa 3-43	17 14 44.0	−29 58 34		7.9
V503 Her	17 34 41	23 20 00	14.4	
W16-312	17 47 03	−30 56 45		7.6
Th 4-4	17 47 25.9	−19 52 56		7.9
Ve 2-57	18 05 17.9	−24 34 13		7.1
V1148 Sgr	18 06 00	−26 00 12	>16.0	
V1988 Sgr	18 24 49	−27 39 24	13.0	
BL Tel	19 02 43.9	−51 29 42	7.7	4.8
SS Sge	19 36 52	16 35 42	11	
V1290 Aql	19 44 09	10 39 30	14	

Table A.3 *Selected properties of symbiotic stars*

Star	F_x	UV	V	Spectral type	K	IR type	$F_{6\,cm}$
EG And		y	7.5	M2	2.6	S	
SMC N60				C	13.0	S	
SMC Ln 358				C	11.5	S	
SMC N73				C	11.5	S	
AX Per	<0.09	y	12	M	5.5	S	<0.71
V741 Per		y		G2	9.8	D'	<0.49
UV Aur		y	7.9	C	2.2	S	<3.00
LMC ---		y			12.9	D	
LMC S63		y		C	11.3	S	
BX Mon		y	12	M4	5.7	S	<0.47
Wra 157				G	9.4	D'	<13
RX Pup		y		M	2.0V	D	20V
Hen 160				M7	7.5	S	<17
AS 201	<0.06		11	G	9.9	D'	
He2-38	<0.06	y		M	3.7V	D	<6
SS 29			13	G	10.7	S	<19
SY Mus		y	11.0	M	4.7	S	<12
BI Cru		y	10.7	M	4.7V	D	
TX CVn		y	9.3	M0	6.3	S	<0.90
He2-87				M7	6.0	S	<12
Hen 828			12.5	M6	7.1	S	<16
SS 38			13.8	M	5.7V	D	13
Hen 863			11.0	K4	8.5	S	<12
St2-22				M	8.5	S	
Hen 905			12.5	K4	8.5	S	<13
RW Hya		y	10	M2	4.7	S	0.38
Hen 916			12.5	M6	7.8	S	<20
V835 Cen	<0.11			M	4.3V	D	20
BD-21°3873	<0.07			G	7.2V	S	<3.00
He2-127	<0.11			M	7.9V	D	<12
Hen 1092	<0.04	y	14	K5	7.8	S	<18
Hen 1103			12.2	M0	8.4	S	<18
He2-139				M	5.3V	D	
T CrB	0.09	y	10.0	M3	4.8	S	<0.88
AG Dra	2.1	y	11.2	K3	6.2	S	<0.41
W16-202				M	6.8	S	
He2-147				M8	5.0	D	<12
UKS-Ce1			15.0	C	11.3	S	<0.37
Wra 1470			12.0	M4	7.8	S	<0.48
He2-171	<0.12	y		M	6.2V	D	2.0V
Hen 1213			10.4	K4	6.7	S	<12
He2-173				M	6.8	S	0.51
He2-176				M7	5.7	D	14
KX TrA	<0.04	y	12.4	M6	6.0	S	<22

Table A.3 (*cont.*)

Star	F_x	UV	V	Spectral type	K	IR type	$F_{6\ cm}$
AS 210			11.5	G:	6.7	D	1.78
HK Sco		y	15	M	7.9	S	<0.36
CL Sco		y	12	M	7.9	S	<0.42
V455 Sco	<0.08		15	M	5.9	S	2.36
Hen 1341			12.0	M0	7.6	S	<1.56
Hen 1342			12.0	M2	8.5	S	<0.42
AS 221			11.0	M4	7.6	S	<0.39
H2-5				M	5.5	S	<0.48
Th3-7				M	8.1	S	
Draco C-1			17.0	C1,2			
Th3-17				M3	8.2	S	
Th3-18				M2	8.2	S	
Hen 1410			12.5	M	8.5	S	<12
V2116 Oph	8000V		19	M	8.1	S	<0.78
Th3-30				K5	8.3	S	
Th3-31				M5	7.6	S	
M1-21			14	M2	7.2	S	<0.60
Pt-1			15	M1	8.6	S	
RT Ser			13	M	7.0	S	1.32
AE Ara	<0.05	y	12.5	M2	6.3	S	<18
SS 96			12	M2	6.4	S	2.57
UU Ser			16	M	9.1	S	<0.48
V2110 Oph			19	M8	7.5V	D	<0.60
SSM 1				M	8.3	S	
V917 Sco			13.0	M7	8.0	S	<25
H1-36		y		M	6.8V	D	40
RS Oph	<0.15	y	11.5	M2	6.5	S	<0.32
AS 245			11.0	M6	7.2	S	0.87
He2-294				M3	8.4	S	
Bl3-14			14.4	M6	8.7	S	
Bl3-6							
Bl L			16.0	M6	7.8	S	
AS 255			12.0	K3	8.4	D	<0.36
V2416 Sgr			13	M	4.6	S	3.52
SS 117			12.5	M6	7.7	S	<0.39
Ap 1-8				M0	7.9	S	<0.43
SS 122			12.0	M7	6.6	S:	1.64
AS 270			12.5	M1	5.5	S	
H2-38				M8	6.7	D	13
SS 129			12.5	K	8.0	S	<12
V615 Sgr			12.5	M	7.6	S	<20
Hen 1591			13.0	G	9.0	S	<0.45
AS 276			11.5	M4	8.1	D	<12
Ap 1-9			12.5	K4	8.8	S	<12
AS 281			12.0	M5	7.0	S	<18

Table A.3 (*cont.*)

Star	F_x	UV	V	Spectral type	K	IR type	$F_{6\,cm}$
V2506 Sgr			12.0	M	8.4	S	<0.47
SS 141			12.5	M	9.0	S	<12
AS 289		y	10.5	M3	5.0	S	
Y CrA	<0.06	y	12	M	6.6	S	<20
V2756 Sgr			11.5	M	7.8	S	<22
HD 319167			12.5	M3	7.5	S	<23
YY Her		y	12	M2	8.0	S	<0.35
He2-374			12.0	M	6.5	S	<0.35
AS 296		y	10.5	M5	4.5	S	0.63
AS 295B	<0.12	y	11.5	M			
AR Pav		y	11	M	7.2	S	<14
Hen 1674			12.5	M5	7.7	S	<0.46
He2-390			12	M	7.0V	D	0.73
V3804 Sgr			12.0	M	7.3	S	<14
V443 Her		y	11.5	M3	5.4	S	<0.44
V3811 Sgr			14	M	8.5	S	
CD-28°14567			11.5	M4	7.6	S	<19
V1017 Sgr	0.56		14	G5	10.6	S	
V2601 Sgr			15	M	8.0	S	<0.50
AS 316			12.0	M	7.8	S	<16
MWC 960			12.0	M0	7.8	S	<0.50
AS 327			11.5	M	8.5	S	<0.48
FN Sgr			11	M	7.9	S	<0.66
Pe 2-16			16.0	M5	8.1	S	<0.42
V919 Sgr			12.5	M	7.2	S	<12
CM Aql			15	M4	7.6	S	<0.57
AS 338			11.5	M5	7.5	S	<0.41
Ap3-1					8.7	S	
BF Cyg		y	12	M4	6.3	S	2.05
CH Cyg		y	7	M6	−0.7	S	1.38
Hen 1761			10.4	M3	5.6	S	<12
HM Sge	0.83	y	16	M	3.6V	D	33V
AS 360			11.0	M6	7.1	S	<15
CI Cyg	<0.05	y	11.1	M4	4.5	S	<0.42
V1016 Cyg	0.08	y	16	M6	4.6V	D	40
RR Tel	0.18	y	14	M	3.6V	D	28
PU Vul		y	9	M4-5	5.9	S	
He2-467			13	G	9.4	S	<0.24
He2-468				M	8.0	S	
V1329 Cyg		y	14	M5	6.9	S	2V
CD-43° 14304		y	10.0	K3	7.6	S	<16
V407 Cyg			15	M			
AG Peg		y	9.4	M2	3.6	S	8.15
Z And		y	10.5	M2	5.0	S	1.16
R Aqr	0.09	y	5.8	M7	−1.2V	S	6V

156   *Appendix*

Many symbiotics are known to be periodic variables; reliable ephemerides are available in some instances and have been collected in Table A.4. For most of the ephemerides, the times of minimum (Min) or maximum (Max) have been derived from photoelectric (U, B, V or K), photographic (pg), or photovisual (pv) data. The remaining ephemerides were derived from radial velocity data of non-uniform quality; T_o refers to the derived time of periastron passage, while Conj refers to the epoch of spectroscopic conjunction for T CrB.

Radial velocity data usually are not reliable for symbiotic stars, but orbital elements nevertheless have been derived for the nine systems appearing in Table A.5. The column headings are rather standard, with P representing the binary period, K_1 and K_2 the semi-amplitudes of the primary and secondary components, V_o the systemic velocity, e the eccentricity, ω the longitude of periastron, a_1 the semi-diameter of the orbit, i the inclination, and $f(M)$ the mass function. In some cases, the primary is the cool component (T CrB, CH Cyg, CI Cyg, and AG Peg), while the hot component is considered to be the primary in the other systems.

Very few symbiotic stars have published proper motions, and even

Table A.4 *Ephemerides for symbiotic stars*

Star	Ephemeris	Source
Z And	Min(pv) = JD 2421298 + 756.85*E	Kenyon and Webbink (1984)
R Aqr	Max(V) = JD 2382892.4 + 386.30*E	Willson, Garnavich and Mattei (1981)
UV Aur	Max(pg) = JD 2415016 + 395.2*E	Zakarov (1951)
TX CVn	T_o = JD 2443635 + 70.8*E	Fried (1980b)
V748 Cen	Min(V) = JD 2441917 + 566.5*E	van Genderen, Glass and Feast (1974)
T CrB	Conj = JD 2433687 + 227.5*E	Paczynski (1965)
BF Cyg	Min(pg) = JD 2415065 + 757.3*E	Pucinskas (1970)
CH Cyg	T_o = JD 2440023 + 5750*E	Yamashita and Maehara (1979)
CI Cyg	Min(V) = JD 2411902 + 855.25*E	Aller 1954
V1016 Cyg	Max(K) = JD2444101 + 472*E	Kenyon and Webbink (1984)
V1329 Cyg	Min(V) = JD 2424870 + 950*E	Grygar, et al. (1979)
AG Dra	Max(U) = JD 2438900 + 554*E	Meinunger (1979)
RW Hya	Max(pg) = JD 2421519.2 + 372.45*E	Kenyon and Webbink (1984)
BX Mon	Max(pg) = JD 2430345 + 1374*E	Kukarkin, et al. (1951)
SY Mus	Min(V) = JD 2435175.7 + 627.0*E	Kenyon and Bateson (1984)
AR Pav	Min(V) = JD 2420330 + 604.6*E	Andrews (1974)
AG Peg	Min(B) = JD 2428250 + 827.0*E	Meinunger (1981)
AX Per	Min(pg) = JD 2436679 + 681.6*E	Kenyon (1982)
RX Pup	Max(K) = JD 2442810 + 580*E	Kenyon and Webbink (1984)
V2601 Sgr	Min(pg) = JD 2429850 + 850*E	Hoffleit (1968b)
V2756 Sgr	Min(pg) = JD 2437485 + 243*E	Hoffleit (1970)
CL Sco	Min(pg) = JD 2427020 + 624.7*E	Kenyon and Webbink (1984)
RR Tel	Min(V) = JD 2442551.7 + 374.2*E	Kenyon and Bateson (1984)
BL Tel	T_o = JD 2434703.8 + 778.0*E	Wing (1962)

fewer have published parallaxes (van Maanen 1936, 1938). Those that do are listed in Table A.6; in most instances the proper motions are significant, while the parallaxes are obviously not significant.

The penultimate table of this section (Table A.7) is rather lengthy, and attempts to summarize the spectroscopic data of those systems that are so poorly studied that they did not deserve an entry in the detailed individual summary. The column headings refer to individual emission lines (He I λ5876, He II λ4686, C IV λ1548,1552, [Fe VII] λ6087, and the unidentified λ6830 feature), or groups of absorption

Table A.5 *Orbital elements for symbiotic stars*

Star	P	K_1	K_2	V_o	e	ω	$a_1 \sin i$	$f(M)$
EG And[a]	470	8.6		−94.8	0.0		80	0.032
TX CVn[b]	70.8	9.8		−6.3	0.56	103	11	0.004
T CrB[c]	227.6	24.0	33.5	−27.0	0.06	90	107	5
T CrB[d]	227.5	22.9	31.3	−30	0.0			3.6
CH Cyg[e]	5750	6.8		−58.4	0.29	201		0.16
CI Cyg[f]	855.25	23	16	+20	0.0		390	1.0
V1329 Cyg[g]	950		63.0	−34.0	0.17	40	118	23
AR Pav[h]	605	13.0		−63.6	0.11	127	154	0.140
AG Peg[i]	820	5.1	17-22	−16.3	0.23	222	82	
BL Tel[j]	778	19.3		+91.8	0.31	91	281	0.496

Notes
(a) Oliversen, et al. (1985)
(b) Fried (1980b)
(c) Kraft (1958)
(d) Paczynski (1965)
(e) Yamashita and Maehara (1979)
(f) Iijima (1983
(g) Iijima, Mammano and Margoni (1981)
(h) Thackeray and Hutchings (1974)
(i) Hutchings, Cowley and Redman (1975)
(j) Wing (1962)

Table A.6 *Proper motions and parallaxes for symbiotic stars*

Star	μ_α (s yr^{-1})	μ_δ ('' yr^{-1})	π
Z And	+0.0010	+0.000	−0.021 ± 0.013
EG And	+0.0009	−0.015	−
R Aqr	+0.0020	−0.021	−
UV Aur	−0.0013	+0.005	−
TX CVn	−0.0013	−0.014	−
T CrB	−0.0011	+0.020	−
CH Cyg	−0.0014	−0.022	−
AG Dra	−0.0006	−0.013	−
RW Hya	−0.0025	+0.014	−
AG Peg	−0.0005	−0.013	−0.003 ± 0.009
BL Tel	+0.0024	+0.005	

Table A.7 *Spectroscopic summary of selected symbiotic stars*

Star	He I	He II	C IV	[Fe VII]	6830	TiO	CO	H₂O	References
SMC N60	y	y		n	y	n			63
SMC Ln 358	y		n	n	y	n			63
SMC N73	y	n		n	y	n			63
LMC ---	y	y	y	y	y	n			5,29
LMC S63	y	y	y	n	n	n			5,29
Wra 157	y	y		n	n	n	n	n	68
Hen 160	y	y		y	y	y	y	n	
AS 201	y	n		n	n	n	n	n	52
He2-38	y	y		y	y	y		y	2,38,65
SS 29	y	y		y	n	n	n	n	
He2-87	y	y		y	y	y	y	n	65
Hen 828	y	y		y	y	y	y	n	
SS 38	y	y		y	n	n	y	n	51,64
Hen 863	y	y		n	n	y	y	n	
St 2-22	y	n		y	y	y			
Hen 905	y	y	y	y	y	y	y	n	
Hen 916	y	y		n	y	y	y	n	
V835 Cen	y	y		y	y	n	n	y	14,51,64
BD-21°3873	y	y		n	n	n	n	n	
He2-127	y	y		y	y	y	y	n	65
Hen 1092	y	y		y	y	y	y	n	65
Hen 1103	y	y		n	n	y	y	n	
HD 330036	y	n	y	n	n	n	n	n	11,33,34
He2-139	n	n		n	n	y			
W16-202	y	n		y	y	y			
He2-147	y	y		n	n	y	y	n	51
UKS-Ce1	n	y		n	n	y			32
Wra 1470	y	y		n	n	y	y	n	58
He2-171	y	y		y	n	n	y	n	51,65
Hen 1213	y			n	n	n?	y		
He2-173	y	y			n	y			65
He2-176	y	y		y	y	n	y	n	65
KX TrA	y	y	n	y	y	y	y	n	11,14,66
AS 210	y	y		n	y	n	n	n	51,67
CL Sco	y	y		n	n	y	y		17,38,56
HK Sco	y	y	y	n	y	n	y	n	16,54
V455 Sco	y	y		y	y	y	y	n	55,57
Hen 1341	y	y		y	y	n	y	n	
Hen 1342	y	y		n	y	n	y	n	
AS 221	y	y		y	y	n	y	n	21,47
H 2-5	y	y		y	y	y	y	y	21
Th 3-7	y	y		y	y	y	y	n	4,60
Draco C-1	y	y		n	n	y	y		1
Th 3-17	y	y		n	n	y	y	n	4,60
Th 3-18	y	y		n	y	y	y	n	4,60
Hen 1410	y	y		y	y	y	y	n	
Th 3-30	y	y		y	y	y	y	n	4,60
Th 3-31	y	y		y	y	y	y	n	4,60

Table A.7 (*cont.*)

Star	He I	He II	C IV	[Fe VII]	6830	TiO	CO	H₂O	References
M 1-21	y	y		y	y	y	y	n	31,42,48
Pt-1	y	y		n	n	y			49,61
AE Ara	y	y	y	n	n	y	y	n	12
SS 96	y	y		y	n	y	y	n	
UU Ser	y	y		n	n	y			20,50
V2110 Oph	y	y		y	y	n	y	n	3,18
SSM 1		y							
V917 Sco	y	y		y	y	y	y	n	
AS 245	y	y		y	y	n	y	n	19
He2-294	y	y		n	y	y	y	n	4,65
Bl3-14	y	y		n	y	y	y	n	4
Bl L	y	y		y	y	y	y	n	4,7
AS 255	y	y		n	n	y	y	n	
V2416 Sgr	y	y		y	y	y	y	n	24,43,62
SS 117	y	y		y	n	y	y	n	
Ap 1-8	y	y		y	y	y	y	n	19
SS 122	y	y		y	n	n	y	n	51
AS 270	y	y		n	n	y	y	n	
H 2-38	y	y		y	y	y	y	n	19,21,51
SS 129	y	y		n	y?	y	y	n	24
V615 Sgr	y	y		n	n	y	y	n	
Hen 1591	y	n		n	n	y?	n	n	24,27,33
AS 276	y	y		n	y	y	y	n	
Ap 1-9	y	y		n	n	y	y	n	19,35,47
AS 281	y	y		y	n	y	y	n	19,24,27,47
V2506 Sgr	y	y		n	y	y			31,35
SS 141	y	y		n	n	y			
AS 289	y	y		n	n	y	y	n	6
Y CrA	y	y	y	n	n	y	y	n	44,45
V2756 Sgr	y	y		y	y	y	y	n	24,27
HD 319167	y	y		n	n	y	y	n	13
He2-374	y	y		y	n	y			
AS 296	y	y		n	n	y	y	n	10
AS 295B	y	y		n	n	n			25
Hen 1674	y	y		y	y	y	y	n	24,27
He2-390	y	y		y	n	y	n		51
V3804 Sgr	y	y	n	n	n	n	y	n	8,9,39
V3811 Sgr	y	n		n	n	y			
CD-28°14567	y	y		y	n	y	y	n	
V2601 Sgr	y	y		n	n	y	y	n	24,26
AS 316	y	y		n	y	y			24,27
MWC 960	y	y		n	y	y	y	n	24,27
AS 327	y	y		n	y	y			24
Pe 2-16	y	y		n	n	y	y	n	46,59
V919 Sgr	y	y		n	n	y	y	n	28
CM Aql	y	y		n	n	y	y	n	22,23,53
AS 338	y	y		y	n	y	y	n	30
Ap 3-1	y	y		y	y	y			

Table A.7 (*cont.*)

Star	He I	He II	C IV	[Fe VII]	6830	TiO	CO	H_2O	References
Hen 1761	y	y		n	n	y	y	n	40,58
AS 360	y	y		y	y	y	y	n	
He2-467	y	y		n	n	n			36
He2-468	y	y		y	y	y			
CD-43°14304	y	y		n	y	n	y	n	
V407 Cyg	n	n		n	n	y			23,41

References for Table A.7

1. Aaronson, Liebert, and Stocke 1982.
2. Allen 1976.
3. Allen 1978.
4. Allen 1979b.
5. Allen 1980c.
6. Blair, et al. 1983.
7. Blanco 1961.
8. Brewster 1975.
9. Brewster 1976.
10. Brugel and Wallerstein 1981.
11. Cannon 1921.
12. Cannon 1933
13. Cannon and Mayall 1938.
14. Carlson and Henize 1974.
15. Cohen and Barlow 1980.
16. Elvey 1941.
17. Elvey and Babcock 1943.
18. Feast and Glass 1980.
19. Frantsman 1962.
20. Hanley 1943.
21. Haro 1952.
22. Harwood 1925.
23. Herbig 1960.
24. Herbig 1969.
25. Herbig and Hoffleit 1975.
26. Hoffleit 1968b.
27. Hoffleit 1970.
28. Hoffmeister 1932.
29. Kafatos, Michalitsianos, Allen, and Stencel 1983.
30. Kohoutek 1965.
31. Kohoutek, Pekny, and Perek 1965.
32. Longmore and Allen 1977.
33. Lutz 1977a.
34. Lutz 1984.
35. Lutz and Kaler 1983.
36. Lutz, Lutz, Kaler, Osterbrock, and Gregory 1976.
37. Luyten 1927.
38. MacConnell 1972.
39. McCuskey 1961.
40. Mayall 1952.
41. Meinunger 1966.
42. Minkowski 1946.
43. Minkowski 1948.
44. Payne 1930a.
45. Payne 1930b.
46. Perek 1960.
47. Perek 1962.
48. Perek 1963.
49. Peterson 1977.
50. Reinmuth 1926.
51. Roche, Allen, and Aitken 1983.
52. Sanduleak and Stephenson 1972.
53. Shapley 1950.
54. Swope 1928.
55. Swope 1938.
56. Swope 1941.
57. Swope 1943.
58. Thackeray 1954c.
59. The 1962.
60. The 1964.
61. Torres-Peimbert, Recillas-Cruz, and Peimbert 1980.
62. Velghe 1957.
63. Walker 1983.
64. Weaver 1972.
65. Webster 1966.
66. Webster 1973.
67. Wilde 1965.
68. Wray 1966.

bands (TiO in the red, CO and H_2O in the near-infrared). A "y" in a given column denotes that the feature in question has been observed in a given object, while an "n" shows the feature has not been detected. The relevant spectroscopic data is not available for some objects, so the appropriate entry has been left blank. Hoffleit (1968b, 1970) searched for optical variability in some of these systems; a few (V2601 Sgr and V2756 Sgr) have light curves resembling that of CI Cyg.

Finally, a list of alternative identifications for symbiotic stars appearing in Table A.7 is given in Table A.8. Most of these systems were originally discovered as planetary nebulae, so many of these designations refer to lists of planetary nebulae or other types of emission-line stars.

Table A.8 *Alternate identifications for symbiotic stars*

Hen 160 = SS 9 = Wra 208
AS 201 = MHα382-43 = Hen 172
He2-38 = PK 280-2°1
He2-87 = SS 36 = PK 302-0°1
Hen 828 = SS 37 = Wra 1022
Hen 905 = SS 40 = Wra 1108
Hen 916 = SS 42 = Wra 1123
V835 Cen = He2-106 = PK 312-2°1
He2-127 = SS 45 = PK 325 + 4°2
Hen 1092 = He2-134 = PK 319-9°1
HD 330036 = BD-48°10371 = Cn 1-1 = PK 330 + 4°1
He2-147 = PK 327-4°1
Wra 1470 = Hen 1187 = SS 55
He2-171 = SS 59 = PK 346 + 8°1
He2-173 = SS 61 = Wra 1518 = PK 342 + 5°1
He2-176 = PK 339 + 0°1
KX TrA = Hen 1242 = Cn 1-2 = He2-177 = PK 326-10°1
AS 210 = MHα276-52 = Hen 1265 = SS 66
CL Sco = HV 4035 = AS 213 = MHα71-5 = Hen 1286 = SS 69 = Wra 1564
HK Sco = HV 4493 = AS 212 = MHα71-6 = Hen 1280 = SS 68 = Wra 1563
V455 Sco = HV 7869 = AS 217 = MHα71-15 = Hen 1334 = SS 74 = H2-2 =
    PK 351 + 3°1
Hen 1341 = SS 75
Hen 1342 = SS 77
AS 221 = MHα276-12 = Hen 1348 = SS 79 = H2-4 = PK 352 + 3°1
H2-5 = PK 354 + 4°2
Th3-7 = PK 356 + 4°3
Th3-17 = PK 357 + 3°3
Th3-18 = PK 358 + 3°5
Hen 1410 = Th3-20
Th3-30 = PK 359 + 2°1
Th3-31 = PK 358 + 1°2
M1-21 = SS 90 = He2-247 = PK 6 + 7°1
AE Ara = HV 5491 = MWC 591 = Hen 1451 = SS 95 = Wra 1754 = PK 344-8°1
UU Ser = AS 237 = SS 98
V2110 Oph = AS 239 = MHα276-12 = Hen 1465

Table A.8 (*cont.*)

V917 Sco = Hen 1481 = SS 103
AS 245 = MHα359-110 = Hen 1510 = SS 107 = H2-28 = PK 6 + 2°2
He2-294 = PK 357-3°1
B13-14 = PK 0-1°4
Bl L = PK 359-2°1
AS 255 = MHα363-45 = Hen 1525 = SS 111
V2416 Sgr = M3-18 = SS 112 = PK 7 + 1°2
Ap1-8 = SS 121 = PK 2-3°1
AS 270 = Hen 1581 = SS 126
H2-38 = SS 128 = PK 2-3°4
SS 129 = T 17
Hen 1291 = SS 132 = T 53
AS 276 = Hen 1595 = SS 135
Ap1-9 = SS 137 = PK 3-4°2
AS 281 = MHα208-83 = SS 138 = Ap1-10 = PK 3-4°1
V2506 Sgr = AS 282 = MHα304-13 = He2-358 = SS 139 = Ap1-11 = PK 3-4°6
AS 289 = SS 143 = F1-11
V2756 Sgr = AS 293 = MHα304-122 = He2-370 = SS 145 = PK 2-5°1
HD 319167 = CnMy 17 = He2-373 = SS 146 = PK 1-6°1
He2-374 = SS 147 = PK 9-2°1
AS 296 = D 143-2 = SS 148
AS 295B = MHα304-41 = SS 149
Hen 1674 = SS 153 = T 21 = Wra 1864
He2-390 = SS 154 = CSV 4026? = PK 5-5°2
V3804 Sgr = AS 302 = MHα304-33 = Hen 1676 = SS 155
CD-28°14567 = AS 304 = SS 162
AS 316 = MHα208-58 = He2-417 = SS 172 = PK 12-7°1
MWC 960 = MHα204-22 = Hen 1726 = SS 174
V2601 Sgr = AS 313 = MHα208-51 = SS 171
AS 327 = MHα208-67 = Hen 1730 = SS 176
Pe2-16 = PK 29-2°1
V919 Sgr = AS 337 = MHα277-6 = SS 178
AS 338 = MHα305-6 = Hen 1737 = K4-12 = PK 48 + 4°1
Ap3-1 = PK 37-2°1
AS 360 = MHα80-5 = Hen 1771
He2-467 = PK 63-12°1
CD-43°14304 = Hen 1924
V407 Cyg = AS 453 = MHα289-90

A.3 A   Compendium of Selected Symbiotic Stars

This section summarizes the basic observational material available for well-studied symbiotic stars which could not be placed easily in tabular form. Nearly all of the objects are variable stars, and they are therefore listed in alphabetical order by constellation. Each entry begins with a variable star designation (if available) and a "namelist" of alternative designations (with priority given to the more familiar catalogues and lists of variables and peculiar stars). After a short introductory paragraph, the remaining observational data have been divided into various topics:

(i) Photometry (ultraviolet, optical and infrared),

(ii) Spectroscopy (ultraviolet, optical and infrared),

(iii) Radio Observations, and

(iv) X-ray Observations.

Since earlier Tables presented numerical data for radio flux densities (Table 3.3) and X-ray fluxes (Table A.3), topics (iii) and (iv) are included only if the data are deemed to be of special interest. Aside from a verbal description of each star's behavior, this section also presents various spectra and light curves of typical systems.

Finding charts have not been provided in this section, as an excellent set of uniform charts has been published by Allen (1984). Allen (1984) also includes a number of fine digital spectra for the vast majority of systems that are not discussed in detail in this Appendix.

Z Andromedae

(BD + 48°4093, HD 221650, SAO 053146, HV 193,
AN 41.1901, MWC 416, AGK3 + 48°2087, AOe 25746, BO VI,
AG Bo 17913, Ku 10435)

Z And is traditionally considered to be the prototypical symbiotic star (e.g., Payne-Gaposchkin and Gaposchkin 1938; Payne-Gaposchkin 1957; Swings 1970). Its variable nature was discovered by Hartwig (1901) and Pickering (1901a,b) who noted its Md spectrum and large range (≈ 2 mag; Hartwig 1910; Graff 1914). Plaskett (1923, 1927) was the first to describe the spectrum in detail; he reported a peculiar class A spectrum with strong nebular lines. Hogg (1932) noted TiO bands on these plates, and this discovery began the study of "stars with combination spectra".

Photometry

The early photometric history of Z And (1888-1937) was reviewed by Pickering (1911b), Yamamoto (1924), Parenago (1933, 1936), Greenstein (1937) and Himpel (1940), while Mattei (1978) has discussed recent variations. Although the 2-3 mag outbursts dominate the optical light curve (Figure 1.2), 0.2-0.3 mag quiescent variations have been noted by a number of observers (Campbell 1942; Payne-Gaposchkin 1946; Gessner 1959; Romano 1960; Erleksova 1964; Mianes 1965; Pucinskas 1970; Lukatskaya 1973; Belyakina 1974a; Splittgerber 1975b; Mjalkovskij 1977; Taranova and Yudin 1981a,c, 1983d, 1984; Belyakina 1985b; Klyus 1982; see also Belokon' and Shulov 1974 for a discussion of variations in the amount of polarization during quiescence). The minima are clearly periodic ($P = 756.85$ days; Kenyon and Webbink 1984), and spectroscopic observations suggest that these minima may be the result of an eclipse of the hot component or a surrounding ionized nebula by its giant companion star.

Five outbursts of Z And have been recorded since 1888 (Greenstein 1937; Petit 1960; Weber 1963; Mayall 1969; Mattei 1978); these are shown quite clearly in Figure 1.2. A given outburst generally lasts ≈ 7 years, and commences with a slow increase in brightness (V$\approx$ 10.5-11.0 to V$\approx$9.5), accompanied by periodic 0.5-1.0 mag oscillations. This is followed by a more rapid rise to visual maximum, which resembles a dwarf nova outburst. As V decreases, B-V decreases while U-B increases (Pucinskas 1964; Belyakina 1965a,b, 1970a). Visual maximum is followed by a slow decline with superimposed 0.5-1.0 mag oscillations. Within this train of oscillations, the interval between successive minima is somewhat shorter than when the system is quiescent, with the result that during an entire outburst there is one more minimum present than there is in a corresponding time period during quiescence (Kenyon and Webbink 1984).

The IR continuum of Z And is similar to that observed in a normal M2 giant (once the data are de-reddened), with some evidence for small excesses at 10 μm and 20 μm (Neugebauer and Leighton 1969; Low 1970; Swings and Allen 1972; Woolf 1973; Szkody 1977; Taranova and Yudin 1979, 1983c; Altamore, et al. 1981; Viotti, et al. 1982b; Eiroa, Hefele and Qian 1982; Kenyon and Gallagher 1983). This giant has a very strong CO absorption band at 2.3 μm, suggesting the star is a luminosity class II rather than a luminosity class III giant (Kenyon and Gallagher 1983).

Spectroscopy

As with most symbiotic stars, the quiscent optical spectrum of Z And is dominated by a late-type absorption spectrum and high ionization emission lines (Figure 1.1). The TiO bands are quite strong in the red, and indicate a normal M giant star (Hogg 1932, 1934; Merrill and Burwell 1933; Bloch and Tcheng 1955; Tcheng and Bloch 1955; Schaifers 1960; Bloch 1960, 1961, 1964, 1965a,b; Belyakina, Boyarchuk, and Gershberg 1963; Dossin 1964; Dolidze and Dzhimshelejshvili 1966; Boyarchuk 1968b,c; Bloch, Jousten and Swings 1969; Mammano, Rosino, and Yildizdogdu 1975; Gravina 1981b; Altamore, et al. 1981; Muratorio and Friedjung 1982a; Viotti, et al. 1982b; Williams 1983). The apparent lack of a λ8432 TiO absorption band allows a more precise determination of the giant's spectral type (M2; Andrillat and Houziaux 1982).

Plaskett (1928) was the first to recognize two distinct emission sources in optical spectra of Z And: a stellar component (lines of Mg II, Fe II, and Ti II) and a nebular component (lines of He II, N III, [O III]). On these initial plates, the He I singlet lines rivaled the triplets in intensity. This is not usually true in planetary nebulae, as the singlets typically have 1/3 the intensity of the triplets (Osterbrock 1974). Plaskett's description of the spectrum holds true today (although additional emission lines such as [Ne V], [Fe VII], and the mysterious λ6830 feature have been identified), as both the narrow stellar lines and the wider nebular lines remain strong emission features (Merrill 1927b, 1944, 1947, 1948; Struve and Elvey 1939; Swings and Struve 1940a, 1941a,b, 1942a,b, 1943a,b; Tcheng 1949a,b; Bloch and Tcheng 1951a,b; Aller 1954; Tcheng and Bloch 1955; Mitchell 1956; Boyarchuk 1964b, 1965, 1967a,b, 1970; Dean and van Citters 1970; Altamore, et al. 1981; Oliversen and Anderson 1982a; Blair, et al. 1983; Williams 1983; Netzer, Leibowitz, and Ferland 1983).

Large-scale intensity variations have been observed in many emission lines; Swings and Struve commented that the O III λ3444 line was markedly variable from night to night, while Merrill found variable He I lines with extremely complex line displacements. Aller (1954) noted fluctuations in the [O III] λ4363/(λ4959 + λ5007) intensity ratio and in various [Fe II] lines. He suggested that some of the variability was a result of erratic continuum fluctuations, but detailed spectrophotometry is needed to verify his conclusion. Baratta, et al. (1978, 1979) found Hα to vary by a factor of 4 over a one year timescale, while He II λ4686 remained relatively constant (Altamore, Baratta, and Viotti 1979). It may be that many of these variations are the result of an *eclipse* of the hot component by the cool giant, since spectrophotometric observations show variations in the strong emission lines and the Balmer continuum, and essentially no variability in the red continuum and TiO absorption bands (Kenyon 1983).

The spectroscopic changes that occur in an optical outburst have been followed in detail by Merrill, Swings and Struve, and Tcheng and Bloch (among others). Although the fine details of these changes vary from outburst to outburst, the overall behavior is fairly repeatable and can be easily generalized. During the rise to maximum, the high ionization features fade as the blue continuum brightens. It is not yet clear if the high ionization lines (He II, N III, [O III], [Ne V], and [Fe VII]) actually disappear, or whether they are simply lost in the rapidly rising blue continuum (e.g., Aller 1954). At maximum light, the spectrum is dominated by bright H I lines. These lines all have strong absorption cores, and the higher members of the Balmer series may appear only in absorption at some stages. The He I lines may also have these P Cygni features, but they are always much weaker than those associated with the Balmer lines. The TiO bands are absent in the blue, but may be weakly present in the yellow or red. The P Cygni absorption features fade and the high ionization lines return as the star declines from visual maximum. The He II emission and the TiO absorption lines strengthen considerably as the blue continuum fades, which suggests that they were always present but were overwhelmed by the continuum at maximum visual light.

Ultraviolet spectra of Z And show a strong continuum that rises towards shorter wavelengths and many intense emission lines (Baratta, et al. 1979; Slovak 1980, 1982a; Altamore, et al. 1981; Viotti, et al. 1982b). The strong intercombination lines indicate an electron density of 10^{10} cm^{-3} (Friedjung 1980), which is significantly larger than that calculated by Aller or Boyarchuk. Thus, two distinct line-forming regions may be present in Z And, and Friedjung (1980; Gorbatskij and Ivanova 1973) hypothesizes that some of the high excitation lines might be formed in the wind leaving the giant, rather than in an ionized region near the hot component.

Z And's UV continuum resembles the Rayleigh-Jeans tail of a blackbody, and various analyses of the flux distribution give a temperature of 43,000-100,000 K if the interstellar reddening, E_{B-V}, is 0.3-0.35 (Slovak 1980; Altamore, et al. 1981; Viotti, et al. 1982b; Caputo 1971a estimated $E_{B-V} = 0.64$ from the optical [Fe II] lines). Kenyon and Webbink (1984) discount the low temperature solutions for the temperature, since these cannot produce enough Balmer continuum radiation. An alternative to a hot stellar model for the hot component is a main sequence star accreting material at $\sim 10^{-5}\,M_{\odot}\,\text{yr}^{-1}$ (cf. Yudin 1986).

Attempts to model the continuum of Z And must account for the large variations ($\geq 50\%$) in emission-line intensities and continuum level discussed by Sahade, Brandi and Fontenla (1981,1984). Some of these variations may be a result of the eclipse discussed above. Additional observations are necessary to delineate the nature of intrinsic variations in the light output of the hot component.

EG Andromedae

(*BD + 39° 167, HD 4174, SAO 036 ̣18, GC 880, IRC + 40° 014, AGK3 + 4° 066*)

This star was grouped with the rather unspectacular Ma stars by Cannon in the HD catalogue, with $m_{pg} = 8.88$ and $m_{pv} = 7.5$. EG And appears in Eggen's (1964) catalogue of high velocity stars as an M2ep giant, with V = 7.5 and an absorption line radial velocity of -96.0 km s^{-1}. The emission lines give a somewhat different radial velocity of -92.8 km s^{-1}. EG And is one of the few symbiotics that Babcock (1950) found to have a magnetic field, although recent observations by Slovak (1982b), using a more sensitive detector, could not verify this result.

Photometry

Optical photometry of EG And was discussed by Jarzebowski (1964), who found a period of 40 days and an amplitude of ≈ 0.1 mag. The fluctuations were definitely larger at λ4200 than at λ5600, suggesting, perhaps, a variable blue component. Tempesti (1977; also Belyakina 1974a) commented that the amplitude was nearly 0.3 mag in 1968-70 with a period of either 40 or 80 days. Smith (1980) noted that variations might be present in AAVSO data, with an amplitude of 0.2 mag and a period of ≈ 470 days.

Kaler and Hickey (1983) reported narrow band filter photometry, and found large intensity variations in the near ultraviolet. Significant fluctuations were observed at λ3500 over one month intervals, suggesting intrinsic changes in the radiation emitted by the hot component. These may be associated with similar variations noted in the satellite ultraviolet, and discussed below.

Published infrared photometry suggests that EG And contains a normal, non-variable late-type giant (Neugebauer and Leighton 1969; Low 1970; Swings and Allen 1972; Kenyon and Gallagher 1983), and confirms the M2 spectral type assigned via optical spectra. Kenyon and Gallagher measured the depth of the 2.3 μm absorption band in EG And, and found it to be comparable in strength to those found in normal M2 giant stars.

Spectroscopy

The peculiar nature of EG And was first noted by Wilson (1950) when he found an M2 giant absorption line spectrum with H I, [O III], and [Ne III] emission lines (Andrillat 1982a). He did not detect emission from He II (Figure A.1). The strong continuum of the giant masks emission lines redward of $H\beta$. The blue continuum is very weak in the optical, but rises somewhat at UV wavelengths (Figure A.1; Stencel and Sahade 1980; Slovak 1982a; Stencel 1982, 1984). H I, [O III], and [Ne III] are usually strong emission lines in the optical, while Ca II (H and K) and He I sometimes appear as emission lines with weak absorption components (Martini 1968; Mammano and Martini 1969; Ciatti and Mammano 1975a; Smith 1980, 1981; Gravina 1981b; Oliversen and Anderson 1982a; Kenyon 1983; Oliversen, et al. 1985). Smith (1981) found that the equivalent width and (perhaps) the radial velocity variations in Hα were associated with the 470 day periodicity seen in AAVSO data.

The ultraviolet emission spectrum is much more interesting, as Mg II, Si II, He II, N III, O III, C IV, and N V have been detected as strong emission lines (Figure A.1). The He II λ1640 flux suggests a hot component with $T_h > 60,000$ K, and Kenyon (1983) estimated a temperature of 60,000-80,000 K (cf. Slovak 1980, 1982a). Both Stencel (1982) and Kenyon (1983) have remarked that the UV continuum in EG And appears to be variable, and Stencel (1984) secured additional observations which suggested that ultraviolet fluctuations might be related to the 470 day periodicity noted by Smith (cf. Sahade, Brandi, and Fontenla 1984). By analogy with other symbiotic stars (e.g., Z And, CI Cyg, SY Mus, and AX Per), this might be caused by an eclipse of the hot component by the giant star.

Figure A.1 - Optical/ultraviolet spectrum of EG And. A weak Hβ emission line is visible at λ4861, and very strong lines from He II, C IV, and N V are visible in the ultraviolet.

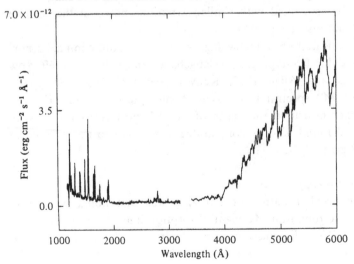

Variations in the profile of the prominent Hα line have been described by Smith (1980), and results from a recently completed program by C. Anderson and N. Oliversen are summarized in Figure A.2. Smith (1980) noted fluctuations in the intensity and the shape of Hα, and suggested a symmetric profile was present when the line was weak. There is some indication for a phase-dependent Hα line in Figure A.2, although both blue and red peaks show large variations in intensity and shape. High resolution ultraviolet observations may be needed to ascertain the nature of these fluctuations, as other optical lines are very weak (Figure A.1).

High resolution observations of the M-type absorption features have confirmed the binary nature of this system, as described by Garcia (1983) and by Oliversen et al. (1985). The orbital solution listed in

Figure A.2 - Time variations of the Hα profile in EG And (C. M. Anderson, private communication). The strong central reversal appears at particular phases in the EG And light curve, and may be associated with binary motion.

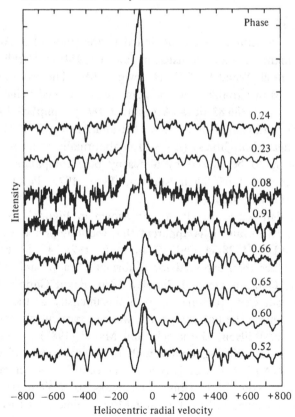

170   *Appendix*

Table A.5 is based on few spectra, and thus the elements are uncertain. Since EG And is a bright target, it should be straightforward to secure additional data and confirm the semi-amplitude and the mass function.

R Aquarii
(BD-16°6352, HD 222800, SAO 165849, HR 8992, GC 32948, IRC-20°642, MWC 100, AFGL 3136, Bo VI, Pu M 3485, 7y 1986, N7y 2718, Q 10618, Ma P 5230, Gou 32068, Du₄ 272b, RC90 6335, AG Wa 8733, Birm 646, Birm Esp 754, Kruger 2118, Boss PGC 6091, Yale 13/I Nr. 8733 G)

R Aqr is one of the more mysterious symbiotic stars (e.g., Sol 1983). It has been detected at virtually every possible wavelength since its discovery as a long period variable star, and its transformations continue to baffle observers and theorists alike. Much of its behavior is very peculiar, and recent activity has been summarized by Michalitsianos (1984).

Photometry

Numerous observers have contributed to the study of R Aqr's light curve, and many of these are listed in Prager (1934; cf. Parkhurst 1890; Valentiner 1900; Wendell 1900; Hartwig 1910). The red star is quite obviously a Mira variable, and the commonly accepted value for its pulsational period is 386.83 days (Argelander 1867; Campbell 1907, 1912, 1918; Sterne and Campbell 1937; Mattei and Allen 1979). However, R Aqr's range of brightness ($\Delta V \sim 5$) is abnormally small for its period; this appears to be caused by a nearby blue companion star with $V \approx 11$ (Campbell 1907; Nicholson and Pettit 1922; Pettit and Nicholson 1933; Hetzler 1940).

The long-term light curve for R Aqr is shown in Figure A.3 (cf. Mattei and Allen 1979), and it is apparent that the long-period variations nearly ceased in 1928-34 and 1974-80. In 1928-34, the color index was much bluer than expected for a normal Mira variable (0.5-0.8; Loreta 1931; Gerasimovic and Shapley 1930; Payne-Gaposchkin and Boyd 1946), but quite reasonable given the intensity of the blue continuum at the time. The hot component was not particularly active in 1974-80; even so, Willson, Garnavich, and Mattei (1981) proposed that the system is an eclipsing binary with a 44 yr period. If this interpretation of R Aqr's behavior is correct, the eclipsing object must be a large, non-luminous body (this is necessary to explain (i) the eclipse depth [$\Delta m \approx 1$ mag] and (ii) the faint Mira minima [$m \approx 10\text{-}11$]).

Insufficient observations are available c. 1846, 1890 to test the eclipse hypothesis - the next eclipse is due in 2020!!

Optical polarimetry might provide a useful probe of the R Aqr binary, but very few measurements are available. Naturally, both the amount and the position angle of the polarization are variable (Serkowski 1970; Shawl 1975; Nikitin and Khudyakova 1979; Dombrovoki and Polyakora 1974; McCall and Hough 1980; Svatos and Solc 1981;

Figure A.3 - Historical visual light curve for R Aqr. The plotted points are ten days means of data accumulated by the AAVSO. Periodic variations of the Mira component are quite obvious, although the amplitude of these pulsations decreased in 1928-34 and in the late 1970's.

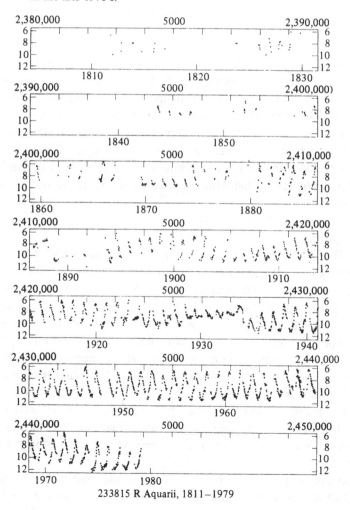

233815 R Aquarii, 1811-1979

Piirola 1983), and there are insufficient data to determine the structure of the gas. Red polarimetry (i.e., $\lambda > 0.6 \mu$ m) is well-correlated with the Mira pulsation, but such a trend was not detected with data secured in the blue.

Infrared photometry of R Aqr was first discussed by Pettit and Nicholson (1933; Nicholson and Pettit 1922), who remarked that the system was similar to other long period variables. More recent observations tend to support this conclusion, as R Aqr shows the features typically associated with Mira variables: 1 mag variability at JHKL, H_2O absorption bands, and 10 μm emission (Moroz 1966; Neugebauer and Leighton 1969; Stein, et al. 1969; Woolf 1969; Low 1970; Simon, Morrison and Cruikshank 1972; Lockwood 1972; Barnes 1973; Morrison and Simon 1973; Merrill and Stein 1976; Noguchi, Maihara, Okuda and Sato 1977; Price and Walker 1976, 1983; Kenyon and Gallagher 1983).

While R Aqr usually looks like a normal long period variable, its IR continuum was noticeably peculiar in 1974-78 (Khozov, and collaborators 1974, 1975, 1976, 1977, 1978; Catchpole, et al. 1979; Whitelock, et al. 1983a). The maxima at K were much fainter than usual during these years, and the color temperature was ≈ 1000 K *cooler*. Whitelock, et al. suggest that this was caused by an opaque dust cloud obscuring the M star, but the observations are insufficient to determine if the dust cloud orbits the Mira or whether it simply formed in our line of sight.

Spectroscopy

The first Harvard spectrum of R Aqr was secured shortly after maximum light (17 August 1893), and showed a normal M6 giant star with Hγ and Hδ as doubly reversed emission lines (Townley, Cannon, and Campbell 1928). Two months later, the M spectrum had vanished, and Hβ had become a strong emission line. While this behavior was unexpected, Merrill (1919, 1920; Bailey 1919b; Moore 1919; Wright 1919) later demonstrated that fairly strong nebular lines (H I, [O III]) were visible when the M star was faint. These lines maintained a fairly constant intensity, and therefore were invisible when the Mira approached maximum light. Normal Mira emission lines (Hγ and Hδ) were visible at maximum, and Merrill noted that the fluctuations in these narrow lines were similar to those observed in other long period variables.

A spectrum of R Aqr near minimum light is displayed in Figure A.4; strong TiO and VO bands dominate in the red and indicate an M7 spectral type (Payne-Gaposchkin 1952). The nebular lines of [O III] and [Ne III] are quite prominent, and their relative intensities indicate an electron density of $\sim 10^5$ cm^{-3} (Herbig 1965; Ilovaisky and Spinrad 1966; Ciatti and Mammano 1975a; Wallerstein and Greenstein 1980; Kaler 1981). Since He II $\lambda 4686$ is usually a weak line, the [Ne III] $\lambda 3869/H\beta$ and [O III] $\lambda 5007/H\beta$ ratios can be used to estimate the temperature of the hot star, $T_h = 50,000$-$60,000$ K. Aside from these strong nebular lines, He I, [N II], [O II], Fe II, [Fe II], Fe III, and [Fe III] are usually present as weak to moderately strong emission features (Merrill 1921, 1922, 1923, 1950b; Ilovaisky and Spinrad 1966; Ciatti and Mammano 1975a; Wallerstein and Greenstein 1980).

High ionization emission lines (He II, N III, C III) have been observed as intense features only during the optical "outburst" in 1924-1933 (Merrill 1927a, 1934, 1935, 1936, 1940, 1941a,b,c,d; Swings and Struve 1940a, 1942a; Beals 1951; however, Zirin [1976] once reported [Fe XIII]). During this period, the spectrum resembled a planetary nebula and its nucleus. The high ionization lines were rather wide, and the H I lines developed absorption components on their blueward edges. The displacements of these absorption wings indicated an expansion velocity of 350 km s^{-1}, which is significantly larger than the 10-20 km s^{-1} velocities commonly observed in red giant winds. This blue companion spectrum disappeared in 1934, when the M star resumed its normal pulsation cycle.

High resolution $H\alpha$ spectra obtained with a Fabry-Perot were

Figure A.4 - Optical spectrum of R Aqr near minimum. Strong TiO bands are visible at wavelengths exceeding 5000 Å, while a weak continuum is present in the blue. A few [O III] and [Ne III] lines are prominent, and the H I lines are also strong.

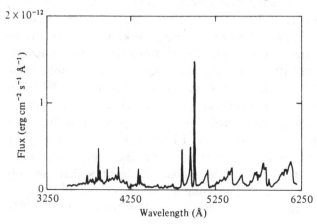

described by Anandarao, Sahu, and Desai (1985). These data appear to indicate that R Aqr possesses two shells of material moving outward at 5 and 15 km s^{-1} from the central source. Observations in other emission lines would be important for verifying these features.

The value of the reddening in R Aqr has been somewhat controversial, since the optical emission lines implied $E_{B-V} > 0.5$ (Wallerstein and Greenstein 1980; Kaler 1981), while the lack of a detectable 0.22 μm absorption dip suggests $E_{B-V} < 0.1$ (Hobbs, Michalitsianos, and Kafatos 1980; Michalitsianos, et al. 1980; Johnson 1980, 1981, 1982). Recent estimates by Brugel, et al. (1984) tend to indicate a decrease in the optical reddening since 1977 ($E_{B-V} = 0.4$ [1979], 0.15 [1982]), implying the dust cloud responsible for the putative eclipses no longer obscures the emission-line region. Nearly all of the reddening in R Aqr must be local, since the amount of interstellar reddening towards the system is negligible (Burstein and Heiles 1982).

Merrill originally reported a 26.7 yr periodicity in the radial velocities of the [O III] and [Ne III] lines, but believed his orbital solution to be of dubious value. Jacobsen and Wallerstein (1975) showed that the periodicity was not evident once additional spectra had been obtained, although Kurochkin (1976) claims to have phased the data with a 24.7 year period. For a reasonable choice of total mass in a possible binary (3 $M_\odot$), a 25-50 yr period implies a maximum radial velocity amplitude of ≈10 km s^{-1}. Given the peculiar "outbursts" experienced every 40+ years, it does not appear likely that a good orbit will be determined for R Aqr in the near future. Zuckerman (1979) suggested that the SiO maser might be used to track the cool star, and this may be the most promising way to measure the period and other orbital parameters.

It was generally expected that spectra secured by the *IUE* satellite would reveal the nature of the hot star in R Aqr. However, the ultraviolet continuum was found to be extremely weak, while various emission lines (C IV, He II, Mg II) were fairly strong (cf. Hobbs, Michalitsianos and Kafatos 1980; Michalitsianos, Kafatos, and Hobbs 1980; Johnson 1980, 1981, 1982; Michalitsianos amd Kafatos 1982a). The weak continuum is very surprising given the 60,000 K temperature estimated from optical data, and Johnson suggested that the large reddening deduced from the optical H I lines could virtually extinguish the ultraviolet continuum. Interestingly, the line spectrum is fairly rich, and analyses of the C III] lines result in densities that are only slightly higher than those derived using optical data ($n_e \sim 10^{5-6}$ cm^{-3}). If the optical reddening is decreasing as noted by Brugel, et al. (1984), ultraviolet observations in the next few years may yield additional information regarding the nature of the hot source.

Radio Observations

R Aquarii has proved to be rather entertaining for the radio observers. The system is a strong, but variable, continuum source, and most of the emission is the result of optically thick thermal bremsstrahlung from a dense central cloud with a size $< 1''$ (Gregory and Seaquist 1974; Schwartz and Spencer 1977; Bowers and Kundu 1979; Johnson 1980; Ghigo and Cohen 1981; Kwok 1982a; Purton, et al. 1982; Knapp 1985). The data is not complete enough to determine if the variability at radio wavelengths is correlated with the optical or IR light curves. Knapp and Morris (1985) were unable to detect the CO 2-1 transition, and estimate an upper limit for the antenna temperature of 0.04 K.

R Aqr is the only symbiotic star known to possess an SiO maser. Both the $v = 1$, $J = 1 - 0$ and the $v = 1$, $J = 2 - 1$ transitions have been detected at a velocity of ≈ 25 km s^{-1} (Lepine, LeSqueren, and Scalise 1978; Zuckerman 1979; Engels 1979). The available data suggest the maser source is variable by at least a factor of 6; if R Aqr behaves like other Miras, the SiO emission should be correlated with the V magnitude.

High resolution *VLA* maps have served only to complicate the R Aqr puzzle (cf. Sopka, et al. 1982; Spergel, Giuliani, and Knapp 1983; Kafatos and Michalitsianos 1982b,c; Kafatos, Hollis, and Michalitsianos 1983; Hollis, et al. 1985). The three maps appearing in Figure A.5 show that the nearly spherical central source has two elongations (labeled A and B), while a fourth feature lies on an extension of the

Figure A.5 - Radio maps of R Aqr as seen by the VLA (Kafatos, Hollis, and Michalitsianos 1983). The central source is surrounded by two elongations (A and A'), while another source, B, lies on a line connecting A with the central object.

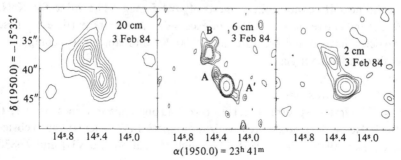

line connecting A with the central source. The central source is coincident with the stellar image of R Aqr, and has the radio spectrum expected from a completely ionized, spherically symmetric, optically thick wind ($S_v \propto v^{0.6}$). Sources A and B are coincident with bright knots in the nebula surrounding R Aqr (Baade 1943; Lampland 1922), and have been interpreted as optically thin emission from a jet accelerated by the hot companion to the Mira variable. If this object is indeed a jet, it suggests that an accretion disk surrounds the hot component, with material being ejected out the poles of the disk (Kafatos 1983; Kafatos and Michalitsianos 1982c). However, Kenyon and Webbink (1984) commented that the UV continuum is much too weak to be explained by accretion onto a main sequence star, and more compact central stars are ruled out by the lack of a substantial hard X-ray flux (unless, of course, the source is hidden by dust).

H1-36 Arae
(PK 353-4°1)
Although H1-36 has not yet been assigned a variable star designation, its 1.3 mag variability at K will one day qualify it to be a full-fledged variable star. It was discovered as a planetary nebula by Haro (1952; Henize 1967), and has been recognized as a symbiotic star only very recently. Observations at optical, infrared, and radio wavelengths have shown this to be a remarkable system that certainly deserves further study.

Photometry
H1-36 is a well-studied infrared source, but no information regarding its optical variability has been published. The variability at K and the strong CO and H_2O absorption bands detected by Allen (1983) imply the system contains a Mira variable with a period of perhaps 3 yr. The depths of these absorption features are comparable to those observed in other Mira-like stars, and a simple de-reddening of the near-IR continuum suggests $A_K \approx 1.5$ mag. The value for K-N, 6.7, is indicative of a large 10 μm excess due to silicate particles. These particles have a temperature near 700 K, and appear to completely enshroud the cool giant.

Spectroscopy
Optical spectra of H1-36 reveal strong emission lines from H I, He I, He II, and other high ionization species (Webster 1974; Webster and Allen 1975; Allen 1983). The high ionization [Fe VII] and λ6830

lines are very strong, since $\lambda 6087/H\beta \approx 1/5$. The Balmer decrement appears normal, and indicates an optical extinction, A_V, $\approx$ 3 mag. Other measures of extinction (including the He II $\lambda 1640/\lambda 4686$ and the [O II]/[S II] ratios) yield $A_V \approx 2$. Since the Balmer line ratios could be modified by collisional processes, Allen adopted $A_V = 2.2$. This results in $A_K = 0.2 - 0.3$, which is well below that derived from the near-IR data. Allen suggested that the hot star and its associated H II region might lie > 40 AU from the cool giant, and outside the dust shroud.

Ultraviolet spectra of H1-36 show few features aside from a weak Balmer continuum and some emission lines. The C III] and C IV lines are very strong, while He II and N III] are somewhat weaker. These lines are commonly observed in symbiotic stars, and imply a fairly large temperature for the hot component.

Radio Observations

H1-36 is the strongest radio source among symbiotic stars, with an observed flux of 91 mJy at 2 cm (Purton, et al. 1977; Wright and Allen 1978; Purton and Feldman 1978). Taylor and Seaquist (1984) obtained additional data with the VLA, and attempted to fit the radio spectrum with a model for the ionized gas (see Chapter 3). For a distance of 4.5 kpc, the derived binary separation is > 2000 AU, implying a binary period of at least 10^4 yr. Although this separation depends on the distance estimate, the predicted Hβ flux (which also depends on the distance) is in good agreement with Allen's observations. The derived period of 10^4 yr is the longest of any symbiotic star.

UV Aurigae
(BD + 32° 957, HD 34842, SAO 57941, HV 3322, AFGL 735, ADS 3934, W$_2$5^h 339, AG Lei 2009, Du$_4$ 569, Ku 2324, Birm Esp 130, Kruger 464)

UV Aurigae was discussed as HV 3322 by Pickering (1911a), who quoted brightness fluctuations between m = 9.2 and m = 10.6, as well as its Md spectral type (see also Doberck 1924). This system is one of a few symbiotic stars known to contain an evolved carbon star, rather than an M giant.

Photometry

Very few optical observations of UV Aurigae have been published, although it is usually very bright. Optical photometry of UV Aur was first discussed by Cannon (1918), who noted a maximum brightness of m = 7.9 and a minimum at m = 10.1 with an unknown period. Zakarov (1951) suggested the variation had a 395.2 day period, which might be due either to a pulsation or an orbital modulation of the light curve. The mean light curve does indeed resemble that of CI Cyg, although the amplitude may be somewhat larger. Khudyakova (1985) reported a 400-day periodicity in polarimetric data, and suggested that the behaviour could be a result of orbital motion in a binary system. Additional UBV observations of this system are very important.

The IR energy distribution of UV Aur is identical to that of other late-type giants, and it has the strong 3.9 μm CS absorption band observed in most carbon stars (Neugebauer and Leighton 1969; Woolf 1973; Bregman, Goebel and Strecker 1978; Hunter, private communication). The excesses at 10 and 20 μm are very strong, and are presumably silicate emission (Woolf; Price and Walker 1983). The system exhibits a fairly strong CO band, confirming the presence of an evolved late-type star (Kenyon and Gallagher 1983).

Spectroscopy

UV Aur is one of a few symbiotic stars that has strong CN absorption bands (Figure 3.10), and the absorption spectrum has been classified as N, R, and S (Wilson 1923; Merrill 1926; Sanford 1935, 1944a; Shapley 1940b; Payne-Gaposchkin 1954; Nassau and Blanco 1954; Cameron and Nassau 1956; Gordon 1968; Richer 1971; Yamashita 1972). Lines from neutral metals are also strong absorption features, and La I and CaCl absorption have been present on occasion (Merrill 1926; Sanford 1944b, 1947b, 1949a, 1950; Rybski 1973). Aside from the interesting absorption lines, the spectrum is unspectacular, since only H I and [O III] are ever obvious emission features (Merrill 1926; Sanford 1950; Ciatti and Mammano 1975b; Oliversen and Anderson 1982a).

The only strong features observed on ultraviolet spectra of UV Aur are those expected from an early-type star. No strong emission lines are obvious, although fairly prominent absorption features are visible at λ1265 (Si II), λ1306 (O I), and λ1336 (C II). The continuous spectrum is well-fit by a 20,000 K stellar atmosphere reddened by $E_{B-V} \sim 0.3$ using the (C_1, C_2) indices discussed in Chapter 2. This temperature suggests that an early B star is the binary companion to the giant. High ionization emission lines have never been reported in UV Aur, so this component is sufficient to produce the emission-line spectrum in Figure 3.10. The observed m_{1700}-K color, corrected for interstellar reddening, implies $R_h/R_g \approx 0.01$; thus the B star must lie below the main sequence if the carbon star has a radius of $\sim 100\ R_\odot$.

TX Canum Venaticorum
(BD + 37°2318, SAO 063173, AGK3 + 37°1208, Lu Gyll 5717)

TX CVn was discovered when Cowley (1956) noted that both its spectrum and its photographic magnitude were variable over a one year timescale. He noted weak H I and possibly weak Ca II emission. TX CVn has been considered as a cataclysmic variable and as a P Cygni star, but its combination spectrum (late B + early M) and frequent outbursts qualify it to be full-fledged symbiotic star.

Photometry

Mumford (1956) discussed the early photographic light curve of TX CVn, and noted three outbursts (1920-25, 1945-48, and 1952). The outbursts generally had similar shapes and maximum light was near V = 9.2 in all three cases. At quiescence, the photographic magnitude is near 12. Bertaud and Weber (1962) reported another outburst in 1962 when m_{pg} rose from 11.4 to 9.3 in about 8 months. The star then declined to $m_{pg} = 9.9$ over the next three months.

The IR energy distribution of TX CVn is consistent with a non-variable late-type star and perhaps a small excess due to hot dust or free-free emission (Swings and Allen 1972; Kenyon and Gallagher 1983; Taranova and Yudin 1983b). The system has a moderately strong 2.3 μm CO absorption band, confirming the M0 spectral type assigned from optical spectra. Observations at longer wavelengths (e.g. 10 μm) are needed to determine if dust is actually present in the system.

Spectroscopy

The optical spectrum of TX CVn (Figure A.6) resembles that of a late B or early A star (weak He I and Mg II absorption), with superimposed emission lines from H I and various singly ionized metals (Pucinskas 1965; Chkhikvadze 1970b; Mammano and Chincarini 1966; Mammano and Taffara 1978; Fried 1980b; Gravina 1981b). Aside from the variable P Cygni absorption components associated with many of the emission lines (indicating mass outflow at a velocity of ≈ 300 km s^{-1}), TX CVn is not much of a symbiotic star, since high excitation emission lines have never been observed in its spectrum (Mammano and Taffara 1978; Gravina 1981b). Both the absorption and the emission lines show variable radial velocities; Fried (1980b)

180   *Appendix*

calculated the preliminary orbit listed in Table A.4 assuming a 70.8
day period. There is a strong indication of TiO absorption bands on
optical spectra of TX CVn (Figure A.6), and Andrillat and Houziaux
(1982) assigned an M1 spectral type to the giant on the basis of the
λ7054 band (cf. Chapter 3).

Ultraviolet spectroscopic observations of TX CVn were discussed by
Slovak (1982a), and he found the UV continuum could be understood
in terms of an early-type stellar atmosphere (A or B). A similar
analysis by Kenyon and Webbink (1984) suggested the hot component
has an effective temperature of 9700 K. No strong emission lines are
visible on the *IUE* spectra, and thus there is no evidence for a component that is hotter than the A or B star. Kenyon and Webbink
(1984) estimated the relative radii of the binary components to be
$R_h/R_g \sim 0.07$, implying $L_h/L_g \sim 0.18$. If the radius of the giant is typical for an M0 III star ($\sim 35 \ R_\odot$), $R_h \sim 2.5 \ R_\odot$ is normal for an A1 V
star (Allen 1973). It is unlikely that this component can be a normal
main sequence dwarf given its photometric behavior, and Kenyon and
Webbink (1984) suggested it might be a low mass white dwarf (~ 0.27
$M_\odot$) with a rejuvenated hydrogen burning shell.

Figure A.6 - Optical spectrum of TX CVn. Only Hα is visible as a
strong emission line, and it has an obvious P Cyg absorption
component. A few TiO absorption bands are visible in the red, while
hydrogen Balmer absorption is prominent in the blue.

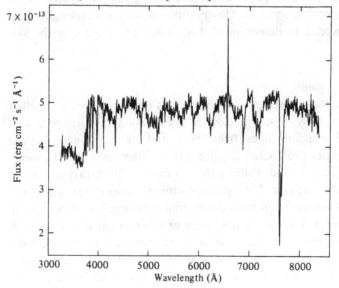

V748 Centauri

(*CoD-32°10517, Hen 1045, S 5003, CSV 2229*)

V748 Cen was suspected to be an R CrB variable by Hoffmeister (1949; cf. Kukarkin, et al, 1951). He found a photographic amplitude of at least 2 mag (m_{pg} = 11 to > 13), and Meinunger (1970) later determined a period of 564.8 days. Early spectra by Eggen and Rodgers (1969) showed hydrogen emission lines, as well as various absorption features (G band, Ca II H and K and H8). On the basis of their spectra, Eggen and Rodgers (1969) identified the X-ray transient Cen X-4 as V748 Cen, but improved X-ray positions later showed that Cen X-4 was not associated with V748 Cen.

Photometry

Optical and infrared photometry discussed by van Genderen, Glass and Feast (1974) confirm that V748 Cen is an eclipsing binary with a 564.8 day period. The system becomes much redder during eclipse, and the J-K and H-K colors at totality imply the eclipsing star is an M4 giant. Aside from the eclipse, van Genderen, Glass and Feast (1974) pointed out irregular brightness variations lasting 5-10 days.

Spectroscopy

A typical optical spectrum of V748 Cen shows an F2 I absorption spectrum with strong H I emission lines (Feast 1970, 1971; van Genderen, Glass and Feast 1974). The Hα emission line has a strong central absorption component, and other low level Balmer lines may also have disk-like emission features. The higher level Balmer lines usually appear only in absorption. The M companion to the F2 star is visible during eclipse; the strength of the TiO bands is consistent with the M4 spectral type assigned from IR photometry.

T Coronae Borealis

(*BD + 26° 2765, HD 143454, SAO 084129,*
Nova Coronae Borealis 1866, HR 5958, AGK3 + 26° 1536,
GC 21491, AG Cbr M. 7433, Oxf ph 26°38034,
Oxf ph 26°38343, Oxf ph 27°34617)

T CrB was first noted as an innocuous 9th magnitude star in the BD catalogue. It captured much attention in 1866 when it rose to 2nd magnitude and declined quite rapidly (Lundmark 1922). A repeat performance of this outburst in 1946 made it one of the few recurrent novae, while Berman's (1932) detection of TiO bands qualified it to be placed amongst the symbiotic stars. T CrB is one of three recurrent novae discussed in this monograph; although many symbiotic binaries display more frequent outbursts, none have the large range exhibited by T CrB. It is also one of the few systems to be detected as an X-ray source; Cordova and Mason (1984; Cordova, Mason, and Nelson 1981) quote a flux of 0.008 ± 0.003 IPC count s^{-1}.

Photometry

On 12 May, 1866, J. Birmingham noticed a 2nd magnitude star in Corona Borealis, and just one day later, Schmidt (1866a) found T CrB roughly equal to Alphecca (α CrB) in brightness. Hind (1866) gave an account of the discovery in Canada by W. Barker on the 4th of May which contradicts negative sightings by Baxendell (1866, 1868) and Schmidt, among others. Lynn (1866a,b,c,d) discussed this discrepancy in great detail, and concluded that Barker had somehow confused his dates. In view of the rapid changes and the number of sightings shortly after the 12th of May (Farquhar 1866; Gould 1866a,b,c; Courbebaisse 1866; Bird 1866; Knott 1866; Schmidt 1866a,b,c; Behrmann 1866a,b; Ern 1866), it is difficult to dispute Lynn's reasoning. The decline from maximum was very fast, as the brightness decreased to $m = 6$ in one week and to $m = 9$ in less than one month (Stone 1866; LeVerrier 1866; Schonfeld 1867; Dawes 1866; Bruhns 1868). The star remained near $m = 10$ for about 4 months, and then rose slowly to $m = 8$ (Campbell and Shapley 1923). T CrB stayed at this "secondary maximum" for a few months, and then declined back to $m = 10$ at a rather leisurely pace.

The 1946 outburst of T CrB was almost identical to that of 1866 (Figure 5.1). Both Knight (1946) and Deutsch (1948) found the star at $m = 3$ on 8 February, and many observers (Pettit 1946a,b,c,d, 1950; Bertaud 1946; Rives 1946; Larsson-Leander 1947; Ashbrook 1946; Davidson 1946a,b; Wright 1946; Groube 1946; Jakovkin 1946; Vandekerkhove 1946; Kharadze 1946; Ernilev 1950; Gilson, Petit, and Koeckelenbergh 1963; Buchnicek 1946; Steavenson 1946; Soloviev 1946a,b, 1949; Isles 1974) tracked its subsequent decline. In each outburst, the rapid decline from $m \approx 2$ to $m \approx 10$ has been followed by a slow rise to $m \approx 8$ and a gradual decline back to $m \approx 10$. This "secondary maximum" is critical to understanding the cause of the outburst, as noted by Webbink (1976; see also Gratton 1951, 1952; Gratton and Kruger 1946, 1949).

Three smaller outbursts of very short duration have been observed by McLaughlin (1939), Ianna (1964), and Radick (Palmer and Africano 1982). Each of these flares (and the two major events as well) occurred near phase zero of the orbital ephemeris derived by Kraft (1964). Palmer and Africano suggest that T CrB be monitored near phase zero to search for other such flares, as these may be associated with periastron passage.

The quiescent light variations of T CrB have been examined by a number of observers (Schmidt 1884, and references therein; Campbell and Shapley 1923; Barnard 1907; Himpel 1938a,b; Bohme 1938; Bertaud 1946; Payne-Gaposchkin and Wright 1946), and Bailey (1975) determined that the low amplitude fluctuations followed a 113.5 day period. This is precisely 1/2 the orbital period discussed by Kraft, and therefore is a result of binary motion. Bailey suggested that ellipsoidal variations in a tidally distorted red giant would account for the photometric periodicity. Rapid flickering on a timescale of minutes has also been observed (Walker 1957; Lawrence, Ostriker, and Hesser 1967; Bianchini and Middleditch 1976; Walker 1977b; Oskanian 1983). This flickering is similar in nature to that observed in the short-period cataclysmic binaries.

Initial infrared observations of T CrB were presented by Geisel (1970), Gehrz and Woolf (1971), and Feast and Glass (1974), and this photometry is consistent with that expected from a normal M star surrounded by a modest amount of 1200 K dust. A 2.3 μm CO absorption band is present on very low resolution spectra of T CrB, and the strength of the band is comparable to that found in CI Cyg, another symbiotic star containing a lobe-filling giant (Kenyon and Gallagher 1983). Hobbs and Marionni (1971) provide an upper limit of 900 mJy to the far-IR flux at 3.5mm; given the lack of detectable radio emission, this is not a particularly useful upper limit.

Spectroscopy

The quiescent optical spectrum of T CrB (Figure A.7) is generally that expected from an M3 giant, with additional H I emission lines and a prominent Balmer jump (Lockyer 1892; Espin 1893, 1900; Adams and Joy 1921; Lundmark 1921b; Berman 1932; Humason 1938; Hachenberg and Wellman 1939; Swings, Elvey, and Struve 1940; Wellman 1940; Swings and Struve 1943b; Gravina 1981b; Andrillat and Houziaux 1982; Blair, et al. 1983; Williams 1983). Occasionally He I, Si I, Ca II, He II, N III, [O III] and [Ne III] have been present as emission lines, but they are never very intense. Minkowski (1943) once observed T CrB to have a B shell spectrum, but this is the only recorded instance of such an occurrence. The Balmer lines usually have the classical centrally reversed profile indicative of a gaseous ring, suggesting they are formed in a disk of material surrounding the hot star.

Spectra secured in an outburst of T CrB are far more entertaining than quiescent spectra. Broad emission lines are characteristic of light maximum, and the absorption components associated with the H I and He I emission lines imply expansion velocities of 4500 km s^{-1} (Huggins 1866a,b,c,d; Morgan and Deutsch 1947; Deutsch 1948; McLaughlin 1947; Bloch, Dufay, Fehrenbach, and Tcheng 1946a,b; Bloch, Fehrenbach, and Tcheng 1946; Bloch 1946; Gregory and Peachey 1946). Strong emission from He II and N III appears a few days after maximum, and weakens as the decline progresses (Morgan and Deutsch; Brahde 1949; Tcheng 1949a,b; Taffara 1949a,b; Belti and Rosino 1952). Several coronal lines brighten markedly as He II begins to fade, and these weaken rapidly once V drops below 7th magnitude (Bloch and Tcheng 1953; McLaughlin 1953; Belti and Rosino 1952; Bochonko and Malville 1968). As the decline progresses, both TiO absorption bands and [Fe II] emission lines grow stronger.

The rise to secondary maximum is characterized by the rapid quenching of He II and N III. During this phase, the spectrum has a strong continuum plus weak lines from H I, He I, Fe II, and [Fe II]. The emission lines, including He II and N III, and the M absorption

Figure A.7 - Optical spectrum of T CrB. The strong TiO bands indicate an M3 giant or bright giant, while H I and He I emission lines are typical of low excitation symbiotic systems.

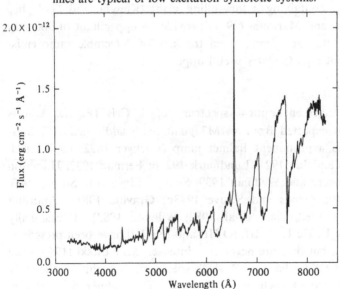

features regain their former prominence as the system declines once again. McLaughlin noted [Fe X] emission was present on spectra secured after secondary maximum, but did not report other coronal features.

Both Sanford (1949b) and Kraft (1958, 1964) noted radial velocity variations in the M absorption features and the Balmer emission lines, and they derived periods of 230.5 and 227 days, respectively. Kraft estimated a mass function of 5 $M_\odot$ and a mass ratio of 1.4 (the M3 component being the more massive star). Paczynski (1965) re-computed Kraft's solution assuming a circular orbit and an inclination of 68°, and obtained masses for the red and blue components of 2.6 $M_\odot$ and 1.9 $M_\odot$, respectively. The elements for each orbital solution are listed in Table A.5.

Low resolution *IUE* spectroscopy of T CrB was discussed by Cassatella, et al. (1982a,b; also Sahade, Brandi, and Fontenla 1984; Figure A.8). Although the system maintained a constant V magnitude over a 2 year period, large variations were observed in the UV. The integrated UV luminosity increased by a factor of 6-7 from March 1979 to February 1981, and then declined slowly. The strong permitted He II and N V lines weakened considerably as the continuum strengthened, while forbidden and intercombination lines intensified. This behavior may be a result of an increase in the density of the

Figure A.8 - Ultraviolet spectrum of T CrB. A few emission lines, notably C III] and C IV, are visible at this epoch, but do not dominate the continuum as in other systems (cf. Figure A.16).

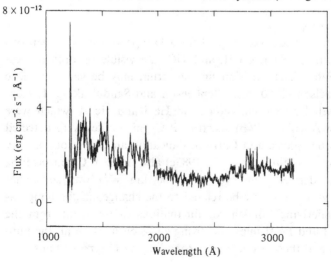

nebular region. Broad wings observed in Si III correspond to a velocity of ≈ 300 km s^{-1}, and such wings may be present on other lines as well. Kenyon and Webbink (1984) suggest the continuum variations in T CrB are a result of transient phenomena in a non-stationary accretion disk surrounding a main sequence star.

BI Crucis
(*Hen 782, Wra 967, SVS 1855*)

Hoffmeister first suspected this star to be variable, and Lee (1973) later associated the emission line star Hen 782 with SVS 1855. Both IR and optical observations have confirmed the variable nature of BI Cru, and it has turned out to be one of the more interesting symbiotic stars

Photometry

BI Cru is a D-type symbiotic, so it is not surprising that the system contains a Mira variable with a 280 day period (Allen 1974b; Whitelock, et al. 1983c). The system has a significant excess at 10 μm (Roche, Allen, and Aitken 1983), but the steam absorption bands usually seen in Mira variables are absent in BI Cru. This may be a result of the short period (and therefore early spectral type) of the Mira-like star. A few strong emission lines (Pβ and Bγ) appear on CVF spectra, and the 2.3 μm CO band may also be a weak emission feature. CO is usually seen in absorption in symbiotic stars; BI Cru is the only system in which CO is in emission.

Spectroscopy

BI Cru is somewhat unusual for a D-type system, since various absorption features (Na I, Ca II, and TiO) are visible on optical spectra (Allen 1974b). The emission line spectrum may be variable, since Henize and Carlson (1980; also Stephenson and Sanduleak 1971; Henize 1976) reported strong emission from He I and He II, which were not present on Allen's (1974b) spectra. P Cygni profiles were detected on a few lines by Henize and Carlson, indicating an expansion velocity of ≈ 450 km s^{-1}. Whitelock, et al. (1983c) also noted P Cygni profiles, although their derived wind velocity (200 km s^{-1}) was somewhat lower. These variations may be related to the changes in the temperature of the underlying hot star or fluctuations in the wind from the Mira. The H I and Fe II lines are strong emission lines on all the published spectra, and these are usually strong lines in D-type systems.

BF Cygni

(AN 112.1914, MWC 315, Hen 1747, Ls II +295)

BF Cygni was discovered by d'Esterre (1915) as an irregular variable, and subsequent observers noted variations between m = 9.4 and m = 11.2 (Hartwig, Kempf and Muller 1919; Hoffmeister 1919). Merrill and Burwell (1933) originally classified the spectrum as Bep, but more detailed spectra revealed BF Cyg's symbiotic nature. This is one of the lower excitation systems, and it presents interesting phenomena (such as the strong [O III] and [Fe III] lines) not visible in many other objects.

Photometry

The early photometric history of BF Cyg was reviewed by Campbell (1940) and Jacchia (1941). Semi-regular variations (P = 757 days; Pucinskas 1970) are apparent in the optical light curve, and these sometimes have the regularity expected of eclipses. At other times, this regularity disappears altogether, so the eclipses may be partial or grazing. A few small flares have been observed in BF Cyg (Romano 1966; Splittgerber 1976; Shaganian 1979), but these have not been as well-studied as those in CI Cyg and Z And. The star reaches m ≈ 9-10 at maximum, and returns to minimum in ≈ 300 days.

BF Cyg has an IR energy distribution that is similar to other S-type symbiotics, and there is, as yet, no evidence for low-amplitude variability (Swings and Allen 1972; Szkody 1977; Taranova and Yudin 1979, 1982a, 1983c; Tamura 1983; Kenyon and Gallagher 1983). A fairly strong 2.3 μm CO band is visible in low resolution spectra, and the band is about as deep as bands found in most M giants.

Spectroscopy

The M-type absorption spectrum (Figure A.9) is not as prominent in BF Cyg as in some other symbiotic stars, but TiO bands in the red suggest an M4 spectral type (Merrill 1943, 1950a; Swings and Struve 1945; Tcheng and Bloch 1954,1956; Payne-Gaposchkin 1954; Belyakina, Boyarchuk, and Gershberg 1963; Boyarchuk 1969b; Gravina 1981b; Andrillat and Houziaux 1982). Bright H I, He I, [O III], and [Ne III] emission lines dominate the optical spectrum, while lines of He II and N III are rather weak (Merrill, Humason, and Burwell 1932; Merrill and Burwell 1933; Shapley 1940a; Merrill 1941b; Aller 1952, 1954; Dossin 1959,1964; Kolotilov 1971; Merlin 1972, 1973; Henize 1976; Oliversen and Anderson 1982a; Oliversen and Anderson 1983;

Kenyon 1983).   Both the [O III] and [Ne III] lines are markedly variable from night to night, and the Balmer lines sometimes develop P Cygni absorption components (Swings 1970).   These fluctuations are not obviously correlated with the visual light curve, but sufficient spectral data are not available to determine the nature of either long-term or short-term variations.

BF Cyg has a fairly simple UV spectrum, with a strong continuum that rises towards short wavelengths and a few intense emission lines (Slovak 1980; 1982a).   Kenyon and Webbink (1984) found the UV continuum is best understood as a hot stellar source ($T_h = 40{,}000 - 45{,}000$ K, with $E_{B-V} = 0.49$) surrounded by a gaseous nebula.   The relatively cool effective temperature of the hot star is consistent with the weakness of the He II lines, and accurately predicts the absolute intensities of the lower level Balmer lines.

CH Cygni
(BD + 49°2999, HD 182917, SAO 031632, IRC + 50°294,
AN 24.1924, AGK3 + 50°1370, Pd 10683, Ag Bo 12955,
Boss 4966)

This low excitation system was first classified as an ordinary Mb star at Harvard, and was not recognized as a symbiotic star until

Figure A.9 - Optical spectrum of BF Cyg (C.M. Anderson, private communication).   The M-type absorption spectrum is rather weak in this object, although the H I and [O III] emission lines are very prominent.

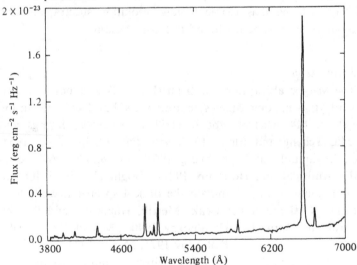

its 1964 outburst. While various single-star models for its behavior have been proposed (e.g., Deutsch, et al. 1974; Wdowiak 1977), recent observations suggest the object is truly a binary star (Kaler, Kenyon, and Hickey 1983). Some of the single star models suggested magnetic activity might be responsible for CH Cyg's peculiarities, but Slovak (1982b) failed to detect the large-scale magnetic field needed in such a model. The binary period is, at best, uncertain. Yamashita and Maehara (1979; cf. Tomov and Luud 1984) proposed an exceptionally long period, 5750 days, based on radial velocity data. Given the complex transformations observed in the emission lines (see below), this interpretation should be accepted with caution. If the giant component is normal, the binary period must be $\geq$ 1-4 yr.

Photometry

Early observations of CH Cyg showed the system to be a red irregular variable with a period of 97-101 days (Figure A.10; Muller and Kempf 1907; Wilson 1942; Model and Lochel 1942; Gaposchkin 1952; Payne-Gaposchkin 1954; Smak 1964, 1966). The variations over the past two decades have been rather erratic, as the star has displayed a number of outbursts (Gusev 1969a, 1976; Isles 1972; Quester 1972; Hopp and Witzigmann 1978, 1981; Taranova, Yudin and Shenavrin 1982; Hubscher 1983; Vennik, Tukivene, and Luud 1984). Generally, CH Cyg has been somewhat bluer than normal during its recent bright maximum (B-V$\approx$0.6, U-B$\approx$-0.4, as opposed to B-V$\approx$1.4, U-B $\approx$0.7 at minimum), although some maxima have been red (B-V$\approx$1.0; Luud 1979). It seems likely that the peculiar behavior of the light curve is the result of irregular variations in a cool giant and a hot companion star (Luud 1979).

CH Cyg also displays rapid photometric variations (Δm$\approx$0.1 mag, $\Delta t \approx$ minutes; Wallerstein 1968; Cester 1968, 1969; Johnson and Golson 1969; Walker, Morris, and Younger 1969; Luud 1970, 1973; Luud, Ruusalepp, and Kuusk 1971; Shao and Liller 1971; Belyakina and Brodskaya 1974; Luud, Ruusalepp, and Vennik 1977; Slovak and Africano 1978; Mikolajewska and Mikolajewski 1980b; Taranova and Shenavrin 1981, 1983; Doroshenko and Magnitsky 1982; Nakagiri and Yamashita 1980; 1982a; Spiesman 1984). This flickering is very small in the red ($\approx$2% at 0.7 μm), and increases dramatically towards the blue ($\geq$30% at 0.32 μm). Larger flares (Δm = 0.4) are sometimes visible in the blue, but these appear to be rare events (e.g., Taranova and Shenavrin 1981). The wavelength dependence of the amplitude (and also the polarization; Piirola 1982, 1983) is similar in many respects to that observed in cataclysmic variables, and may indicate the presence of a compact accreting star in CH Cyg (Luud 1979).

CH Cyg is one of the brightest symbiotic stars at K (-0.6 to -0.8), and is therefore a relatively easy target in the infrared. The broadband energy distribution is that of a normal M6-7 giant star with large excesses at 10 µm and 20 µm (Figure 3.3; Neugebauer and Leighton 1969; Low 1970; Gillett, Merrill, and Stein 1971; Swings and Allen 1972; Morrison and Simon 1973; Noguchi, Maihara, Okuda, and Sato

Figure A.10 - Optical light curve of CH Cyg before its eruptive activity of the 1960's and 1970's. The plotted points are ten day means of observations collected by the AAVSO.

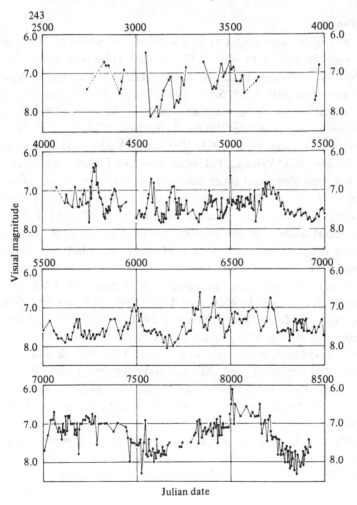

Julian date

1977; Taranova and Yudin 1979, 1982c; Price and Walker 1983; Roche, Allen and Aitken 1983; Kenyon and Gallagher 1983; Taranova and Shenavrin 1983). Very strong absorption bands observed in the near-IR (TiO, VO, and H_2O) and the IR (CO; Figure 3.11) also indicate a very late-type giant (Gusev 1969b,c, 1977; Gusev and Caramish 1970; Kenyon and Gallagher 1983). Gusev estimated a total luminosity of $\approx 4000\ L_\odot$ for the giant (based on the distance modulus, $m - M = 6.5$, estimated by Deutsch, et al. 1974), which would make it only slightly less luminous than the giant in CI Cygni. Note that this distance modulus is inconsistent with the 28 day period proposed by Deutsch, et al. and recently adopted by Duschl (1983). If the red star is to be contained within its Roche lobe, the 28 day period implies a distance ≤ 40 pc (assuming (i) $K = -0.7$, (ii) a bolometric correction to K for an M6 star [Frogel, Persson and Cohen 1981], (iii) an effective temperature for an M6 star [Ridgway, et al. 1980], and (iv) a total binary mass $\leq 10\ M_\odot$), which is ruled out by the small proper motion ($\mu \approx 0''.02 - 0''.03\ yr^{-1}$; AGK3, SAO) and lack of a detectable parallax.

Although the giant is a low amplitude variable in the infrared ($\Delta K = 0.2$; Ipatov, Taranova, and Yudin 1984), it is not responsible for the optical outbursts. Luud, Vennik, and Pehk (1978a,b) noted that K was nearly constant as the system brightened by 1.5 mag at V and 2.3 mag at U.

Spectroscopy

Initial optical spectra of CH Cyg revealed an M star, and later observations have revised the Mb spectral type to M6 III or M7 III (Stebbins and Whitford 1945; Stebbins and Kron 1956; Smak 1964, 1966; Galatola 1973; Maehara and Yamashita 1978; Gravina 1981b; Andrillat 1982a). Joy (1942) estimated a radial velocity of -53 km s^{-1} from the absorption lines, and did not report the appearance of any emission lines. Thus, it appears that CH Cyg did not display a combination spectrum before the recent series of outbursts.

The spectrum of CH Cygni in outburst seems to be constantly evolving, and it is not quite clear from recent observations if there is ever a quiescent state (Figure 3.2; Figure 5.4; Faraggiana 1967, 1980b; Swings and Swings 1967; Swings 1968; Faraggiana and Hack 1969a,b, 1971, 1980; Dzhimshelejshvili 1968, 1970, 1971; Dolidze and Dzhimshelejshvili 1969; Ivanova 1969; Gusev 1970, 1971, 1977; Doan 1970; Luud 1970, 1980a; Luud, Vennik and Pehk 1978a,b; Smith 1979; Ichimura, et al. 1979; Anderson, Oliversen and Nordsieck 1980; Oliversen and

Anderson 1982a; Hack and Selvelli 1982a,b; Hack, et al. 1982; Chochol and Hric 1982; Luud, et al. 1982; Wallerstein 1983; Blair, et al. 1983; Boehm and Hack 1983; Galkina 1983; Ipatov and Yudin 1983; Kaler, Kenyon, and Hickey 1983; Welty 1983; Luud and Tomov 1984; Persic, Hack, and Selvelli 1984). TiO absorption bands and H I emission lines are always strong features, while shell-like lines are present when the system is very bright (Wallerstein 1983; Yoo and Yamashita 1982, 1984; Yoo 1984). High excitation optical lines (He II, N III) have never been observed in CH Cyg, but low excitation emission from He I, [O III], Fe II and [Fe II] have been visible as weak lines (e.g., Blair, et al. 1983; Kenyon 1983; Rodriguez 1985; Tomov 1986).

The line profiles of many emission lines are very complicated, and fluctuate in a seemingly random fashion (e.g. Anderson, Oliversen and Nordsieck 1980; Swings and Andrillat 1981; Galkina 1983; Wallerstein 1983; Welty 1983; Tomov 1984). The lower-level Balmer lines are generally double-peaked emission features, while the upper levels often appear only in absorption. Inverse P Cygni profiles have been observed in a number of lines (Ti II, Fe II; Faraggiana and Hack 1980; Wallerstein 1983); this might indicate that material is flowing onto the hot component. Wallerstein (1983; Faraggiana and Hack 1980) suggested that multiple absorption layers surround the hot component at velocities of -44 and -63 km s^{-1}. While this may indeed be true, many observations will be needed in the future to attempt to correlate the complex variations in the lines with continuum fluctuations.

The ultraviolet spectrum of CH Cyg resembles that of an A star, although no model atmosphere gives an acceptable fit to the UV continuum (Hack 1979; Faraggiana, Hack, and Selvelli 1979; Faraggiana 1980a; Wing and Carpenter 1980; Hack and Selvelli 1982a,b; Hack, Persic and Selvelli 1982; Slovak 1982c; Persic, Hack, and Selvelli 1984). Most of the observed emission lines can be attributed to Fe II, although lines from Mg I, Mg II, O I, C II, and Al II have also been identified. A few high excitation lines (C IV and Si IV) are also visible, but they are not as strong as in some symbiotic stars. The Mg II line is cut by two distinct absorption components, and one of these is interstellar. The other component has a velocity of -50 km s^{-1}, and may correspond to the -44 km s^{-1} shell identified by Wallerstein (1983).

The ultraviolet continuum in CH Cyg is highly variable, having increased by a factor of 5 at $\lambda 1400$ during a one-year period. Kenyon and Webbink (1984) were unable to find a simple combination of a hot stellar photosphere and Balmer continuum radiation that satisfactorily reproduced any single *IUE* observation. The sharp continuum break at λ1700 in CH Cyg is also observed in A5-F0 stars, but is not usually seen in earlier type stars (Wu, et al. 1983). Since the system is so weak in the UV, one would not expect it to be an X-ray source. However, Luud (1980b) suggested CH Cyg was the X-ray source H1926+503, which has yet to be confirmed.

CI Cygni
(HV 3625, AN 10.1922, MWC 415, Hen 1784, VV 172)

A. Cannon discovered CI Cygni on a Harvard objective prism plate as a bright continuum source with superimposed H I and He II emission lines (Shapley 1922b). Closer examination of the plate revealed several absorption features and a bright band at λ4650. Merrill and Humason (1932) re-discovered it, and found very strong TiO bands in addition to the bright lines discussed by Miss Cannon. Thus, CI Cyg is one of the first "stars with combination spectra". A brief, but useful review of CI Cyg's historical behavior appears in Miller (1967).

Photometry

The visual light curve of CI Cyg is shown in Figure 5.2, and both the irregular outbursts and the periodic eclipses are fairly distinct. Whitney (see Aller 1954) first noted the periodic nature of the long-term variations, and derived the ephemeris listed in Table A.4. Both Hoffleit (1968a) and Pucinskas (1972) suggested that the variation is the result of an eclipse, which was verified in a series of UBV observations obtained by Belyakina (1968a, 1974b, 1975, 1976, 1979a,b, 1980, 1981, 1983, 1984; Kenyon, et al. 1982). The long duration of the total eclipse (~60-70 days) requires that the occulting star fill its tidal lobe.

The depth of a quiescent eclipse is about 0.3 mag in V and is somewhat larger at U and B. The duration of totality is longer in V than in either U or B; however, the length of the eclipse (first contact to fourth contact) may be slightly longer in U than in B or V (Bauer 1975, 1977; Tempesti 1976; Splittgerber 1975a; Schweitzer 1976; Burchi, et al. 1980; Huruhata 1980; Gevorkian 1981; Slovak 1982d). The B-V and U-B colors become redder during eclipse, suggesting that a late-type star eclipses a hotter component. The brightness at mid-eclipse (V=11.1) is independent of the periodic outbursts. A few eclipses have occurred in the middle of an outburst, and Figure 5.2 shows eclipses observed following the 1975 outburst. The length of the eclipse ingress implies the radius of the eclipsed object is $\sim 5 \times 10^{12}$ cm.

There have been five outbursts in the recorded history of CI Cyg, and each one has been a unique event. The 1911 outburst lasted less than 100 days, and the star reached $m_{pg} = 10.6$. The next outburst occurred in 1937; the star achieved $m_{pg} = 10.2$ at maximum, and returned to minimum after only 50 days (Greenstein 1937). Both of these outbursts occurred near $\phi \sim 0.6$ on Whitney's ephemeris. The remaining three outbursts have been of somewhat longer duration (see Figure 5.2), and all have commenced slightly before or during an eclipse (e.g., Stienon 1973; Belyakina 1979b). Although detailed color information is not available for the first two outbursts, B-V decreased from its normal quiescent value of 1.1 to 0.6 in each of the last 3 outbursts. The behavior of B-V in outburst implies an M star was replaced by a predominantly F or G continuum, and this has been confirmed by contemporaneous spectroscopic observations.

Burchi, et al. (1983) claim to have observed night-to-night UBV variations in CI Cyg with amplitudes of 0.35 mag in B and V and 0.90 mag in U. However, significant variations (0.1-0.15 mag) were visible in their comparison stars, and this casts some doubt on their conclusions. No variations were observed by Belyakina (1984) during the 1982 eclipse. Such "flickering" is expected given the spectroscopic variability reported by Merrill and others, and additional observations are needed to check Burchi, et al.'s data.

IR photometry of CI Cyg is consistent with a non-variable M4-5 giant (Swings and Allen 1972; Szkody 1977; Taranova and Yudin 1979, 1981b), and CVF photometry shows that this giant has a very strong 2.3 μm CO absorption band (Kenyon and Gallagher 1983). The depth of this band is comparable to that found in some supergiant stars (e.g., BC Cyg), and thus the giant may have supergiant dimensions and fill its tidal lobe. Additional evidence for a very evolved red giant is provided by the 10 μm excess reported by Yudin (1981c; Taranova and Yudin 1983c; Mikolajewska 1983; although Low's [1970] data had previously indicated no 10 μm excess). This dust may explain the polarization detected by Szkody, Michalsky, and Stockes (1982).

Spectroscopy

The quiescent optical spectrum of CI Cyg is characterized by strong TiO absorption bands and intense high ionization emission lines, (Figure 3.9). The spectral type of the cool star has been alternatively classified as M4 III or M5 III (Merrill 1944, 1950a; Aller 1954; Tcheng and Bloch 1954; Boyarchuk 1969a; Andrillat and Houziaux 1982), but recent analyses suggest M4 is more appropriate (Chapter 3).

Much of the optical light is obviously emitted by this component, but H I, He I, He II, [Ne V], and [Fe VII] are usually quite intense emission lines (Merrill 1933; Merrill, Humason, and Burwell 1932; Merrill and Burwell 1933; Payne-Gaposchkin and Gaposchkin 1938; Himpel 1940; Swings and Struve 1943b; Aller 1952, 1954; Tcheng and Bloch 1954, 1956; Boyarchuk 1969a; Henize 1976; Gravina 1980, 1981a,b; Iijima 1981a,b; Swings and Andrillat 1981; Baratta, Mengoli, and Viotti 1982; Oliversen, Anderson, and Nordsieck 1982; Andrillat 1982a; Fehrenbach 1982; Oliversen and Anderson 1982a, 1983; Huang 1982a; Blair, et al. 1983). Other high ionization lines have been visible at some epochs (e.g., [Fe X]; Swings and Struve 1940a), but they are not usually very strong. The [O III] and [Ne III] lines are generally very weak in CI Cyg, which is in stark contrast to V1016 Cyg where these lines dominate the spectrum.

Many authors (Merrill 1944, 1950a; Swings and Struve 1943b; Aller 1954; Tcheng and Bloch 1954, 1956; Boyarchuk 1969a) have noted spectrum variability in CI Cyg, and much of this may be caused by the periodic eclipses discussed earlier. CI Cyg's spectrum becomes much redder in an eclipse, as the Balmer continuum disappears and many emission lines weaken considerably in intensity (Tcheng and Bloch 1956; Kenyon 1983; Oliversen and Anderson 1983). The [O III] and [Ne III] lines are *not* eclipsed, so these lines must be formed in a low density region surrounding the binary (Oliversen and Anderson 1983). The H I and He II lines are eclipsed, and their behavior implies that the bulk of the H II region has a radius of ≈ 0.5 AU (Mikolajewska 1983; Mikolajewska and Mikolajewski 1982, 1983; Mikolajewska, Mikolajewski, and Krelowski 1983; Oliversen and Anderson 1983; Kenyon 1983). The level of the continuum at wavelengths greater than $\lambda 5000$ is not affected by the eclipse (Kenyon 1983), and thus the giant is not significantly variable (however, the data do not exclude variations at the 0.2-0.3 mag level).

Merrill first described fluctuations in the line displacements, and noted that the H I and He II lines varied together, and that the He I, [O III], and [Ne III] lines also varied together. The Fe II lines may have followed the nebular lines, but these lines were weak and difficult to measure accurately. Iijima (1982; also Chochol, et al. 1984) analyzed the radial velocity variations of the TiO bands and the H I emission lines to estimate a mass ratio, q ($= M_{hot}/M_{cool}) = 1.5$; the orbital elements derived in his study are listed in Table A.5.

The interstellar reddening to CI Cyg was first estimated by Boyarchuk (1969a), and he derived a visual extinction of 1.2 magnitudes.

Mikolajewska and Mikolajewski (1980a) examined stars in the field
surrounding CI Cyg, and estimated the line-of-sight reddening to CI
Cyg was slightly larger, with $E_{B-V} \approx 0.45$. The Balmer line ratios have
also been used to estimate the amount of interstellar reddening to CI
Cyg, and results have ranged from $E_{B-V} = 0.35 - 0.47$ (Iijima 1982;
Oliversen and Anderson 1983; Kenyon 1983; Blair, et al. 1983). Similar results are obtained from the He I and He II lines (Kenyon 1983);
the preferred value for E_{B-V} is now 0.45 based on a combination of
optical and ultraviolet estimates.

The optical spectrum of CI Cyg in outburst is radically different
from that found in quiescence, and the observed changes are well-
documented. The rise to visual maximum is characterized by a large
increase in the intensity of the blue continuum and a considerable
weakening in the *relative* intensity of the high ionization emission lines
and the TiO absorption bands (Mammano, Rosino, and Yildizdogdu
1975; Yildizdogdu 1981; Belyakina 1979a; Ivanova and Khudyakova
1980; Audouze, et al. 1981). Detailed spectrophotometry is needed to
confirm if the *absolute* intensities of the emission lines decrease in an
outburst.

At maximum light, the optical spectrum resembles that of an F
supergiant with superimposed H I and He I emission lines (e.g.,
Belyakina 1979a). Strong central absorption cores are visible in the H
lines, but are much weaker in the He I lines (Mammano, Rosino, and
Yildizdogdu 1975). The F-type absorption features (Ca II H and K,
Na I D, and other metallic lines) are usually quite numerous, and there
is no evidence that these lines ever have emission components (Belyakina 1979a).

The high ionization lines and the TiO absorption bands return as the
system declines from visual maximum and the blue continuum dimin-
ishes in intensity (Mammano, Rosino, and Yildizdogu 1975; Belyakina
1979a). The central absorption cores in the Balmer lines remain as the
decline progresses (Fehrenbach and Huang 1981), although He I lines
with associated absorption components have not been observed in
quiescent spectra. A few eclipse cycles are required for the spectrum to
evolve from recent light maxima to a normal quiescent state (Belyakina
1979a; Gravina 1980, 1981a,b).

The ultraviolet spectrum of CI Cyg (Figure 3.9) shows that the system is a bright continuum source (once the spectra are de-reddened)
with many strong emission lines (Stencel, et al. 1980, 1982; Michalitsianos, et al. 1982c; Baratta, et al. 1982; Lambert, et al. 1980; Stencel,
Michalitsianos, and Kafatos 1982). The bright UV continuum and

some emission lines disappear during an eclipse, although a few lines (e.g., Mg II, C IV, and N V) are not eclipsed. These data suggest that a stream is present in the system which feeds a large accretion disk ($R \sim 10^{13}$ cm). Slovak (1980, 1982a,c) modelled the observed UV continuum with a subdwarf A star reddened by $E_{B-V} = 0.2$, but this is insufficient to provide the high energy photons needed for the intense emission lines.

Kenyon and Webbink (1984) found that the observed continuum is understood best as an accretion disk surrounding a main sequence star. Their derived accretion rate, $\sim 10^{-5} M_{\odot}$ yr^{-1}, requires a lobe-filling giant companion. Indeed, the available IR and optical observations suggest the giant is an evolved asymptotic branch giant that does fill its Roche lobe (see above). The observed line fluxes from the Balmer lines agree very well with those predicted from the accretion model; although the He II line flux remains underpredicted.

V1016 Cygni
($MH\alpha$ 328-116, AS 373)

The "peculiar emission star in Cygnus" was reported by FitzGerald, et al. (1966) to have undergone a nova-like outburst in 1965. Previous to this outburst, it had been catalogued as an Hα source (Merrill and Burwell 1950), and as a long period variable (Nassau and Cameron 1954). V1016 Cyg is one of a handful of very slow or symbiotic novae which have undergone protracted outbursts of ≥ 20 yrs, and is also a weak X-ray source (Allen 1982b). These stars represent a unique chance to understand nova-like outbursts in general, as noted by Kenyon and Truran (1983).

Photometry

The early photometric history of V1016 Cyg was reviewed by FitzGerald, et al. (1966; cf. Ahern, et al. 1977, 1978; Cassatella 1982). Before the outburst, the object was red with a B-V of about 1.5 (Nassau and Cameron 1954; also Palomar Sky Survey and Maria Mitchell Observatory plates). Two post-outburst UBV measurements made by Mumford (1966) suggest a large variability in U, although B and V were relatively constant (V = 11.32, U-B = -0.55, B-V = 0.02 on JD 2439385.65 and V = 11.31, U-B = -0.38, B-V = 0.05 on JD 2439386.76). The post-outburst light curve has been discussed by Ciatti, Mammano, and Vittone (1978a,b, 1979); Figure A.11 shows visual observations obtained by the AAVSO and sources summarized by Ciatti, Mammano, and Vittone (1979). The rise from minimum was fairly slow, and lasted ∼2-3 years. Since the system reached maximum, it has remained relatively constant at V = 10.5 (cf. Arhipova 1983).

The strong emission lines observed in the optical require a very intense source of UV radiation. However, the molecular absorption lines of AlH, TiO, and VO that also have been observed in optical spectra of V1016 Cyg (see below) imply a much cooler region. A cool IR source was detected by Knacke (1972) and Swings and Allen (1972), who found large excesses at 3 μm and at 10 μm. Both excesses originally were thought to be the result of thermal radiation by optically thick dust (T = 200 K and T = 1000 K); recent observers have suggested the excess at 3 μm might be caused by an enormous amount of circumstellar reddening. The 10 μm excess in V1016 Cyg is similar in shape to excesses observed in other late-type stars, as noted by Aitken, Roche, and Spenser (1980). They found evidence for a three-fold *decrease* in the 10 μm flux over 1972-78, which was confirmed by Yudin (1982c). This feature stabilized for a few years, since observations from 1979-81 are consistent with 1978 data (Taranova and Yudin 1980, 1983a). However, Ipatov, Taranova, and Yudin (1985) have noted the presence of additional dust at 3 μm in 1983, which could be indicative of new dust formation in this system.

Whatever the true nature of the IR continuum might be, the source is definitely variable with a period of ≈470 days (Harvey 1974; Feast,

Figure A.11 - Optical light curve of V1016 Cyg during its recent outburst. Before this system rose to V = 10 in 1964, it was classified as an M-type giant with V ∼ 14 and a strong Hα emission line. The optical luminosity has been nearly constant since 1965, although its spectrum has increased towards higher excitation.

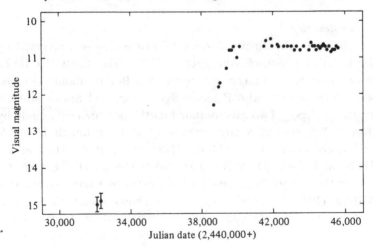

Robertson, and Catchpole 1977; Taranova and Yudin 1980, 1983a; Yudin 1982c; Kenyon and Gallagher 1983; Lorenzetti, Saraceno, and Strafella 1985). The amplitude of these fluctuations ($\Delta K \sim 1$) is similar to that of Mira variables, and higher resolution observations detected the CO and H_2O absorption bands commonly seen in evolved red giants (Puetter, et al. 1978; Kenyon and Gallagher 1983). The presence of VO and TiO absorption bands in optical spectra (and the 470 day period) imply the Mira has an M6-7 spectral type (Ciatti, Mammano, and Vittone 1979).

Spectroscopy

The optical spectrum of V1016 Cyg (Figure A.12) is currently dominated by extremely intense H I and [O III] emission lines superimposed on a very weak continuous source (Boyarchuk 1968a; O'Dell 1967; FitzGerald and Houk 1970; Ciatti, Mammano, and Rosino 1971, 1975; FitzGerald 1973; Baratta, Cassatella, and Viotti 1974; FitzGerald and Pilavaki 1974; Mammano and Ciatti 1975; Ahern 1975, 1978; Ciatti, Mammano and Vittone 1978a,b; Strafella 1980; Swings 1981; Gravina 1981b; Yudin 1981a; Muratorio and Friedjung 1982b; Oliversen and Anderson 1982a, 1983; Stauffer 1984). Higher ionization lines (He II, [Ne V], and [Fe VII]) have appeared since the outburst began,

Figure A.12 - Optical/ultraviolet spectrum of V1016 Cyg. The weak continuum and emission lines have been expanded in this spectrum; the 2200 Å absorption feature is quite pronounced, and implies $E_{B-V} \sim 0.25$.

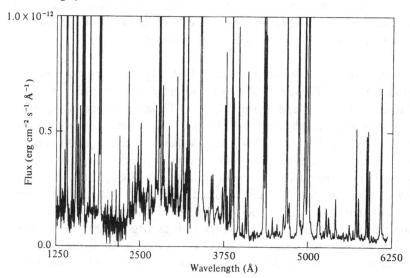

and are slowly increasing in intensity (Boyarchuk 1968a; Ciatti, Mammano, and Vittone 1979; FitzGerald and Pilavaki 1974; Ipatov and Yudin 1981; Andrillat, Ciatti, and Swings 1982; Blair, et al. 1983; Kenyon 1983). The evolution of the spectrum has paralleled the development of the very slow nova RR Tel; a simple interpretation of the data suggests the ionizing source may be getting hotter. The nebular conditions appear to have remained relatively constant, as the strong [O III] and [Ne III] lines continue to give $n_e = 10^{5-6}$ cm^{-3} and $T_e = 8,000\text{-}30,000$ K (Boyarchuk 1968a; Ahern 1978; Kenyon 1983; Oliversen and Anderson 1983).

FitzGerald and Pilavaki (1974) detected AlH absorption in V1016 Cyg, and this sparked a search for a cool M star. Both TiO and VO absorption bands occasionally have been observed since 1974, and the appearance of these bands might be correlated with maxima in the IR light curve (Mammano and Ciatti 1975; Ciatti, Mammano, and Vittone 1979; Andrillat, Ciatti, and Swings 1982). The absorption lines are typical of late-type Miras, strengthening the case for V1016 Cyg being a symbiotic star, rather than a proto-planetary nebula.

Ultraviolet spectra of V1016 Cyg are dominated by very intense emission lines (C IV, He II, C III, O III, N V, etc.) superimposed on a weak continuum (Figure 5.5; Figure A.12; Flower, Nussbaumer, and Schild 1979; Nussbaumer and Schild 1981; Feibelman 1982a). The relative intensities of the [Ne V] $\lambda\lambda$ 1575, 2973 and the He II $\lambda\lambda$ 1640, 2385, 2511, 2733, 3203 lines imply $E_{B-V} = 0.2 - 0.3$, which is close to that derived by Ahern (1978; $E_{B-V} = 0.26$) using the hydrogen Balmer lines. Nussbaumer and Schild adopted $E_{B-V} = 0.28$, and developed a nebular model for the UV emission lines. Their value for the electron density $(3 \times 10^6$ cm$^{-3})$ is comparable to that determined from optical data.

High resolution optical spectra of V1016 Cyg were first discussed by FitzGerald and Pilavaki (1974), who found double-peaked profiles in the strong H I lines and in He II λ4686. These observations suggested a shell of gas expanding at 30-40 km s^{-1}. More recent data described by Solf (1983), Stauffer (1984), and Wallerstein et al. (1984) confirm this basic feature, as double-peaked structure was observed in [N II], [O III], and the λ6830 emission feature (cf. Swings and Andrillat 1981). Similar structure was reported in C IV λ1550 by Feibelman (1982b), but this was later refuted by Deuel and Nussbaumer (1983). Some high ionization ultraviolet lines such as [Mg V] λ2926 might possess multiple components, but most lines appear to be symmetric with broad wings (Nussbaumer 1982a,b; Kindl, Marxer, and Nussbaumer

1982). It is curious that the ultraviolet lines have different profiles than those of the strong optical lines.

Radio Observations

V1016 Cyg is a fairly strong radio source, although not as intense as its sister object HM Sge (Seaquist and Gregory 1973; Marsh 1976; Purton, Feldman, and Marsh 1973; Marsh, Purton, and Feldman 1976; Kwok 1982a). The slope of the radio spectrum ($\alpha \sim 1$) can be explained by a constant velocity outflow in an inverse-square density law (Chapter 3), with a mass loss rate of $\sim 10^{-5} \, M_\odot \, \text{yr}^{-1}$ if the distance is 1 kpc.

Harris and Scott (1976) were the first to resolve the shell of V1016 Cyg, and thought the nebula had a bipolar appearance. This was confirmed by Hjellming and Bignell (1982; Newell and Hjellming 1981; Becker and White 1985), as their VLA data clearly show two blobs of material on opposite sides of the stellar source. Hjellming and Bignell (1982) suggested that this structure resembled the appearance of classical novae, which provides additional support for a model in which the 1965 outburst is due to a thermonuclear runaway on the surface of a white dwarf (Chapters 4,5).

V1329 Cygni
(*HBV 475*)

V1329 Cyg was discovered as HBV 475 by Kohoutek (1969) on the basis of its peculiar emission spectrum; a search of previous plates revealed the system had undergone a nova-like outburst in 1964 and was still at maximum. This symbiotic star is one of a handful of objects that have been called very slow or symbiotic novae (Allen 1980a; Kenyon and Truran 1983). In that regard, it is similar to AG Peg (and distinct from V1016 Cyg, HM Sge, and RR Tel) in having an obvious TiO absorption spectrum, a dust-free infrared continuum and no evidence for X-ray emission.

Photometry

The outburst of V1329 Cyg began in 1956, as m_{pg} decreased from 15.5 to 14.2 in 7 years (Figure A.13; Kohoutek and Bossen 1970; Bossen 1972; Richter 1972; Arhipova and Mandel 1973, 1975; Hicks 1973; Grygar and Chochol 1982). This was followed by a dramatic rise to $m_{pg} = 12.5$ in 1964 and a very gradual fading. The decline has been accompanied by 1 mag oscillations that are clearly periodic. Current estimates place the period at 950-960 days (Arhipova 1977;

Grygar, Hric, and Chochol 1977; Hric, Chochol, and Grygar 1978), and it is usually interpreted as an eclipse of the outbursting star by its M companion.

Pre-outburst observations of V1329 Cyg indicate the system once contained a late-type star (B-V = 1.6), and infrared photometry confirms that this star is an M giant (Knacke 1972; Allen 1974a; Tamura 1983; Lorenzetti, Saraceno, and Strafella 1985). The H-K color is similar to that observed in most late-M giants, and there is no evidence for a substantial circumstellar dust shell (Knacke 1972). Significant variability has been observed at J and H ($\Delta H \sim 0.7$; Yudin 1981b), but a low resolution search for CO or H_2O absorption has not been attempted.

Figure A.13 - Optical light curve and spectroscopic development for the recent outburst of V1329 Cyg. The upper panel shows the B light curve since 1960; the tic marks refer to times of minima from the photometric ephemeris listed in Table A.4. The lower panel indicates the evolution of the spectrum as a function of time.

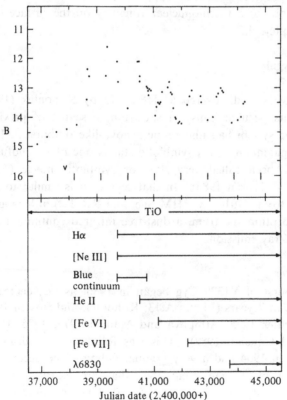

Spectroscopy

Only one spectrum of V1329 Cyg was obtained before its 1964 rise in brightness, and it shows an M star with no trace of any emission lines (Stienon, Chartrand and Shao 1974). Spectra secured a few years after outburst have strong H I emission, weak He I and [Ne III] lines, and a diffuse λ4650 band (Stienon, Chartrand and Shao 1974; Mammano and Righini 1973). Many more emission lines were visible in 1969, when the system was discovered by Kohoutek (Crampton, et al. 1970; Dean and van Citters 1970; Dumortier and Stram 1970; Andrillat 1970, 1973; Mammano and Righini 1973; Ciatti, D'Odorico, and Mammano 1974). V1329 Cyg has slowly evolved towards higher excitation; [Fe VII], [Ca V], and the λ6830 band appeared in the mid-1970's and have continued to increase in intensity (Andrillat and Houziaux 1976; Andrillat 1982b; Tamura 1977, 1978, 1981a,b; Ahern 1978; Gravina 1981b; Iijima and Mammano 1981; Oliversen and Anderson 1982a). Wide He II and N III lines were visible in 1969, and these Wolf-Rayet features disappeared as the high ionization forbidden lines grew stronger (Andrillat and Houziaux 1976).

Although Dean and van Citters (1970) claimed the cool star could not be later than K5, most spectra of V1329 Cyg show TiO bands quite clearly (Andrillat 1982b; Andrillat and Houziaux 1982; Mammano and Righini 1973; Oliversen and Anderson 1983). The presence of the λ8432 TiO band implies the giant is later than M3; if the VO band reported by Mammano and Righini (1973) is truly present, the spectral type is M6 or later.

The line profiles of V1329 Cyg are very complex, and the strongest lines have many sharp features (Tamura 1977, 1978, 1981a,b; Oliversen and Anderson 1982a). These may be caused by individual expanding gas clouds, or perhaps expanding shells as in classical novae. Two blue-shifted components in [Fe VII] have velocities of -200 and -350 km s^{-1}; the asymmetrical shape of these lines is reminiscent of [Fe IV] and [Fe VI] lines in RR Tel. The interpretation of the line profiles is complicated by peculiar variations. Observations of Hα at times reveal a wide line with a deep central absorption, while a narrow, symmetrical line is visible on other occasions (Oliversen and Anderson 1982a).

V1329 Cyg has a binary period of 950 days, and various investigators have attempted to derive an orbital solution (Grygar, et al. 1979; Chochol, Grygar, and Hric 1980; Iijima and Mammano 1981; Iijima, Mammano, and Margoni 1980, 1981). These solutions give an enormous mass function ($f(M) = 23\ M_\odot$), and such a large mass is inconsistent with other optical and infrared data. The radial velocity curve for V1329 Cyg has been derived by combining velocities from H I, He I, He II, Fe II, and [O III] emission lines, and it may be that some of these lines are not associated with the hot star. By analogy with other symbiotics (e.g., AG Peg), the Fe II and [O III] lines should not be associated with the hot star. At the very least, the He^{+2} zone should not be coincident with the O^{+2} zone (Osterbrock 1974). Until these ambiguities are sorted out, the mass function is, at best, uncertain.

Ultraviolet spectroscopic observations of V1329 Cyg have been discussed by Nussbaumer and Schmutz (1983; also Kindl and Nussbaumer 1982a; Feibelman 1982a). The UV continuum level and the emission line intensities follow the 950 day optical variation, and could probably be used to verify that the hot star is eclipsed by the giant. The radial velocity variations also appear to be correlated with the 950 day period, and these should be used to determine the orbit more accurately. As in a few other symbiotics, the UV continuum is very weak, while the lines are very intense. The gaseous nebula in V1329 Cyg appears to be similar to those of V1016 Cyg and HM Sge, with $n_e = 1-5 \times 10^6$ cm^{-3} and $T_e = 15,000$ K.

Radio Observations

The behavior of V1329 Cyg in the radio parallels the optically thin decline phase of classical novae (Kwok 1982a). Early observations reported a mean flux density of 14 ± 5 mJy during 1973-78 (Altenhoff and Wendker 1973; Terzian and Dickey 1973; Purton, et al. 1982) and a later observation placed the system at 5 mJy (Hjellming 1981). Kwok, Purton, and Keenan (1981) failed to detect the system with the *VLA*, although Seaquist, Taylor and Button (1984) later measured a flux of 2.16 ± 0.44 mJy. The source therefore may be variable, and continued monitoring may reveal the source of the fluctuations.

AG Draconis
(*BD + 67° 922, SAO 016931, AGK3 + 66° 715, CΠZ 1155,*
Do 35139, BV 48)

AG Draconis was discovered by Janssen and Vyssotsky (1943) on objective prism plates. They found a weak continuous spectrum with superimposed H I and He II emission lines. Wenzel (1955a,b) reviewed the discovery paper, and included a short description of the light curve, a finding chart, and a list of comparison stars. Based on Wilson's spectra, Eggen (1964) listed AG Dra as a high velocity star with a G7pe spectral type and V=9.44, B-V=0.88 and U-B=-0.68.

The absorption line radial velocity was measured at -137.6 km s^{-1}, while that of the emission lines was -146 km s^{-1}.

Photometry

The photometric variations of AG Dra were first described by Sharov (1954) and Beriev (1955; also Tsesevich 1956; Badalyan 1958), both of whom noted 2.5 mag variations with no obvious periodicity. Historically, the quiescent photographic magnitude of AG Dra is ≈ 11, with small amplitude variations (≈ 0.6 mag; Rathman 1961; Robinson 1964, 1969; Belyakina 1965a,b, 1966, 1969; Gessner 1966; Splittgerber 1975a; Klyus 1978; Meinunger 1979; Burchi 1980; Oliversen, et al. 1980; Taranova and Yudin 1982a). These fluctuations are largest at ultraviolet wavelengths (e.g., $\Delta U \approx 1$ mag), and are negligible at optical wavelengths ($\Delta V \approx 0.15$; Belyakina 1969; Meinunger 1979). The variations have been found to be regular only recently, and Meinunger (1979) determined a period of 554 days.

Like Z And and AX Per, AG Dra has undergone a number of 2 mag outbursts, during which time the brightness fluctuations may be quite large (Sharov 1960; Robinson 1964, 1969). The most recent outburst occurred in 1980, when the visual magnitude rose from $V = 10$ to $V = 8$ over a period of weeks. According to Kaler (private communication), both the blue continuum and the strong emission lines increased in intensity during the outburst. However, the He II $\lambda 4686$/Hβ ratio remained relatively constant, suggesting the temperature of the hot source did not vary significantly during the six-fold increase in visual brightness. This was later confirmed by *IUE* observations, as discussed below. This behavior is dramatically different from that experienced by CI Cyg and AX Per, as the relative intensities of the emission lines *fade* during the outbursts of these systems.

As noted below, the spectral type of the cool star in AG Dra is somewhat uncertain. Infrared photometry points to a K-type continuum (Swings and Allen 1972; Bopp 1981; Kenyon and Gallagher 1983; Taranova and Yudin 1983c), but does not shed any light on the controversy regarding the luminosity class of this component. It is evident that the K-type star is not strongly variable in the IR, as various estimates suggest $\Delta L < 0.3$ mag (Taranova and Yudin 1983c).

Spectroscopy

Wilson (1943,1945) obtained the first high dispersion optical spectra of AG Dra, and found strong emission lines of He I in addition to those found by Janssen and Vyssotsky. An example of a low dispersion spectrum is given in Figure A.14, and it shows the strong emission lines and red continuum expected of a symbiotic star. The absorption spectrum of the cool component has been classified as either K1 II (Roman 1953), G5 (Mirzoyan and Bartaya 1960), K3 III (Arakelyan and Ivanova 1958; Boyarchuk 1966b; Dolidze and Dzhimshelejshvili 1970), G-type (Dolidze and Pugach 1962; Chkhikvadze 1970a), K5 III (Belyakina 1969), or K0 I (Huang 1982b). Obviously, the nature of the cool star is not well-known (some of the differences represent a poor choice in the spectral region covered, since blue spectra give systematically earlier spectral types than red spectra), but K3 is probably a reasonable compromise regarding the spectral type. The luminosity class is, at best, ambiguous for the present.

Generally only H I, He I, and He II emission lines are observed in AG Dra (Mirzoyan 1955, 1956, 1958a,b; Boyarchuk 1966b; Boyarchuk and Gershberg 1966; Doroshenko and Nikolov 1967; Johnson and Golson 1969; Chkhikvadze 1970a; Aslanov and Rustamov 1970; Huang 1982b; Oliversen and Anderson 1982b), but recent spectra have found weak emission from O III, [Fe V], [Fe VI], and the ubiquitous

Figure A.14 - Optical spectrum of AG Dra. The relatively featureless continuum is typical of symbiotic stars with G or K giants. Strong emission lines from H I and He II dominate this spectrum, and additional lines from He I and the λ6830 band are also visible.

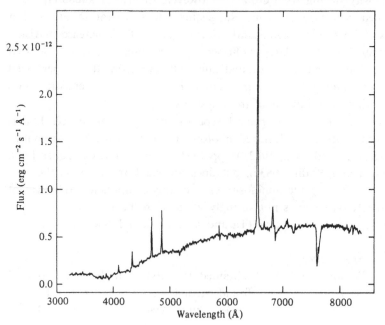

λ6830 band (Smith and Bopp 1981; Gravina 1981b; Faloma 1983; Blair, et al. 1983). Boyarchuk (1966b) reported a weak [O III] line which is not present on later spectra. Given the variability in the line spectrum reported by Mirzoyan (1955, 1958a,b), Doroshenko and Nikolov (1967) and Faloma (1983), the disappearance of [O III] is no cause for alarm.

Ultraviolet observations of AG Dra were obtained both before and after the recent outburst, and these data represent a unique opportunity to test various models for the outburst phase of a symbiotic star. All the IUE spectra are characterized by a strong continuum (which resembles the Rayleigh-Jeans tail of a blackbody), intense permitted emission lines and weak intercombination lines (Lutz and Lutz 1981; Altamore, et al. 1982; Viotti, et al. 1982a, 1983, 1984). The strongest emission line is He II λ1640, which is composed of a sharp, central emission superimposed on a weak, broad line (implying $v \sim 1000$ km s^{-1}). The intercombination lines (O IV], N III]) suggest a very dense nebula ($n_e \lesssim 10^{10}$ cm^{-3}), with the He II formed in a somewhat denser region.

The ultraviolet continuum observed before the recent outburst can be fit by a simple blackbody at a temperature of $\approx 130,000$ K (Kenyon and Webbink 1984; Viotti et al. 1983, 1984), providing the flux distribution has not been affected by interstellar reddening. Various independent reddening estimates give essentially no reddening towards AG Dra as discussed by Kenyon and Webbink (1984). If the cool companion to this hot source is a normal K giant, then the hot star is definitely a white dwarf. However, it seems more likely that the K star is a Pop II giant, in which case the hot star is ≈ 5 times larger than a white dwarf.

Ultraviolet observations secured during the decline from the 1980 maximum show that the UV continuum of AG Dra was ≈ 2.5 mag brighter with little change in overall shape (Altamore, et al. 1982). Most of the high ionization lines increased in intensity as well, although their *relative* intensities showed little evidence of a change. As noted above, the optical spectrum behaved in a similar fashion, and together with the UV data, this implies that the hot star increased dramatically in luminosity while undergoing only a modest increase in its effective temperature (Kenyon and Webbink 1984). Kenyon and Webbink (1984) suggest these exploits are good evidence that the outbursts of AG Dra are powered by thermonuclear runaways (Chapters 4,5).

X-ray Observations

AG Dra is one of the few symbiotic stars detected by *Einstein*. The total X-ray luminosity is $\leq 1\ L_\odot$ (depending on the distance), and the temperature of the source is $\approx 150,000\text{-}250,000$ K (Anderson, Cassinelli, and Sanders 1981; Anderson, et al. 1982). The sole observation of AG Dra was made $\approx 1/4$ cycle before maximum (and before the recent outburst); new observations should concentrate on correlating the X-ray flux with the photometric period.

YY Herculis
(*MH* α 352-34, AS 297)

YY Her is a classical symbiotic star, having been identified by Herbig in the late 1940's. Unfortunately, the system is ≈ 1 mag fainter than more prominent systems such as Z And and CI Cyg, and has therefore suffered from a lack of exposure.

Photometry

YY Her was first observed by Wolf (1919), who found optical fluctuations from 11.0 to 13.0. Similar variations were observed by Plaut (1932; also Bohme 1939); he included a finding chart and a list of comparison stars. Belyakina (1974a) found 1 mag variations at UBV; and noted that the U-B and B-V colors are similar to those of other symbiotic stars.

YY Her is a fairly simple IR source, containing a normal M giant with little or no evidence for large-amplitude variability (Swings and Allen 1972; Kenyon and Gallagher 1983; Taranova and Yudin 1983c). The strong CO band commonly observed in symbiotic stars at 2.3 μm has not been searched for in this system.

Spectroscopy

Initial spectroscopic observations of YY Her were discussed by Merrill and Burwell (1950) and Herbig (1950), who found a red continuum with moderately strong TiO bands (Figure A.15). Strong emission from lines of H I, He II, and [O III] was visible on these early spectra, and He I lines were weakly detected. Herbig noted rapid variability in the radial velocities (≈ 20 km s^{-1} in 2 weeks), which may have been indicative of activity in the hot component. More recent optical spectra show prominent H I and He II emission, as well as strong TiO absorption bands in the red (Gravina 1981b; Blair, et al. 1983; Huang 1984). The Hα emission line has a central absorption feature (Welty 1983), which may be indicative of an accretion disk.

The ultraviolet spectrum of YY Her is reminiscent of CI Cygni (Figure 3.9), as lines of He II, C IV, and N V are all very prominent emission features (Michalitsianos, et al. 1982a). Various other lines (O III], N IV], Si IV], O IV], and O III) are also strong. The continuum is relatively flat, and is not easily understood as a simple combination of a stellar continuum with additional nebular radiation. Kenyon and Webbink (1984) suggest that the UV continuum is produced in an accretion disk surrounding a main sequence star companion to the late-type giant. The derived accretion rate of $\sim 10^{-5} \, M_\odot \, \mathrm{yr}^{-1}$, along with the deduced interstellar reddening, $E_{B-V} = 0.13$, are consistent with independent measurements of the reddening (Burstein and Heiles 1982) and luminosity (Kenyon and Webbink 1984).

V443 Herculis

(*MWC 609*)

V443 Her was discovered by Tifft and Greenstein (1958a,b), who remarked on its resemblance to the classical symbiotic stars. This

Figure A.15 - Optical spectrum of YY Her. Infrared photometry confirms the early M-type spectrum indicated by the strong TiO bands in YY Her, although low resolution IR spectra of the 2.3 μm CO absorption band have not been acquired. The strong line at the left edge of this spectrum is [Ne V] λ3426; other strong lines are due to H I and He II.

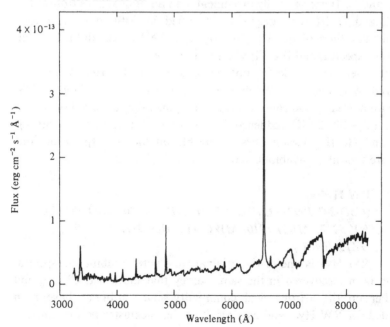

object is a fairly easy target for northern observers, and has much in common with RW Hya.

Photometry

Only Belyakina (1968a, 1974a) has observed V443 Her with UBV photometry, and she found variability in U-B and B-V as well as in V. Periodic variations were not reported by Belyakina, so additional observations are needed to determine adequately the nature of the optical fluctuations. Given the luminosity of the hot source, it would not be surprising if V443 Her exhibited the reflection effect as in AG Peg.

Infrared photometry confirms the middle-M spectral type originally assigned by Tifft and Greenstein (Swings and Allen 1972; Kenyon and Gallagher 1983). The moderately strong CO absorption band detected by Kenyon and Gallagher (1983) indicates the cool star is either a normal Pop I giant or a metal poor Pop II giant.

Spectroscopy

Modern spectra of V443 Her (Figure A.16) show a typical combination spectrum. The TiO bands are well-marked, and the lack of TiO λ8432 indicates the spectral type of the giant is M2-M4 (Andrillat 1982a; Chapter 3). Sharp emission lines from H I, He I, He II, and Fe II are generally prominent, and additional lines from C III, N III, and O III were originally identified by Tifft and Greenstein. The nebular lines of [O III], [Ne III], and [Fe V] were diffuse on the discovery spectra, and [Fe VI] was possibly present.

Ultraviolet spectroscopic observations of V443 Her were first analyzed by Kenyon and Webbink (1984), and the shape of the far UV continuum clearly indicates the hot component is a very hot stellar source ($T_h \sim 80,000$ K) reddened by $E_{B-V} = 0.31$ (Chapter 3). Strong C IV and He II emission lines are visible on the *IUE* spectrum, and these are typical of symbiotic stars.

RW Hydrae
(*CoD-24° 10977, CpD-24° 5101, HD 117970, SAO 181760, HV 3237, AN 51.1910, MWC 412, Hen 917, AW 1062, GZ13°1648*)

RW Hya is one of the original stars with combination spectra, having been discovered in the same survey that unleashed CI Cyg and AX Per on the astronomical community. No outbursts have been recorded for RW Hya, and as a result of this lackluster performance, it generally has been overlooked. This system is the best example of a symbiotic star with a hot stellar source, and has a rather simple emission spectrum. Its short period, 372 days, makes it a prime candidate for detailed monitoring projects at a variety of wavelengths.

Photometry

RW Hya varies from $m_{pg} \sim 10$ to $m_{pg} \sim 11.2$ over a period of 372.45 days (Pickering 1910; Townley, Cannon, and Campbell 1928; Shapley 1924b; Kenyon and Webbink 1984). If this period is orbital in origin, the radius of the Roche lobe is $\sim 110\ R_{\odot}$ for a reasonable choice of masses. An M2 giant (see below) has a radius of $\approx 90\ R_{\odot}$, so the cool component in this system may fill its tidal lobe.

Infrared photometry shows that the cool component is a normal M giant, with some evidence for an excess at 10 μm (Gehrz and Woolf 1971; Glass and Webster 1973; Feast, Robertson, and Catchpole 1977; Kenyon and Gallagher 1983). There is no evidence, as yet, for large amplitude variability in the IR.

Figure A.16 - Optical/ultraviolet spectrum of V443 Her. This system contains a middle-M giant star and a hot component with $T_h \sim 80,000$ K (Kenyon and Webbink 1984). Emmission lines from He II, C IV, and N V are fairly prominent in the ultraviolet shortwards of the λ2200 interstellar absorption feature, while H I and He I lines dominate the optical.

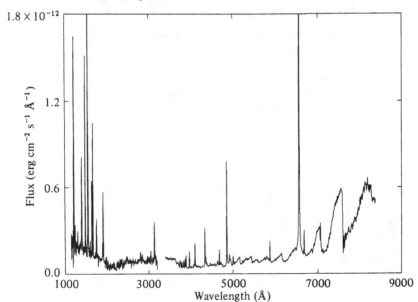

212   *Appendix*

Spectroscopy

Cannon secured the first spectra of RW Hya, and found an M-type absorption spectrum with H I emission lines. Merrill and Humason (1932; also Merrill and Burwell 1933) detected a He II emission line and TiO bands on their objective prism plates, and estimated a photographic magnitude of 10. Subsequent observations at higher dispersion revealed variable TiO bands, along with H I, He I, He II, and [O III] emission lines (Merrill 1933, 1940; Swings and Struve 1941b; Bidelman and MacConnell 1973; Henize 1976). All the emission lines had a variable radial velocity, with (i) H I, He I, and N III, (ii) [O III] and [Ne III], and (iii) He II following separate radial velocity curves.

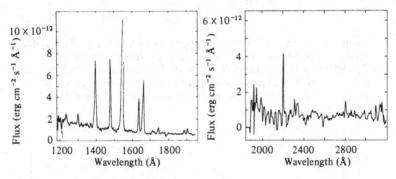

Figure A.17 - Ultraviolet spectrum of RW Hya (Kafatos, Michalitsianos, and Hobbs 1980a,b). The slope of the continuum is characteristic of S-type systems containing hot stellar sources ($T_h \sim 10^5$ K).

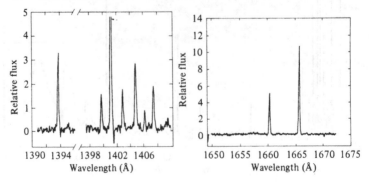

Figure A.18 - High resolution ultraviolet spectra of RW Hya (Kafatos, Michalitsianos, and Hobbs 1980b). The lines show no evidence for structure at the 10-20 km s^{-1} level, and have widths ~ 100 km s^{-1}.

Ultraviolet spectra of RW Hya have been discussed by Slovak (1980, 1982a) and Kafatos, Michalitsianos, and Hobbs (1980a,b). The *IUE* cameras detected RW Hya rather easily, and found a strong UV continuum and many intense emission lines (Figure A.17). The temperature of the hot star has been estimated to be 50,000-100,000 K (Slovak 1982; Kafatos, Michalitsianos, and Hobbs 1980b; Kenyon and Webbink 1984); the higher temperature is preferred since He II is observed as a fairly strong emission line.

High dispersion *IUE* observations of RW Hya were discussed by Kafatos, Michalitsianos, and Hobbs (1980), who determined $n_e \sim 10^8 - 10^9$ cm^{-3} and $T_e \sim 10,000$-15,000 K from an analysis of strong intercombination lines (Chapter 3). Most emission lines are very symmetric, with full widths at zero intensity of 100-200 km s^{-1} (Figure A.18). P Cygni absorption components do not accompany any emission lines, and thus there is no evidence for a large-scale stellar wind. The velocity widths of the ultraviolet lines are ~ 10 times smaller than the inferred escape velocity from the surface of the hot component (Kenyon and Webbink 1984); most of the emission therefore must originate in an extended nebula surrounding the binary system.

BX Monocerotis
(*HV 10446, AS 150, MH* α 61-12, SS 4)

BX Mon has the longest measured period among symbiotic stars ($P \approx 1380$ days; CH Cyg may have a 5750 day period, but it is not yet convincing). Although this system is rather faint in the optical, it is a fairly bright IR source. Since it has a known period, BX Mon deserves careful study.

Photometry

BX Mon was discovered by Mayall (1940) as a very long period variable ($P = 1380$ days). The mean light curve has a standstill from $\phi \approx 0.4$ to $\phi \approx 0.6$. Regular optical photometry of this system has not been acquired since the 1940's, although amateurs in the AAVSO monitor its brightness. Small-amplitude variations are apparent on the AAVSO visual light curve, but it is difficult to assign a period to the fluctuations.

The infrared energy distribution of BX Mon is appropriate for a late-type giant (Whitelock and Catchpole 1983), and the system has a prominent CO absorption band at 2.3 μm. Emission from Pβ is also visible in the near-IR, and thus FTS observations of BX Mon might be profitable.

Spectroscopy

Optical spectra of BX Mon show a typical symbiotic star, with strong H I and weak He I emission lines (Figure 3.10; Merrill and Burwell 1950; Bidelman 1954; Sanduleak and Stephenson 1973; Iijima 1984). Iijima (1984) noted λ4686 emission on one of his spectra, suggesting a variable hot source or perhaps an eclipse. Many TiO bands are prominent on optical spectra, and the giant has an M4 spectral type (Chapter 3).

The ultraviolet spectrum of BX Mon discussed by Michalitsianos, et al. (1982a) is similar to SY Mus near minimum, strengthening the possibility that this object is indeed an eclipsing binary. Strong absorption lines were visible in the λλ2300-3000 region, while various emission lines (C IV, N III], O III], Si III], and C III]; He II was not detected) were prominent at shorter wavelengths (Michalitsianos, Hobbs, and Kafatos 1980). The absorption features (mainly due to Fe II and Mn II) suggest that the near-UV was dominated by a bright stellar component, and indeed a late A or early F line-blanketed model atmosphere reproduces the observed continuum rather well (Michalitsianos, et al. 1982a). Such models do not provide enough high energy photons to explain the emission spectrum.

SY Muscae
(*HD 100336, HV 3376, AN 118.1914, Hen 667, Wra 824, SS 32*)

This interesting system was discovered by Cannon (Pickering 1914b) as a peculiar emission line star with a 1 mag range in photographic brightness. The periodic phenomena observed at optical and ultraviolet wavelengths may be the result of an eclipse, in which case the orbital period of SY Mus would be nearly identical to that of AR Pav.

Photometry

SY Mus is a well-known periodic variable, as both Uitterdijk (1934) and Greenstein (1937) estimated periods of ≈625 days. Recent visual observations have been made by the Variable Star Section of the Royal Astronomical Society of New Zealand (Bateson, Jones, and Menzies 1975), and a recent analysis of these data by Kenyon and Bateson (1984; Figure A.19) suggest the system is an eclipsing binary with $P = 627$ days. Large amplitude variations are not seen in the infrared (Glass and Webster 1973; Feast, Robertson, and Catchpole 1977), and the J-K and H-K colors are consistent with a moderately reddened M giant.

Spectroscopy

Optical spectra of SY Mus show strong TiO bands and prominent emission lines from H I, He I, and He II (Henize 1952; Allen 1984; Kenyon, et al. 1985). Lines from Fe II, [Fe II], [O III], [Ne V], and [Fe VII] are usually rather weak emission features. The strong H I and He II lines do not disappear in an eclipse, although they appear to weaken considerably (Kenyon, et al. 1985).

Ultraviolet data for SY Mus were first obtained by the *TD-1* satellite, although positive detections were made in only two bandpasses (λ1965 and λ1565; Thompson, et al. 1978). Multiple *IUE* observations have been discussed by Michalitsianos, et al. (1982a,b) and by Michalitsianos and Kafatos (1984), and reveal large-scale variations in both the continuum level and the strengths of various emission lines. Kenyon and Webbink (1984; also Michalitsianos and Kafatos 1982b) suggested that these variations were consistent with an eclipse, and subsequent *IUE* observations demonstrated that the hot source is indeed occulted during primary minimum (Kenyon, et al. 1985). The UV continuum outside eclipse can be fit by a 120,000 K blackbody in combination with Balmer continuum radiation if $E_{B-V} = 0.38$ (Kenyon and Webbink 1984). This value for the reddening is consistent with

Figure A.19 - Mean visual light curve of SY Mus from observations made by amateurs of the Variable Star Section, Royal Astronomical Society of New Zealand (Kenyon and Bateson 1984).

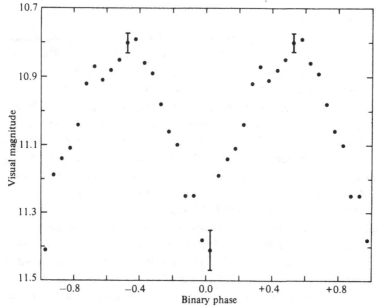

that derived from the He II λ1640/λ4686 intensity ratio (Kenyon, et al. 1985).

Strong intercombination lines are visible on ultraviolet spectra of SY Mus, including O III], O IV], N IV], and C III]. These lines decline in intensity by factors of 3-7 during eclipse, comparable to the decline experienced by the He II λ1640 and C IV λ548,1552 lines. Michalitsianos, et al. (1982a,b) estimated an electron density of $\sim 10^{10}$ cm^{-3} from the intensity ratios of these lines, and such large values are typical of symbiotic stars (Chapter 3).

RS Ophiuchi
(BD-6° 4661, HD 166214, MWC 414, SS 106)

The recurrent nova RS Ophiuchi was discovered by Fleming (1907; Pickering 1901b), although the range and spectral type were not noted. The discovery was announced by Hartwig (1901), and subsequent studies showed that the star had undergone an outburst in 1898 (Pickering 1904,1905). The spectrum was observed by Cannon to have strong hydrogen Balmer lines in addition to emission from N III and He II. In this respect, the spectrum resembled the spectra of other novae. Objective prism plates taken before the 1898 outburst showed a K-type spectrum. The available photographic data are rather incomplete in 1898, but the star reached at least a magnitude of 7.7, and quickly declined back to its normal value of 11-12.

Photometry

The optical outbursts of RS Oph are well-documented, although the quiescent variations are poorly understood. Each outburst has been followed by numerous groups (Loreta 1933, 1934a,b,c; Peltier 1933; Campbell 1933; Bloch 1934; Brydon 1933; Bloch, Ellsworth and Liau 1933a,b; Dermul 1933, 1958; Plakidis 1933; Pearce 1933; de Roy 1933, 1934; Flammarion and Quenisset 1934; Jacchia 1934; Roumens 1934; Vandekerkhove 1934; Viaro 1934; Guerrieri 1934; Prager 1940; Bertaud 1949, 1958; Antal and Tremko 1959; Cragg 1959; Connelley and Sandage 1958; Abrami and Cester 1959; Cousins 1958; Andrews 1959, 1960; Weber 1958; Kurochkin and Starikova 1958; Starikova 1960; Filin 1959; Magalashvili and Kumsishvili 1958; Mayall 1960; Eskioglu 1963; Bartolini 1965; Svovpoulos 1966; Isles 1974); the mean light curve in Figure A.20 has been compiled from many of these sources. Since the rise and decline are so very rapid, it is not clear if the *actual* visual maximum has ever been observed. The three most recent outbursts are nearly identical, and do not show an obvious secondary maximum as observed in each outburst of T CrB. Since RS Oph declines somewhat more slowly than T CrB, secondary maxima might well have been missed.

Variations following the 1898 outburst were discussed by Campbell (1912, 1918) and Townley (1906), and are similar to those displayed by most symbiotic systems. Few quiescent UBV observations have been secured recently, but the available data are sufficient to show 0.5 mag variations with no obvious periodicity (Tempesti 1975, 1977). The B light is polarized (3-6% at a position angle of 37-48°), but it is not clear if variations in the polarization are associated with variability in B or in radial velocity (Serkowski 1970).

A few infrared observations are available (Geisel, Kleinmann and Low 1970; Swings and Allen 1972; Feast and Glass 1974; Szkody 1977; Kenyon and Gallagher 1983), and these data are consistent with an M giant surrounded by 1200 K dust.

Figure A.20 - Optical light curve and spectral development for a typical outburst of RS Oph. The decline from a maximum in RS Oph is not as rapid as in T CrB, but follows a trend towards increasing excitation as a function of time.

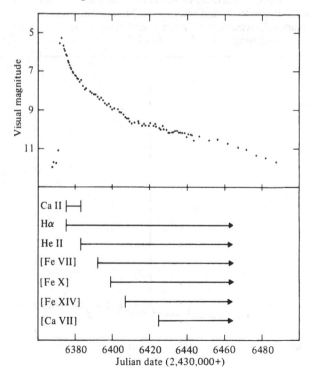

Spectroscopy

The quiescent spectrum of RS Oph is more or less typical of a symbiotic star, having a late-type absorption spectrum and various emission lines (Figure A.21). The absorption spectrum has been classified as G5, K0, and M2 (Pickering 1904,1905; Adams, Humason and Joy 1927; Sanduleak and Stephenson 1973); and early M is currently favored (Chapter 3). Strong H I and numerous Fe II emission lines usually are visible in RS Oph, and many other lines have been reported occasionally (Gaposchkin 1946; Humason 1938; Swings and Struve 1941b, 1943a; Sanford 1947a; Merrill and Bowen 1951; Barbon, Mammano, and Rosino 1969; Wallerstein 1963; Wallerstein and Cassinelli 1968; Mustel, Boyarchuk, and Bartash 1963; Sanduleak and Stephenson 1973; Williams 1983). In 1942, the high ionization lines were especially pronounced, and [Fe X] λ6374 was detected (Swings and Struve 1943a). Later, Joy and Swings (1945) reported a bright band at λ6830, which appears in many highly excited symbiotic stars.

As with T CrB, the spectrum of RS Oph transforms dramatically in an outburst. Although the 1898 outburst was not well-covered, the

Figure A.21 - Optical spectrum of RS Oph. The red TiO bands are fairly weak in this system, and indicate an M0 spectral type (Chapter 3). The Balmer decrement is very steep, and reflects the large amount of interstellar reddening in the direction of RS Oph.

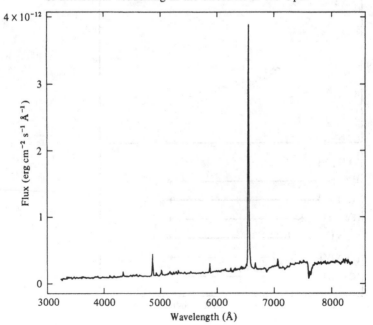

spectral evolution during the 1933, 1958, and 1967 outbursts has been documented by a number of observers (Adams and Joy 1933a,b, 1934; McLaughlin 1933; Colacevich 1933; Westgate 1933; Wright and Neubauer 1933; Wilson and Williams 1934; Sayer 1938; Wallerstein 1958, 1969; Rosino, Taffara and Pinto 1960; Joy 1961; Dolidze and Alanya 1962; Bloch and Dufay 1958; Dufay, Bloch, Bertaud and Dufay 1958, 1964, 1965a,b; Griffin and Thackeray 1958; Folkart, Pecker and Pottasch 1964a,b,1965a,b; Folkart, et al. 1961; Pottasch 1967; Tolbert, Pecker and Pottasch 1967; Dufay and Bloch 1964, 1965; Dufay, et al 1964, 1965a,b; Code 1968; Barbon, Mammano and Rosino 1968; Friedjung 1972). A diagram accompanying the light curve (Figure A.20) highlights the major developments common to each outburst.

At maximum visual light, the H I emission lines are wide (25-30 Å) and fairly strong. Narrow absorption and emission components are superimposed on these bands, and a weak absorption is visible at the violet end of the wide band. The He I and Fe II lines display a similar appearance, but are generally much weaker features. The continuum at maximum is also quite strong, and has a color temperature of $\approx 10,000$ K.

Both the sharp lines and the continuum fade during the decline, while the broad bands increase in intensity. The [O III] lines emerge as sharp lines initially, but these weaken rapidly and are replaced by wide bands. Broad [O I], He II, and N III lines appear as the decline continues. The $\lambda 4686$ line is a doubled emission line, while the $\lambda 4640$ band has multiple emission components.

The late decline phase begins with the arrival of very strong coronal lines. The [Fe X] $\lambda 6374$ line appears first as a strong feature, with [Fe XI] $\lambda 3987$ following as a weak line. Very strong [Fe XIV], [A X], and [Ni XII] lines surpass their lower ionization counterparts over a period of weeks, and these lines remain strong for many weeks. Wallerstein measured [Fe X] $\lambda 6374$ to be more intense than $H\beta$ during the 1967 outburst, so a large amount of energy is radiated by the coronal lines (Grewing and Habing 1972).

Given the large intrinsic reddening of RS Oph, one would not expect it to be a strong UV source. Indeed, the far-UV continuum is very weak, and only Mg II, O III, N IV], Ne IV, and C IV appear as relatively strong emission lines. Shortwards of 2400 Å the spectrum is extremely weak, and it is not possible to identify the continuum level accurately.

V2116 Ophiuchi

(GX1 + 4, GX2 + 5, 3U 1728-24)

V2116 Oph is the only symbiotic star that was originally discovered as an X-ray source (Lewin, Ricker, and McClintock 1971). With a time-averaged flux of $4.5 - 22 \times 10^{-9}$ erg cm^{-2} s^{-1} (Doty, Hoffman, and Lewin 1981), this system is one of the strongest X-ray sources in the sky. It is located in the direction of the galactic center, and if its distance is 9 kpc, its total X-ray luminosity is $\sim 20,000 \, L_{\odot}$.

X-ray Observations

The X-ray spectrum of V2116 Oph can be fit either by a power law ($F_E \propto E^{-\alpha}$, with $\alpha = 1.4 - 1.6$), or a Planck law ($F_E \propto E^3/[e^{(E/kT)} - 1]$, with $kT = 18 - 28$ keV; Ricker, et al. 1973, 1976; Hawkins, Mason, and Sanford 1973; Thomas, et al. 1975; Becker, et al. 1976; White, et al. 1976; Doxsey, et al. 1977; Doty, Hoffman, and Lewin 1981; Coe, et al. 1981). Obviously the spectrum is quite hard, suggesting the X-ray source might be an accreting neutron star. Becker, et al. (1976) defined the hardness ratio, h, to be the ratio of the number of X-rays with $E > 4.6$ keV to those with $E < 4.6$ keV. They found the system exhibited a high state with $h = 1.03$, while the low state had $h = 0.79$.

Early observations of V2116 Oph revealed that the X-ray emission was pulsed at a period of ≈ 2 min (Ricker, et al. 1976; Hawkins, Mason, and Sanford 1973; Thomas, et al. 1975; Becker, et al. 1976; White. et al. 1976; Doxsey, et al. 1977; Mason 1977; Koo and Haymes 1980; Strickman, Johnson, and Kurfess 1980; Doty, Hoffman, and Lewin 1981; Coe, et al. 1981). Subsequent observations showed that the mean light curve exhibited two distinct maxima if a period of ≈ 4 min was adopted. A 4 min period is easily understood if the system contains a rotating, magnetized neutron star, since the two maxima can be ascribed to the two magnetic poles. Whatever the true period may be (either 2 or 4 min), it is decreasing quite rapidly. The value for $\dot{P}/P$ derived by Doty, Hoffman, and Lewin (1981) implies the neutron star is spinning up; the rate of spin-up is consistent with that expected if the neutron star accretes the material necessary to produce the total X-ray flux.

Optical and Infrared Observations

V2116 Oph has had an exciting history as an X-ray source, and has not been neglected by observers at other wavelengths. Glass and Feast (1973) found the object to be a typical red giant in the IR, and their optical spectra showed a strong Hα emission line and weak emission from Hβ and [O III] λ5007 (cf. MacConnell 1978). Many high excitation lines (including He I, [Fe VII], and possibly [Fe X], [A X], and [A XI]) were visible on higher quality spectra, as well as a weak blue continuum that rose rapidly into the red (Davidsen, Malina, and Bowyer 1976, 1977). The Hα/Hβ intensity ratio, and the V-K color, imply a very large reddening ($E_{B-V} \approx 4-6$). Given the high excitation lines associated with the optical source, it seems quite likely that this system is indeed the X-ray source.

AR Pavonis
(CpD-66° 3307, HV 7860, MWC 600, Hen 1649, P 4603)

AR Pavonis was discovered by Mayall (1937) to be an unusual eclipsing binary with a 605 day period. This is an interesting symbiotic system, as it combines some of the features seen in quiescent objects (e.g., He II emission) with those usually found in erupting systems (e.g., F-type shell spectrum). Since it is an eclipsing binary, an understanding of this system is critically important for determining the nature of other symbiotic variables.

Photometry

AR Pav is one of a few symbiotic stars known to be an eclipsing binary, and Andrews (1974) derived a period of 604.6 days. The depth of an average AR Pav eclipse is ≈ 2 mag at V, and a smaller, secondary, eclipse might be present at $\phi = 0.54$ with a depth of 0.5 mag. The duration of totality suggests the radius of the eclipsing star is $\sim 100\ R_{\odot}$, while that of the eclipsed star is $\sim$ 45-75 $R_{\odot}$. Kenyon and Webbink (1984) noted that the rather shallow eclipse depth (considering the projected radii and the effective temperatures of the two components) requires the eclipsed object to be an accretion disk rather than a spherical star. The irregular variations observed outside eclipse then might be a result of instabilities in either the accretion stream or perhaps the disk itself.

The available IR and near-IR photometry show that AR Pav contains a normal M3 giant star (Glass and Webster 1973; Menzies, et al 1982). This component is slightly variable, and definitely fills its Roche lobe. Observations at 10 μm are needed to determine if the system has a silicate excess similar to that found in CI Cyg.

Spectroscopy

The optical spectrum of AR Pav is rather complicated, and Thackeray (1954b) recognized 4 distinct components. High excitation emission lines (e.g. He II λ4686) and a cF absorption spectrum (Ti II and other singly ionized metals) are distinctly visible outside of eclipse (Merrill and Burwell 1943; Sahade 1949; Beals 1951; Thackeray 1959; Thackeray and Hutchings 1974; Henize 1976), and are associated with the hot component. Late-type TiO absorption features and low excitation emission lines (e.g. Fe II) are prominent when the cool star eclipses the hot component. The cF spectrum was very strong after the 1954 eclipse, suggesting that some type of outburst had taken place. Thackeray estimated $m = 9$ from his spectra, which is significantly brighter than normal. The cF spectrum faded somewhat during the next year, and has never returned to this high state.

The line profiles are very interesting, although they greatly complicate radial velocity measurements (Thackeray 1959; Thackeray and Hutchings 1974). The H I and He II lines have strong central absorption cores, and are probably formed in a disk surrounding the hot component. The separation of the twin emission peaks in He II $\lambda4686$ is 137 km s^{-1}, and the He II emission is highly concentrated towards the center of the disk. The He I lines do not appear to be double peaked, and are presumably formed in a semi-spherical region surrounding the disk.

The ultraviolet spectrum of AR Pav (Figure A.22) resembles that of CI Cyg (Figure 3.9), and can be interpreted in terms of an accretion disk surrounding a low mass main sequence star (Kenyon and Web-

Figure A.22 - Ultraviolet spectrum of AR Pav obtained by *IUE*. The weak continuum and strong emission lines are characteristic of symbiotic stars whose hot component is an accreting main sequence star.

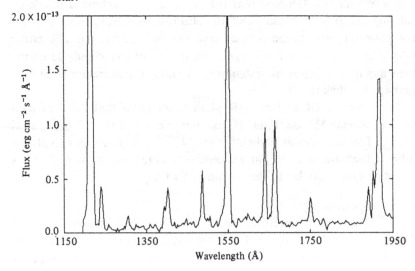

bink 1984). The cF optical spectrum and the high excitation lines imply an accretion rate of $\sim 10^{-4} M_\odot \, yr^{-1}$ (Chapter 3). Although Andrews estimated $E_{B-V} = 0.1$ from optical photometry, Kenyon and Webbink (also Slovak 1982a,d) derive $E_{B-V} = 0.3$ from the 0.22 μm feature. Additional observations are needed to resolve this ambiguity.

A series of *IUE* spectra obtained by Hutchings and Cowley (1982) and Hutchings, et al. (1983) suggest that only the outer portion of the disk is eclipsed during primary maximum. A fair fraction of the ultra-violet continuum and many of the strong emission lines do not show significant variations at primary minimum, and thus the central portions of the disk may never be obscured by the giant star. Hutchings, et al. (1983) noted variations in the duration of the eclipse from cycle to cycle, implying that the size of the disk and/or the accretion stream are not constant.

AG Pegasi

(BD + 11° 4673, HD 207757, SAO 107436, AGK3 + 12° 2570, Zi 2046, AG Lpz I 8693)

A. Cannon discovered this remarkable symbiotic star, and noted strong H I emission lines with P Cygni absorption components. AG Peg was classified therefore as a peculiar Be star, but its symbiotic nature was discerned by Merrill in the 1920's. It is one of the few symbiotic stars believed to have a magnetic field (Babcock and Cowling 1953 reported 1-4 kG), although Slovak (1982b) was unable to verify these results. The system is one of the brightest symbiotics (V = 8.6, K = 3.4), and is therefore one of the best-studied. However, its behavior should not be considered as "typical", since it is the only symbiotic star known to require > 100 years to decline from a major outburst.

Photometry

Lundmark (1921a) reviewed the historical variability of AG Peg, and chronicled the nova-like outburst that began in the 1850's. Before c. 1850, AG Peg was an inconspicuous 9th magnitude star; its slow brightness rise concluded in 1871 when m ~ 6. The slow decline from visual maximum has been well-observed, and presently V ≈ 8.6 (see Figure 5.3). Shorter-term photometric variations were discussed by Rigollet (1947, 1948; also Stebbins, Huffer, and Whitford 1940), who noted irregular 0.3 mag fluctuations in V. These were studied in more detail by Belyakina (1968a,b, 1970b), and she found an 800 day photometric period with an amplitude of 0.3 mag. The maximum in V is correlated with a maximum in the number of Fe II emission lines observed in the spectrum, and she concluded that the variations are due to orbital motion (Figure 3.1). This interpretation requires that the hemisphere of the giant facing towards the hot star be significantly hotter (and brighter) than the hemisphere facing away from the hot star, as discussed in Chapter 3. Meinunger (1981; Kenyon 1983) refined Belyakina's period estimate to 827.0 days; this is likely to be the orbital period of the AG Peg binary (cf. Luthardt 1984; Slovak 1982d; Belyakina 1985a; Fernie 1985).

Infrared observations show that AG Peg contains a normal M giant, with little or no trace of a 10 μm excess (Low 1970; Mendoza 1972; Swings and Allen 1972; Kolotilov and Shenavrin 1976; Feast, Robertson, and Catchpole 1977; Szkody 1977; Feast, et al. 1983a; Roche, Allen, and Aitken 1983; Kenyon and Gallagher 1983). Feast, Robertson, and Catchpole found no evidence for long period variability, although Kolotilov and Shenavrin (1976) reported 0.3 mag fluctuations at L. Such variations are smaller than those expected from a normal Mira variable, so it is unlikely that the giant in AG Peg is a Mira-like star (the detection of a seemingly normal 2.3 μm CO absorption band by Kenyon and Gallagher supports this conclusion).

Spectroscopy

In his initial description of AG Peg's optical spectrum, Merrill (1916; see also Shapley 1922a; Merrill and Burwell 1933) reported strong H I emission lines with violet absorption borders and faint emission lines of singly ionized metals. Numerous absorption lines without bright emission components were also present; these were predominantly He I lines, although the Ca II K line was also visible. The appearance of these first spectra resemble those of P Cygni, and thus Merrill concluded AG Peg was a P Cygni star.

The optical spectrum of AG Peg (Figure A.23) has transformed remarkably since 1915, and these alterations have been faithfully reported by Merrill (1929a,b, 1932, 1941a, 1942, 1951a,b, 1959) and countless others (Swings and Struve 1940a; Struve 1944; Tcheng and Bloch 1946; Tcheng 1950, 1952a,b; Beals 1951; Payne-Gaposchkin 1952; Bloch and Tcheng 1952a,b, 1958; Burbidge and Burbidge 1954; Grant and Aller 1954; Pagel 1958; Dokuchaeva 1958; Ivanova 1960, 1963; Arhipova, Dokuchaeva, and Vorontsov-Velyaminov 1961; Schaifers 1961; Arhipova and Dokuchaeva 1962; Dolidze and Pugach 1962; Arhipova 1963; Hang and Mo 1963; Luud and Il'mas 1965, 1966; Boyarchuk 1966c, 1967c; Boyarchuk, Esipov, and Moroz 1966;

Lopez and Sahade 1968; Johnson and Golson 1969; Chkhikvadze 1970b; Caputo 1971b; Kolotilov 1971; Ciatti, D'Odorico and Mammano 1974; Oliversen and Anderson 1982a; Welty 1983). The evolution of the spectrum has been characterized by a tediously slow development of high ionization emission lines and TiO absorption features. In many instances (e.g. He I and Si IV), a bright emission line was heralded by the appearance of a strong absorption line, which slowly faded as the bright line developed. Early in his study, Merrill noted the line intensities were modulated with an 800 day period. This phenomenon disappeared as the lines grew stronger and wider in the late 1930's.

As noted above, AG Peg was originally classified as a P Cygni or peculiar Be star, since it had He I absorption and P Cygni-like H I lines. The He I lines developed emission components in 1920, while the absorption features remained strong. The relative displacement of the absorption and emission lines implied a low expansion velocity in 1920, but their separation increased to ≈ 250 km s^{-1} in 1950. As this terminal velocity increased, the He I lines (especially $\lambda 3888$) developed *multiple* absorption components at velocities of ≈ 72, 125, 180, 230, 300, and 382 km s^{-1}. The 72 km s^{-1} component was persistent at

Figure A.23 - Optical spectrum of AG Peg.
The strong TiO bands present in AG Peg are typical of early M giants, while the strong emission lines from H I and He II indicate the presence of a hot companion star ($T_h > 30,000$ K; cf. Figure 3.14). The Balmer jump at $\lambda 3646$ is very pronounced in AG Peg, which suggests the electron temperature in the ionized nebula is $\sim 20,000$ K.

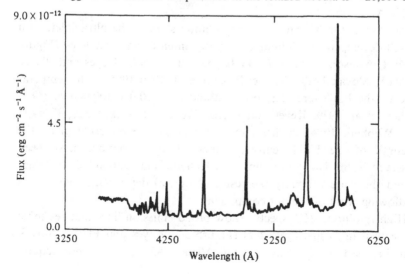

λ3888, but was weak and seldom seen in other He I lines. P Cygni profiles remain present in recent spectra of AG Peg, with velocities similar to those found by Merrill (Kolotilov 1975). The behavior of the He I lines in AG Peg is analogous to the development of multiple velocity systems in classical novae (Payne-Gaposchkin 1957).

Traces of TiO bands were noted on 1921 spectra, but their strength fluctuated erratically. By the 1940's, the bands were fairly prominent, and indicated an M1 giant. The TiO bands have become more conspicuous as the blue continuum has faded, and other late-type absorption lines (Ca I, Na I, Fe I, and Ca II) are now easily visible. The inferred spectral type of the giant is M3 (Andrillat 1982a), and this is confirmed by IR data.

After a few years of photographic plates had been accumulated, Merrill reported that the radial velocities of many lines displayed an 800 day period. These variations persisted throughout his study, although he could not define the value of the period more precisely. Recent spectra (Hutchings and Redman 1972; Cowley and Stencel 1973; Hutchings, Cowley, and Redman 1975) have confirmed that the velocity changes reflect the orbital motion of an M giant and a hot, compact star, and place the period somewhere between 820 and 830 days. An orbital solution is complicated by changes in the relative phasing of the emission lines (He I is an obvious culprit: early in the outburst these lines were ≈160 days out of phase with the H I lines, but by 1950 the two sets of lines were nearly in phase!). In spite of these difficulties, the giant is believed to be 3-4 times more massive than its hot companion; the most recent orbital solution is given in Table A.5.

AG Peg is the brightest of all symbiotic stars in the ultraviolet, with a well-developed continuum and many intense emission lines (Figure 3.14; Thompson, et al. 1978; Gallagher, et al. 1979; Keyes and Plavec 1980a,b; Slovak 1980, 1982a,c; Penston and Allen 1985). The temperature of the hot source has been estimated at 30,000-100,000 K (Gallagher, et al. 1979; Keyes and Plavec 1980a,b; Slovak 1982a; Kenyon and Webbink 1984); higher temperatures may be preferred given the strength of the He II emission lines. Large variations have been observed in the UV continuum, and Kenyon and Webbink (1984) suggested these are probably the result of optical depth variations in the outflowing wind of the hot component.

High resolution *IUE* spectra show broad emission lines that resemble those seen in Wolf-Rayet stars (Figure 5.7; Keyes and Plavec; Slovak and Lambert 1982; Penston and Allen 1985). These lines require expansion velocities > 1000 km s^{-1}, and are probably formed in the hot component's wind. If the hot source is indeed a Wolf-Rayet star, its temperature may be closer to 30,000 K than 100,000 K (Keyes and Plavec 1980a,b).

Radio Observations

AG Peg was one of the first symbiotic stars to be detected in the radio, and its radio spectrum is consistent with free-free emission in an outflowing stellar wind (Gregory, Kwok, and Seaquist 1977; Altenhoff, et al. 1976; Wendker, et al. 1973; Woodsworth and Hughes 1977; Seaquist, Taylor, and Button 1984). The mass loss rate derived from a fit to the observed spectral energy distribution is $\sim 10^{-6}\ M_\odot$ yr^{-1} (cf. Friedjung 1967), which is somewhat lower than that observed in Wolf-Rayet stars. *VLA* observations imply the source has a linear dimension of ≈ 100 AU (Ghigo and Cohen 1981), which is larger than the size implied by the total Hβ flux (< 10 AU; Kenyon 1983). More recent *VLA* data show three sources in AG Peg: a nebula with a diameter of 1″.5, a point source, and a condensation of material inside the large nebula (Hjellming 1985). The central source can be associated with the wind from the WN star, while the large nebula is the remnant wind of the cool giant which has recently been ionized by the erupting hot component. The condensation is probably a density enhancement in the outflow, as indicated by the optical line profiles.

AX Persei
(*HV 5488, MWC 411, MW 228*)

AX Per was discovered by Merrill and Humason (1932) during the same objective prism survey that discovered CI Cygni. AX Per has a classical symbiotic spectrum, with strong TiO bands and intense high ionization emission lines. Recent observations suggest the system is an eclipsing binary; its relatively short period (682 days) makes it a promising candidate for synoptic monitoring projects.

Photometry

The photographic light curve of AX Per from 1887-1931 has been discussed by Lindsay (1932), and shows two major outbursts during which AX Per reached a visual magnitude of ≈ 9. The system was quite active in 1950-51 (Wenzel 1956; Seidel 1956) and more recently in 1980, with outbursts resembling those experienced by Z And and CI Cyg. The rise to visual maximum is rather rapid, and B-V decreases from ≈ 1 to ≈ 0.6. The colors at maximum light resemble those of an F star, and this has been confirmed by spectroscopic measurements.

The periodic oscillations observed in quiescence are much more pronounced during the decline from visual maximum, as in Z And.

Low amplitude oscillations in the quiescent phases of AX Per were first discussed by Payne-Gaposchkin (1946; also Seidel 1956; Romano 1960; Belyakina 1968a), who suggested a period of 675 days. Mjalkovskij (1977) noted variability in the photographic and the photovisual ($\Delta m_{pg} \approx 0.7$, $\Delta m_{pv} \approx 0.35$; Figure 3.1), and Kenyon (1982) used these data to revise the period to 681.6 days. Kenyon concluded that the periodicity reflected orbital motion in a binary system; spectroscopic observations later showed that the variation was caused by an eclipse of the hot component by the cool giant.

Little infrared photometry is available for AX Per, but these data do show that the system contains a late-type giant star, with a small 10 μm excess (Low 1970). The early IR photometry suggested that the giant component was strongly variable, since Swings and Allen (1972) measured $K = 6.45$, and Szkody (1977) found $K = 5.53$. More recent observations show no large-scale variations over a 4 month timescale ($K = 5.45 \pm 0.02$; Kenyon and Gallagher 1983). As in other symbiotic stars, AX Per has the strong 2.3 μm CO absorption band that is characteristic of a late-type giant star (Kenyon and Gallagher 1983).

Spectroscopy

AX Per was discovered spectroscopically, as objective prism plates revealed the star to be an M giant with bright H I, He II, [O III], [Ne III], and Fe II lines (Figure A.24; Merrill and Humason 1933; Merrill 1933; Merrill and Burwell 1932; Merrill, Humason, and Burwell 1932). Merrill (1944; also Boyarchuk 1966a) originally classified the cool star as an M3 giant, but subsequent classifications appear to prefer a somewhat later spectral type for the giant (later than M4; Boyarchuk 1969b; Andrillat and Houziaux 1982). The emission lines are always quite strong (except in outburst), and H I, He II, [Ne V], and [Fe VII] are usually detected as intense emission features (Swings and Struve 1940a, 1941b, 1942a,b, 1943a,b; Gauzit 1955a,b,c; Tcheng and Bloch 1954, 1957; Bloch and Tcheng 1958; Boyarchuk 1966a, 1969a; Gravina 1981b; Oliversen and Anderson 1982a, 1983; Blair, et al. 1983).

Early spectra of AX Per showed variations in the intensity of many emission lines (Merrill 1944; Swings and Struve 1940a), and Oliversen and Anderson (1982a, 1983) showed that the H I and He II line fluxes vary in phase with the ephemeris derived by Kenyon (1982). The behavior of these lines is suggestive of an eclipse (as in CI Cyg), and additional monitoring is needed to determine accurately the size of the two binary components in AX Per.

The ultraviolet spectrum of AX Per is similar to that of CI Cyg, in that it has a bright, flat UV continuum with many strong emission lines. Slovak (1980, 1982a) has noted that the continuum has the same shape as a reddened A star, but this type of hot component cannot ionize a gaseous nebula to the degree observed in AX Per. Kenyon (1983) determined that an accreting main sequence star will produce a flat UV continuum; a detailed fit to observations of AX Per suggests an accretion rate of a few $\times 10^{-5} M_\odot$ yr^{-1}.

V741 Per
(*M*1-2, *VV* 8, *PK* 133-8°1)

This object was discovered independently by Minkowski (1946; M1-2) and Vorontsov-Velyaminov (1946; VV 8) as a stellar planetary nebula. While M1-2 may still be considered by some observers as a

Figure A.24 - Optical spectrum of AX Per (C.M. Anderson, private communication). This system is nearly identical to CI Cyg. Strong TiO bands are obvious in the red, and the prominent emission lines are associated with transitions in H I, He I, and He II. Emission from [Fe VII] is also visible, and the lack of the λ6830 feature is characteristic of symbiotic systems containing accreting main sequence stars.

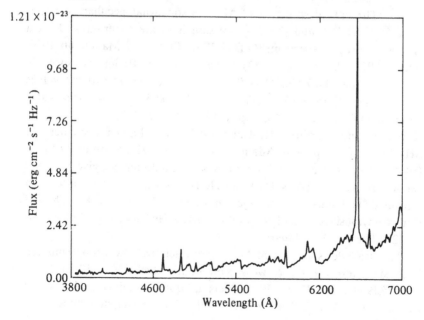

planetary nebula, it is generally recognized to be a "yellow" symbiotic star consisting of a G-type giant or supergiant and a hot companion. There remains some controversy concerning the luminosity class of the G-type star: heavy element absorption lines indicate a supergiant, which places the star outside the galaxy. If M1-2 is an interacting binary, the heavy elements in the photosphere might be enhanced by various mechanisms as noted by Grauer and Bond (1981).

Photometry

Although M1-2 is designated V741 Per, accounts of its variable nature are premature. Early UBV observations detected a constant G-type star with a UV excess (Arhipova, Noskova, and Gapbarov 1974), and various infrared data have attributed the large nonvariable IR excess to dust emission (Gillett, Knacke, and Stein 1971; Gillett, Merrill, and Stein 1972; Allen 1973; Cohen and Barlow 1974; Persson and Frogel 1973). Drummond (1980) proposed M1-2 as a possible binary planetary nebula nucleus, on the basis of periodic (4 hr), 0.06 mag variations in V. Grauer and Bond (1981) were unable to verify this periodicity, using a larger telescope than Drummond did, and discarded the system as a short-period binary.

Spectroscopy

The optical spectrum of M1-2 is somewhat peculiar, since intense H I, [O III], and [Ne III] emission lines are superimposed on a G-type absorption spectrum (O'Dell 1963; Ciatti and Mammano 1976; Barker 1978; Blair, et al. 1983). The $\lambda4363$ [O III] line is especially prominent, and the $\lambda4363/(\lambda4959+\lambda5007)$ ratio implies extreme values for n_e and T_e ($n_e \sim 10^{6-7}$ cm^{-3}; $T_e \sim 10,000$-$30,000$ K; Razmadze 1960a,b; O'Dell 1963, 1966; Zipoy 1975; Adams 1975; Lutz 1977a,b; Kohoutek and Martin 1981; Kaler 1983). The Balmer decrement of M1-2 is very steep, and Adams reported that Hα is optically thick ($\tau \approx 15$ at line center); combining this with the decrement gives a visual extinction, $A_V = 2$. Weak He I and He II emission lines have also been measured, and thus the hot object must have $T_h \geq 60,000$ K. The G2 star is obviously too cool to produce many He$^+$-ionizing photons, so M1-2 is likely to be a binary star.

The ultraviolet spectrum of M1-2 is characterized by strong emission lines superimposed on a weak continuum (Feibelman 1983). The C III] $\lambda1909$ and C IV $\lambda1549$ doublets are quite prominent, while He II $\lambda1640$ and N V $\lambda1240$ are rather weak. Strong emission lines dominate a weak UV continuum, and thus it is not clear if the continuum is produced by a hot star as in SY Mus. Although the excitation is lower in M1-2, its UV spectrum resembles those observed in the symbiotic novae V1016 Cyg, V1329 Cyg, and HM Sge.

Radio Continuum Observations

Radio observers have been quite busy with M1-2, although they have yet to detect it. Various upper limits have been set (Higgs 1971; Johnson, Balick and Thompson 1979; Kwok, Purton, and Keenan 1981), and the most stringent is 0.49 mJy at 6 cm (Seaquist, Taylor, and Button 1984).

RX Puppis
(CoD-41° 3911, CpD-41° 2287, HD 69190, Hen 138, SS 8)

RX Pup is an interesting symbiotic star that was originally thought to be related to η Carinae or R Coronae Borealis (Pickering 1897, 1914a; Fleming 1912; Payne-Gaposchkin and Gaposchkin 1938). M-type absorption features were not detected until the 1970's, although the peculiar emission-line spectrum had been common knowledge since 1900. This is one of ≈ 20 symbiotics that contains a Mira variable, and this component may be the source of RX Pup's baffling transformations.

Photometry

RX Pup has been badly neglected by optical photometrists, although Cannon noted irregular variability between $m = 11.5$ and $m = 14.1$. Fortunately the infrared observers have not overlooked this system, and there is a wealth of broad-band and narrow-band photometry available. It is now well-established that the system contains a long period variable ($P \approx 580$ days; Figure A.25) and a rather large 3.5 μm excess which is due to a combination of free-free and hot dust emission (Glass and Webster 1973; Feast, Robertson, and Catchpole 1977; Allen, et al. 1982; Whitelock and Catchpole 1982; Roche, Allen and Aitken 1983; Whitelock, et al. 1983b; Whitelock, et al. 1984). The behavior of the JHKL light curves in the 1970's is more complicated than usually seen in a normal Mira, and may be a result of a reduction in the free-free emission and a contemporaneous increase in the dust emission (Whitelock, et al. 1983b). Steam absorption was detected by Barton, Phillips and Allen (1979), while Whitelock, et al. (1983b) noted additional CO absorption and Bγ emission (Figure A.26).

Spectroscopy

The optical spectrum of RX Pup is comprised of a variable M-type giant and a plentiful supply of emission lines (Sanduleak and Stephenson 1973; Andrillat 1982a). The behavior of the emission line

Figure A.25 - Infrared light curves for RX Pup (Whitelock, et al. 1983b). The variations in J and L are clearly periodic ($P \sim 580$ days), and are consistent with those expected of a Mira variable.

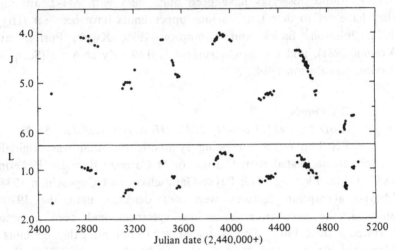

Figure A.26 - Infrared spectra of RX Pup (Whitelock, et al. 1983b). Strong water bands are visible at 1.3 and 1.9 μm, and a weak CO band is present at 2.3 μm. D-type systems typically have a Pβ emission line, and this transition is fairly prominent in RX Pup.

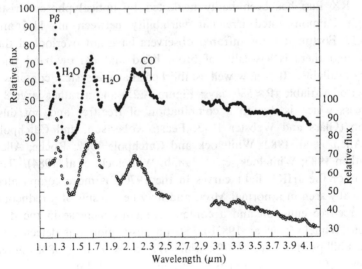

spectrum points to a slow evolution from high to low to high excitation in 1975-81. Generally, H I, He II, N III, C IV, and [Fe VII] are very strong emission lines (Swings and Struve 1941b; Henize 1976; Swings and Klutz 1976), but these were absent in 1977-78 when the spectrum resembled that of a Be or shell star (Klutz, Simonetto and Swings 1978; Klutz 1979). The H I lines lost their P Cygni components in Dec. 1979, when the He II and N III lines re-appeared with strengths similar to those seen in WN7-8 stars (Klutz and Swings 1981).

RX Pup's ultraviolet spectrum has been investigated in some detail by Kafatos, Michalitsianos, and Feibelman (1982; Kafatos and Michalitsianos 1982a). The continuum is fairly flat from $\lambda 1200$ to $\lambda 2000$ (Figure A.27), and rises towards longer wavelengths. The continuum resembles that of an A0 II star, but such a star is incapable of producing the high energy photons need to explain the emission-line spectrum (Kenyon and Webbink 1984). High dispersion *IUE* spectra show that many of the strong emission lines are comprised of multiple components (Figure 3.19; Figure A.28). These features suggest ionized material in RX Pup is moving at velocities ≤ 500 km s^{-1}. The two major components of He II $\lambda 1640$ are separated by ≈ 200 km s^{-1},

Figure A.27 - Ultraviolet spectrum of RX Pup (Kafatos, Michalitsianos, and Feibelman 1982). The flat continuum and strong emission lines are usually observed in D-type systems, and this appearance resembles that of dusty planetary nebulae.

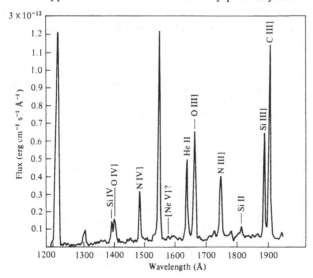

which may indicate that either an accretion disk or an expanding shell surrounds the hot component in RX Pup.

Radio Observations
RX Pup is a fairly strong radio source, with an intensity of ~30 mJy at 4 cm (Seaquist 1977; Wendker 1978; Wright and Allen 1978; Purton, et al. 1982). This emission is most likely produced in the optically thin wind of the Mira component (which has been ionized by the hot star). A straightforward interpretation of the radio spectrum is complicated by rapid variations in the observed flux density (factors of 2-3 in 24 hr; Seaquist 1977).

HM Sagittae
HM Sge is one of a select group of symbiotic stars known to have undergone a slow nova-like outburst. It has much in common with its more northerly counterpart, V1016 Cyg, and with the southern system RR Tel. HM Sge is one of a few symbiotics that has been resolved with the *VLA*, although higher resolution observations are needed to resolve the central source completely. HM Sge is one of a half-dozen systems detected with the *Einstein* satellite, and is the most luminous symbiotic nova at X-ray wavelengths (Allen 1982b).

Photometry
Various patrol plates place HM Sge at $m_{pg} > 17$ before it rose to $V = 11$ in 1975 (Dokuchaeva 1976; Dokuchaeva and Balazs 1976; Noskova, et al. 1979). The system was fairly red prior to outburst $(B - V > 1)$, and evolved to the blue $(B - V \approx 0.4)$ during the rise to

Figure A.28 - High resolution ultraviolet spectra of C IV and O III] in RX Pup (Kafatos, Michalitsianos, and Feibelman 1982). Multiple components separated by 40-60 km s^{-1} are visible on most lines in RX Pup, and vary in velocity over timescales of months.

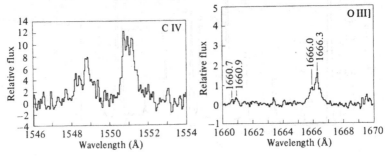

visual maximum (Wenzel 1976). Since 1975, HM Sge has varied between V = 11 and V = 12 with no obvious periodicity (Wenzel 1976; Ciatti, Mammano, and Vittone 1978a,b, 1979). As noted by Ciatti, Mammano, and Vittone (1979), this variability suggests that the emission line intensities vary, since HM Sge has such a weak optical continuum.

The infrared behavior of HM Sge is similar to that of V1016 Cyg, although its amplitude may be somewhat smaller at K (Slovak 1978). There is a strong 10 μm excess, as well as CO and H_2O absorption, all of which suggest that a Mira variable lurks somewhere in the system (Davidson, Humphreys and Merrill 1978; Puetter, et al. 1978; Bregman 1982; Yudin 1982a,b; Taranova and Yudin 1980, 1982b, 1983a; Roche, Allen and Aitken 1983; Kenyon and Gallagher 1983; Ipatov, Taranova, and Yudin 1985; Lorenzetti, Saraceno, and Strafella 1985).

High resolution FTS spectrograms of HM Sge show strong H I Brackett lines, which appear to be reddened by 1.5 magnitudes (implying an optical extinction of 12 mag!!; Thronson and Harvey 1981). This reddening is much larger than that indicated by the Balmer lines ($E_{B-V} = 0.4 - 0.7$; Davidson, Humphreys and Merrill 1978; Blair, et al. 1981, 1983; Kenyon 1983; Taranova and Yudin 1983a). Thronson and Harvey (1981) suggest the dust may be abnormal or patchy (which might be supported by polarization data, as noted by Efimov 1979); an alternative interpretation is that the emission-line region lies outside a dust cocoon surrounding the late-type star (Allen 1983).

Spectroscopy

The optical spectrum of HM Sge could be mistaken for that of a planetary nebula (Figure A.29), given its weak continuum and strong emission lines (Bopp 1977; Stover and Sivertsen 1977; Arhipova and Dokuchaeva 1978; Blair, et al. 1981, 1983; Taranova and Yudin 1983a; Kenyon 1983). The evolution of this spectrum since Arhipova and Esipov (1976) reported nebular emission from [O I], [O II], [O III], [N II], [S III], and [A III] has been characterized by a slow increase in ionization at roughly constant visual luminosity (which is supplied mainly by emission lines; Shanin 1978; Andrillat, Ciatti, and Swings 1982; Blair, et al. 1981, 1983). Wolf-Rayet features were observed in 1977; as these lines faded, He II and [Fe VII] emerged as very intense emission lines (Belyakina, Gershberg, and Shakhovskaya 1978; 1979; Brown, et al. 1978; Wallerstein 1978; Ciatti, Mammano, and Vittone 1979; Arhipova, Dokuchaeva, and Esipov 1979; Feibelman, et al. 1980; Blair, et al. 1981, 1983; Welty 1983; Stauffer 1984). Late-type absorption bands are occasionally visible in the red, and their periodic appearances may be correlated with the IR variability (Andrillat, Ciatti, and Swings 1982; Taranova and Yudin 1983a).

High resolution optical observations of HM Sge show features similar to those found in V1016 Cyg (Figure 5.6; Stauffer 1984; Wallerstein, et al. 1984; Willson, et al. 1984; Solf 1984). The lines tend to be fairly broad, consisting of two components separated by 50-75 km s^{-1} in velocity and $\sim 1''.5$ in space. The elongation of the optical nebula in [N II] agrees with that observed in the radio, although the twin radio knots do not spatially coincide with those identified by Solf (1984) in the optical. It remains to be determined if the material detected by Solf (i) was ejected in the 1975 outburst (in which case the inferred distance, ~ 400 pc, disagrees with that estimated from IR data, or (ii) was ejected by the Mira prior to 1975 and is now becoming ionized by high energy photons emitted by the hot component. High ionization features such as the $\lambda 6830$ feature tend to have broader wings than the low excitation [N II] and [O I] lines, and may be formed closer to the hot component.

Ultraviolet spectroscopy of HM Sge has been discussed by Mueller and Nussbaumer (1985; Kindl and Nussbaumer 1982b), Feibelman (1982a) and Kenyon (1983). The UV continuum (Figure A.29) is very

Figure A.29 - Optical/ultraviolet spectrum of HM Sge. This spectrum is nearly identical to that of V1016 Cyg (Figure 5.5), with a very weak continuum and intense emission lines of H I and [O III].

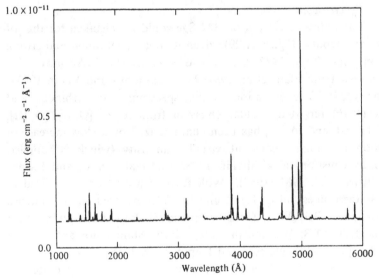

similar to that of V1016 Cyg, and the (C_1, C_2) colors are nearly identical to those of the dusty planetary nebula IC 4997 (Kenyon 1983; Kenyon and Webbink 1984). Various reddening indicators, including the 2200 Å feature and the strong He II emission lines, suggest $E_{B-V} \sim 0.6$, which is comparable to the reddening derived from optical H I, He I, and He II lines ($E_{B-V} \sim 0.65$; Kenyon 1983; Blair, et al. 1981, 1983).

The fluxes of the strong UV emission lines are clearly variable, having increased by a factor of 2 since 1979 (Mueller and Nussbaumer 1985). This evolution was accompanied by a dramatic increase in the $\lambda 4686/H\beta$ intensity ratio, implying a variation in the temperature of the underlying hot source (Chapter 2; Stauffer 1984). The physical conditions within the nebula appear not to have significantly evolved, as the C III], N III], and Si III] doublets consistently yield $n_e \sim 10^6$ cm^{-3} and $T_e < 60,000$ K at all epochs. These values are similar to those found in V1016 Cyg and RR Tel.

Radio Observations

The radio history of HM Sge has been discussed in detail by Purton, Kwok, and Feldman (1983) and Purton, et al. (1982; cf. Feldman 1977; Kwok and Purton 1979; Kwok 1982a,b; Kwok, Bignell, and Purton 1984). The system has steadily brightened at radio wavelengths since 1977 (Figure 5.8), and this qualitatively parallels the optically-thick expansion phase observed in classical novae. The radio spectrum has remained roughly constant in shape during the ten-fold increase in flux density, with the spectral index, α, lying between 0.9 and 1.2 at all epochs. More rapid variations (e.g., days) have been reported at low frequencies (cf. Purton, et al.), and these are apparently uncorrelated with the overall rise in the flux density since 1977.

VLA observations of HM Sge were discussed by Purton, Feldman, and Kwok as well as by Purton, et al. and by Becker and White (1985). The source is clearly resolved by the *VLA* (Figure 5.9), and model fits to the data imply an angular size of $\approx 0''.2 - 0''.3$. The ionized region is elliptical, with the major axis lying at a position angle of $\approx 135°$.

Attempts to fit the radio data with simple models have not been entirely successful, although the interacting winds model best reproduces the features observed in HM Sge (see Chapter 5). Given the complexity of the gas distribution expected in a binary system in which both components lose mass, it may be necessary to supplement the radio data with simultaneous UV and optical data.

FN Sagittarii

(*AS 329 = SS 177*)

FN Sgr is a typical symbiotic star that has been overlooked by most observers. Two outbursts were recorded in the 1920's and 1930's, and these appear to have been similar in many respects to those recently experienced by Z And and CI Cyg.

Photometry

Ross (1926) discovered FN Sgr as a peculiar variable star with 2 mag fluctuations in brightness over a 21 year period. Subsequent observations (Beliavsky 1927; Payne 1928) revealed the system had undergone a small nova-like outburst in 1924-26 ($\Delta m \approx 4$); a similar outburst took place in 1936-41 (GCVS). Recent visual observations have been discussed by Bateson (1971), who found 2 mag variations ($P \approx 400\text{-}1000$ days) superimposed on smaller random fluctuations.

Infrared photometry of FN Sgr confirms the M spectral type assigned from optical spectra (Swings and Allen 1972; Glass and Webster 1973). Allen (1980a) noted a strong CO absorption band at 2.3 μm which is common to M giants and supergiants. If FN Sgr contains a normal giant, the strength of this band suggests an M4 spectral type.

Spectroscopy

Optical spectra of FN Sgr show the strong TiO bands common to most symbiotic stars. Emission lines of H I, He II, N III, [O III], and [Ne III] are usually prominent, but there have been no reports of high excitation lines such as [Ne V] and [Fe VII] (Herbig 1950; Merrill and Burwell 1950; Bidelman and MacConnell 1973; Sanduleak and Stephenson 1973). Spectra have not been obtained during the outburst state of FN Sgr, and thus it is not known if its outbursts are similar in nature to those of other symbiotic stars.

V1017 Sagittarii

(*Nova Sagittarii 1919*)

V1017 Sgr is one of three recurrent novae included in this discussion. It is among the handful of symbiotic stars known to contain a G giant, and usually displays a low-excitation emission-line spectrum. Since V1017 Sgr is also an X-ray source ($F_X \approx 0.022 \pm 0.005$ IPC count s^{-1}; Cordova and Mason 1984), it merits more attention than it has hitherto received.

Photometry

Three outbursts have been documented in V1017 Sgr (1901, 1919, 1973), but only the 1973 outburst was well-observed (Figure A.30). The range is ≈3-7 mag in B; although the rise to visual maximum is quite rapid, the decline requires many months (Bailey 1919a; Mattei 1974; Vidal and Rodgers 1974; Landolt 1975). These outbursts are similar to those of T CrB and RS Oph, but they evolve on a longer timescale.

Infrared photometry shows that V1017 Sgr does contain a G giant (H-K≈0.1; J-K≈0.6). Since the system is faint at K (10.6), no searches have been made for a 10 μm excess.

Spectroscopy

Optical spectra of V1017 Sgr typically show a well-developed G-type absorption spectrum, and Kraft (1964) classified the cool star as G5 III (G9 according to Bruck 1935). Both H I and He I are usually detected as emission lines, and He II is sometimes present as a weak line (McLaughlin 1946; Williams 1983). As with most symbiotics, the spectrum is variable: Humason (1938) once recorded a strong continuous spectrum with no trace of absorption or emission lines.

Figure A.30 - Optical light curve of a recent eruption in V1017 Sgr, as collected by members of the Variable Star Section, Royal Astronomical Society of New Zealand.

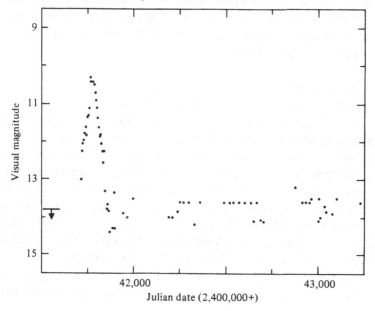

The high ionization emission lines (e.g. He II) increase in intensity during the rise to visual maximum, while the absorption lines fade rapidly (McLaughlin 1946; Vidal and Rodgers 1974). It is not known if any of the lines develop P Cygni profiles at maximum or during the decline. Very high excitation lines (e.g. [Fe VII] or [Fe XIV]) have not been observed in V1017 Sgr; in this respect, it differs dramatically from T CrB and RS Oph.

RT Serpentis
(MWC 265, Hen 1452, SS 94)

RT Ser is the prototypical symbiotic nova. Payne-Gaposchkin (1957) listed this object as one of the slowest of all classical novae, and recent spectra show its symbiotic nature quite clearly.

Photometry

Leavitt and Wolf discovered RT Ser during its 5300 day rise from $m > 16$ to $m \approx 9$ (cf. Shapley 1919; Barnard 1919). The nova remained at maximum visual light for a few years before beginning a tediously slow decline (Shapley 1923; Lacchini 1933; Payne-Gaposchkin 1957; Glebocki 1970). Recent photometry suggests the system has only recently returned to its pre-outburst brightness.

Although visual data for RT Ser are lacking, infrared observations show that this symbiotic star contains a normal M giant with a moderately strong CO absorption band (Feast and Glass 1974; Allen 1982a). Glass and Feast suggested the M component might be a Mira variable, but recent photometry has given no indication of large amplitude variability in the infrared.

Spectroscopy

Optical spectra of RT Ser were obtained at maximum visual light, and showed the A8-F0 supergiant spectrum usually observed in outbursting symbiotic stars and classical novae (cf. Shapley 1919; Adams and Joy 1920; Bailey 1921). Hydrogen emission lines were detected early in the decline, and eventually these were accompanied by lines of Fe II and other singly ionized metals (Adams, Joy and Sanford 1924; Adams and Joy 1928; Merrill and Burwell 1933). Higher excitation lines (e.g., [O III] and He II) were identified in the 1930's (Joy 1931), and very high ionization lines of [Ne V] and [Fe VII] were noted by Swings and Struve (1940b, 1942c; also Grandjean 1952).

More recent optical spectra show RT Ser to be a typical symbiotic star, with strong high ionization emission lines and prominent TiO absorption bands (Sanduleak and Stephenson 1973; Henize 1976; Allen 1980a; Fried 1980a; Mochnacki and Starkman 1985). Allen assigned an M6 spectral type to the giant component on the basis of the 2.3 μm CO absorption band. The strong [Fe VII] and λ6830 emission lines in RT Ser are observed in all symbiotic novae, including V1016 Cyg and RR Tel.

RR Telescopii
(HV 3181)
RR Tel ranks as one of the slowest of all the classical novae (Payne-Gaposchkin 1957), and only recently has it been classified as a symbiotic star. It could be considered as the prototypical D-type system: Mira-like pulsations were observed well before its 1944 outburst and recently have been recovered in the optical and in the infrared.

Photometry
RR Tel was discovered as a semi-regular variable with a 386.73 day period and an unusually small range (Fleming 1907, 1908; Payne 1928; Gaposchkin 1945; Whitelock, Catchpole, and Feast 1982). The amplitude increased in 1940-44, as if the star anticipated the nova-like outburst that began in 1944 (Mayall 1949; Glebocki 1970; Robinson 1975). The rise from $m_{pg} = 14$ to $m_{pg} = 7$ took less than 1 yr, and the color index changed from ≈ 1 to ≈ -0.2 (de Kock 1948). The system remained nearly constant at $m_{pg} \approx 7$ for 5 years, and then began a slow decline (Mayall 1949; Heck and Manfroid 1982; Kenyon and Bateson 1984). The early phases of the decline were slow, as the star faded by ≈ 2 mag in 1000 days (Figure A.31). However, subsequent decreases in brightness were extremely slow: ≈ 2500 days were needed to reach $V = 10$, and the system has not yet returned to its pre-outburst brightness.

The long period oscillations observed before outburst became visible once RR Tel had faded to $V \approx 10$ (Kenyon and Bateson 1984; Heck and Manfroid 1982, 1985). These variations do not repeat precisely, but a 374 day period fits the times of minima rather well (Figure A.31). The mean light curve is similar to that of R Aqr, implying that the 374 day variation is a pulsation. More rapid flickering has also been observed in RR Tel (Penston, et al. 1983), but additional observations are needed to determine if the variations are in the lines or in the continuum.

Infrared observations show that the cool component in RR Tel is indeed a Mira variable with a period of ≈ 387 days (Figure 3.5; Feast,

Robertson, and Catchpole 1977; Whitelock, Catchpole, and Feast 1982; Feast, et al. 1983b). Strong H_2O and weak CO absorption bands are present on CVF photometry, and $P\beta$ is observed as a strong emission line (Figure A.32; Allen, et al. 1978; Feast, et al. 1983b). Dust emission is prominent when the Mira is faint (Roche, Allen and Aitken 1983 [10 μm]; Feast, et al. 1983b), and this dust is probably heated by radiation from the hot nova-like companion.

Figure A.31 - Optical light curve of the slow nova RR Tel. The observations in (a) show the slow decline from maximum, as witnessed by members of the Variable Star Section, Royal Astronomical Society of New Zealand. The mean light curve in (b) is a result of binning the data-set in (a) on a 374 day period, as described in the main text.

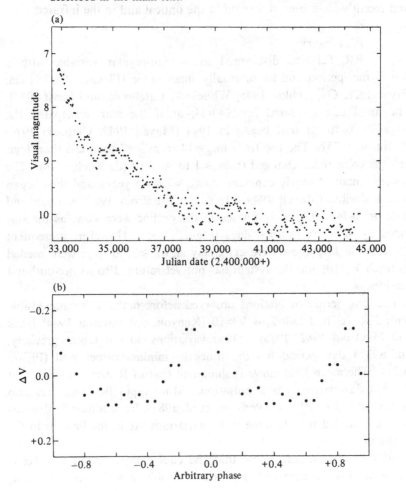

Spectroscopy

The evolution of the optical spectrum of RR Tel since its 1944 outburst is "typical" for a symbiotic nova, as similar behavior has been observed in a few other systems (e.g. AG Peg, RT Ser). At maximum light, the spectrum resembled that of an F supergiant (Shapley 1949a,b; Thackeray 1950; Henize and McLaughlin 1951; Pottasch and Varsavsky 1960). Most of the Balmer lines were present as strong absorption features, although Hβ was absent. This line evolved into a strong emission feature as Hγ disappeared, leading Thackeray to suggest that *all* the H I lines were present as emission and absorption features. As the emission lines grew stronger, the absorption lines were filled in and became fainter. Eventually, all the absorption lines had vanished, and the spectrum began to resemble a declining nova.

As RR Tel has declined from light maximum, the optical spectrum has transformed remarkably (Thackeray 1950, 1954a, 1969, 1977; Thackeray and Webster 1974; Webster 1974; Pottasch and Varsavsky 1960; Friedjung 1966, 1974; Aller, et al. 1973; Aller and Keyes 1974; Walker 1977a). High ionization lines developed slowly, and some of them now rival Hβ in intensity. The [O III] lines are particularly strong, and their relative intensities imply a density of $\sim 10^6$ cm^{-3} pro-

Figure A.32 - Infrared spectrum of RR Tel (Whitelock, et al. 1983b). As with RX Pup discussed earlier, strong water absorption bands dominate the near-IR spectrum of this symbiotic system. The prominent emission line is due to Pβ.

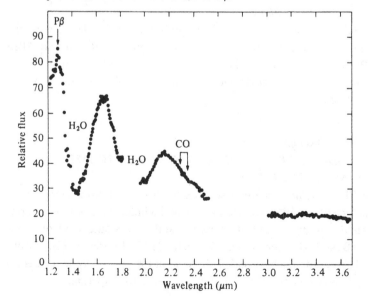

viding $T_e = 10,000$-$20,000$ K. The continuum has always been very weak, although TiO bands have been identified on some spectra (Webster 1974).

The emission line spectrum of RR Tel is extremely rich, and many exotic forbidden transitions were first identified in its spectrum (Thackeray 1977; Lambert and Thackeray 1969). The identification of lines from [Fe IV], [Fe V], [Fe VI], [Fe VII], [Ni IV], [Ni V], [Ni VI], and [Ni VII] has greatly expanded our knowledge of atomic physics, as the observed wavelengths have been used to refine the calculations of energy levels (Edlen 1969; Garstang 1968; Heinrichs 1975; Thackeray 1977; Raassen and Hansen 1981; Johansson 1983; Raassen 1985). In spite of this success, strong bands at $\lambda\lambda6830,7088$ remain unidentified.

The ultraviolet spectrum of RR Tel is a forest of emission lines superimposed on a weak continuum source (Heck, et al. 1978; Penston, et al. 1980, 1983; Ponz, Cassatella, and Viotti 1982). The ultraviolet color indices (as defined in Chapter 2) are nearly identical to those of IC 4997, a planetary nebula with a large amount of circumstellar dust (Kenyon and Webbink 1984). Thus, it is not surprising that Penston, et al. (1983) were unable to find a combination of hot blackbody and Balmer continuum radiation that satisfactorily fits the observed continuum. The identification of [Fe XI] $\lambda2649$ by Raassen (1985) is very important for understanding the slow evolution of this system, and should be confirmed by observations of the [Fe X] $\lambda6374$ transition.

Radio Observations

RR Tel was one of the first symbiotics to be detected in the radio, and it remains as one of the strongest sources (Wright and Allen 1978; Purton, et al. 1982). The spectral index, $\alpha = 0.52$, suggests the matter lost by the Mira component has been completely ionized by the hot nova-like star.

BL Telescopii
(CoD-51°11917, HD 177300, SAO 245923, HV 9587,
AN 950.1935, P 4963)

BL Tel is not usually associated with the symbiotic stars, but it seems sufficiently similar to AR Pav to be included in this monograph. The system was discovered at Harvard, and it was suspected of being an R CrB star (Luyten 1935; Howarth 1937; Mattei 1972). More recent observations have revealed the true nature of the system: it is an eclipsing binary composed of an M giant and an F supergiant.

Photometry

A number of eclipses have now been observed in BL Tel, and the best value for the period is currently 778.6 days (Cousins and Feast 1954; Knipe 1968; Gaposchkin 1970; Walraven and Walraven 1970; van Genderen 1977, 1980, 1982, 1983). The mean light curve is strikingly similar to that of AR Pav. A major difference between the two systems is that the F star in BL Tel has been shown to have a 65 day variation in its V magnitude (Wing 1962; van Genderen 1983, and references therein).

Spectroscopy

The optical spectrum of BL Tel usually resembles that of an F8 supergiant, with no evidence for high ionization emission lines (Cousins and Feast 1954; Feast 1967; van Genderen, Glass and Feast 1974). Many of the F-type absorptions (Cr II, Fe II, Sc II, Sr II, and Ti II) are strong and sharp; these were used by Wing (1962) to derive the binary elements listed in Table A.5. The M companion to the F star is visible only during eclipse, and the overall weakness of its TiO absorption bands suggests this star is an M0 giant.

The ultraviolet spectrum of BL Tel is that of an F star, with a strong absorption feature due to the Mg II doublet at λ2800 (Wing and Carpenter 1980). Other weaker absorption features are visible farther towards the blue, although the continuum is rather weak. Most of these features appear to be Fe II lines, which is consistent with other ultraviolet spectra of F stars. There is no evidence for the C IV or He II emission lines usually present in symbiotic stars, and emission lines have never been observed in the visual.

PU Vulpeculae

(*Nova Vulpeculae 1979*)

PU Vul was independently discovered by Honda and Kuwano as a ninth magnitude nova. Liller and Liller (1979) showed the system had undergone a 5-6 mag rise in brightness during 1978-79, which is similar to the eruption experienced by RR Tel. Aside from the deep minimum, the evolution of PU Vul is reminiscent of the early development of RT Ser and RR Tel. Spectrophotometric observations at a variety of wavelengths are important if such nova-like outbursts are to be understood.

Photometry

The pre-outburst light curve of PU Vul has been discussed by Liller and Liller (1979). Prior to 1960, the system preferred B ~ 16.5, and occasionally rose to B = 15. The star was fainter than B = 14 in the late 1960's and early 1970's, before beginning a slow rise in 1978 (Figure A.33). The rise to maximum light required over one year; B reached 9.3 when B − V ∼ 0.55. Adopting $E_{B-V} \sim 0.5$ suggests B − V ∼ 0.05 near maximum; and this agrees with the A-type spectrum expected of a slow nova.

The 1979 maximum of PU Vul was characterized by 0.15 mag oscillations (P ∼ 80 days) superimposed on a very gradual fading (Mahra, et al. 1979; Bruch 1980; Nakagiri and Yamashita 1982b; Chochol, Hric, and Papousek 1981; Purgathofer and Schell 1982, 1983; Hron and Maitzen 1983; Kolotilov and Belyakina 1982). A rapid decline began in early 1980, when the system faded by over 3 mag in 100 days. The color changes were very interesting: B − V increased to ∼ 1 as V fell to 11, and then increased to ∼ 0.5 once V reached 13.5. The brightness remained at V ≈ 13.5 for nearly 100 days, and then slowly increased to V ≈ 8.5. The complete recovery required over 200 days; B − V first increased to 0.9 (when V = 9-9.5) and finally steadied at B − V ≈ 0.8 (Kolotilov 1983a,b).

The evolution of PU Vul in the near-IR has not been as dramatic as in the visual (Bensammar, Friedjung and Assus 1980; Belyakina, et al.

Figure A.33 - Optical light curve of the recent outburst in PU Vul. Pre-eruption observations placed the system at V > 14. Small oscillations with *P* ∼ 80 days are present before the deep minimum, but have not been visible since the system recovered from what may have been an eclipse of the hot component by its M-type giant companion.

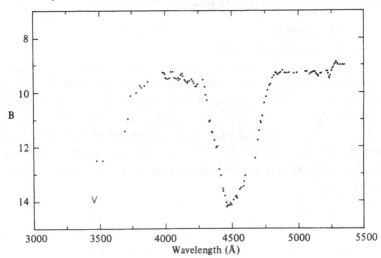

1982a,b,c; Friedjung, et al. 1985). The reduction in the overall brightness at V and in the IR argues against a model in which the minimum is due to the formation of an optically thick dust cloud, unless the dust is very cold (T < 100-200 K). The behavior during minimum is significantly different from the dust minima in classical novae (e.g., FH Ser), in which a significant rise in IR flux accompanies the decline in V.

Infrared observations obtained by the author in October 1984 confirm the presence of a late-type giant star in PU Vul. The H-K color is comparable to H-K in BF Cyg and CI Cyg, while J-K is more indicative of a K giant. It seems likely that the F component contributes to the flux at J. Low resolution spectroscopic data revealed a strong 2.3 μm CO absorption band, and the strength of this band in PU Vul is comparable to the CO band observed in EG And (cf. Kenyon and Gallagher 1983). These data suggest the giant in PU Vul has a spectral type later than M0, and perhaps as late as M5 or M6.

Spectroscopy

As with most symbiotic novae, very few pre-outburst optical spectra of PU Vul are available. An M-type absorption spectrum (gM4) was visible in 1958 and again in September 1978 (cf. Honda, et al. 1979). This suggests that an M giant was present in the system before outburst, in agreement with the IR data discussed above.

The spectrum of PU Vul during its first maximum resembled that of an F supergiant, with no evidence of strong emission lines. There is some indication that the spectral type evolved from A7 II to F5 I during April-October 1979, which would parallel the early evolution of RR Tel (Honda, et al. 1979; Hric, Chochol, and Grygar 1980; Chkhikvadze 1981; Gershberg, et al. 1982; Kolotilov 1983a).

Following this maximum, the visual light faded (Figure A.33) and the spectrum evolved dramatically. An M-type absorption spectrum was visible in April 1980, when V had declined by 1.5 mag (Nakagiri and Yamashita 1982b). Some emission lines (Hα and Hβ) had appeared by June 1980, and Hα remained visible throughout the minimum (Gershberg, et al. 1982). Emission lines from Na I, [N II], and [O III] were strong features near minimum, although Hβ had faded. These emission lines also faded during the 4-5 mag increase in brightness following minimum, although Hα remained bright (Zongli and Xiangliang 1983; Yamashita, Maehara, and Norimoto 1982; Hron and Maitzen 1983; Yamashita, Norimoto, and Yoo 1983). Subsequent spectra (Figure A.34; Zongli and Xiangliang 1983; Zongli 1984) show variable Hβ and Hγ emission, and some indication of a TiO band in the red.

Recent observations reported by Iijima and Ortolani (1984) discovered P Cygni profiles in Hα and various metallic lines. These lines indicate expansion velocities of ~ 100 km s^{-1}, which are comparable to those detected during the early decline of AG Peg. The two absorption components visible in Hα may indicate the ejection of multiple shells, a feature also present in AG Peg, as discussed earlier.

IUE spectra of PU Vul show a late-A star reddened by $E_{B-V} = 0.49$. The decline and recovery of the UV continuum followed the optical quite closely, and may have been of somewhat longer duration (Friedjung, et al. 1985). The behavior of the UV continuum suggests the minimum might have been a result of an obscuration of the A-F supergiant by small dust particles. Similar minima have been observed in classical novae (e.g., DQ Herculis), but this is the first instance of this type of event in a symbiotic nova. Strong emission lines are not obvious in the *IUE* spectra of PU Vul, although lines of Fe II are probably weakly present.

Figure A.34 - Optical spectrum of PU Vul. A weak TiO band is visible at λ7500, while Ca II H and K absorption lines are present in the blue. Both Hα and Hβ are fairly strong emission lines, and Hγ might be weakly present.

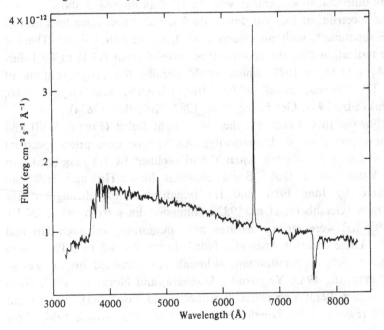

References to the Appendix

Aaronson, M., Liebert, J. and Stocke, J. 1982. *Astrophys. J.*, **254**, 507.

Abrami, A. and Cester, B. 1959. *Mem. Soc. Astr. Ital.*, **3**, 183.

Adams, T.E. 1975. *Astrophys. J.*, **202**, 114.

Adams, W.S., Humason, M.L. and Joy, A.H. 1927.
  Pub. Astr. Soc. Pac., **39**, 365.

Adams, W.S. and Joy, A.H. 1920. *Pop. Astr.*, **28**, 513.

Adams, W.S. and Joy, A.H. 1921. *Pub. Astr. Soc. Pac.*, **33**, 263.

Adams, W.S. and Joy, A.H. 1928. *Pub. Astr. Soc. Pac.*, **40**, 252.

Adams, W.S. and Joy, A.H. 1933a. *Pub. Astr. Soc. Pac.*, **45**, 249.

Adams, W.S. and Joy, A.H. 1933b. *Pub. Astr. Soc. Pac.*, **45**, 301.

Adams, W.S. and Joy, A.H. 1934. *Pub. Astr. Soc. Pac.*, **46**, 223.

Adams, W.S., Joy, A.H. and Sanford, R.F. 1924.
  Pub. Astr. Soc. Pac., **36**, 137.

Ahern, F.J. 1975. *Astrophys. J.*, **197**, 639.

Ahern, F.J. 1978. *Astrophys. J.*, **223**, 901.

Ahern, F.J., FitzGerald, M.P., Marsh, K.A. and Purton, C.R. 1977.
  Astr. Astrophys., **58**, 35.

Ahern, F.J., FitzGerald, M.P., Marsh, K.A. and Purton, C.R. 1978.
  in *IAU Symposium No. 76, Planetary Nebulae: Observations and Theory*,
  ed. Y. Terzian (Dordrecht: Reidel), p. 326.

Aitken, D.K., Roche, P.F. and Spenser, P.M. 1980.
  Mon. Not. Roy. Astr. Soc., **193**, 207.

Allen, D.A. 1973. *Mon. Not. Roy. Astr. Soc.*, **161**, 145.

Allen, D.A. 1974a. *Mon. Not. Roy. Astr. Soc.*, **168**, 1.

Allen, D.A. 1974b. *Inf. Bull. Var. Stars*, No. 911.

Allen, D.A. 1976. *Proc. Austr. Astr. Soc.*, **3**, 78.

Allen, D.A. 1978. *Inf. Bull. Var. Stars*, No. 1399.

Allen, D.A. 1979a. in *IAU Colloquium No. 46*,
  Changing Trends in Variable Star Research,
  ed. F.Bateson, J.Smak and I.Urch (Hamilton, N.Z.: U. Waikato Press), p. 125.

Allen, D.A. 1979b. *Obs.*, **99**, 83.

Allen, D.A. 1980a. *Mon. Not. Roy. Astr. Soc.*, **192**, 521.

Allen, D.A. 1980b. *Mon. Not. Roy. Astr. Soc.*, **190**, 75.

Allen, D.A. 1980c. *Astrophys. Lett.*, **20**, 131.

Allen, D.A. 1981a. in *Proceedings of the North American Workshop on Symbiotic Stars*, ed. R.E. Stencel (Boulder: JILA), p. 12.

Allen, D.A. 1981b. *Mon. Not. Roy. Astr. Soc.*, **197**, 739 [Erratum: *Mon. Not. Roy. Astr. Soc.*, **201**, 1199].

Allen, D.A. 1982a. in *IAU Colloquium No. 70, The Nature of Symbiotic Stars*, ed. M. Friedjung and R. Viotti (Dordrecht: Reidel), p. 27.

Allen, D.A. 1982b. in *IAU Colloquium No. 70, The Nature of Symbiotic Stars*, ed. M. Friedjung and R. Viotti (Dordrecht: Reidel), p. 115.

Allen, D.A. 1983. *Mon. Not. Roy. Astr. Soc.*, **204**, 113.

Allen, D.A. 1984. *Proc. Astr. Soc. Aust.*, **5**, 369.

Allen, D.A., Baines, D.W.T., Blades, J.C. and Whittet, D.C.B. 1982. *Mon. Not. Roy. Astr. Soc.*, **199**, 1017.

Allen, D.A., Beattie, D.H., Lee, T.J., Stewart, J.M. and Williams, P.M. 1978. *Mon. Not. Roy. Astr. Soc.*, **182**, 57P.

Allen, D.A. and Glass, I.S. 1974. *Mon. Not. Roy. Astr. Soc.*, **167**, 337.

Allen, D.A. and Glass, I.S. 1975. *Mon. Not. Roy. Astr. Soc.*, **170**, 579.

Aller, L.H. 1952. *Pub. Astr. Soc. Pac.*, **64**, 228.

Aller, L.H. 1954. *Pub. DAO Victoria*, **9**, 321.

Aller, L.H. and Keyes, C.D. 1974. *Astrophys. Space Sci.*, **30**, 387.

Aller, L.H., Polidan, R.S., Rhodes, E.J. and Wares, G.W. 1973. *Astrophys. Space Sci.*, **20**, 93.

Altamore, A., Baratta, G.B., Cassatella, A., Friedjung, M., Giangrande, A. and Viotti, R. 1981. *Astrophys. J.*, **245**, 630.

Altamore, A., Baratta, G.B., Cassatella, A., Giangrande, A., Ponz, D., Ricciardi, O., and Viotti, R. 1982. in *IAU Colloquium No. 70, The Nature of Symbiotic Stars*, ed. M. Friedjung and R. Viotti (Dordrecht: Reidel), p. 183.

Altamore, A., Baratta, G.B. and Viotti, R. 1979. *Inf. Bull. Var. Stars*, No. 1636.

Altenhoff, W.J., Braes, L.L.E., Olnon, F.M. and Wendker, H.J. 1976. *Astr. Astrophys.*, **46**, 11.

Altenhoff, W.J. and Wendker, H.J. 1973. *Nature*, **241**, 37.

Anandarao, B.G., Sahu, K.C. and Desai, J.N. 1985. *Astrophys. Space Sci.*, **114**, 351.

Anderson, C.M., Cassinelli, J.P., Oliversen, N.G., Myers, R.V. and Sanders, W.T. 1982. in *IAU Colloquium No. 70, The Nature of Symbiotic Stars*, ed. M. Friedjung and R. Viotti (Dordrecht: Reidel), p. 117.

Anderson, C.M., Cassinelli, J.P. and Sanders, W.T. 1981. *Astrophys. J. (Letters)*, **247**, L127.

Anderson, C.M., Oliversen, N.A. and Nordsieck, K.H. 1980. *Astrophys. J.*, **242**, 188.

Andrews, P.J. 1974. *Mon. Not. Roy. Astr. Soc.*, **167**, 635.

Andrews, R.G. 1959. *J. Brit. Astr. Assoc.*, **69**, 114.

Andrews, R.G. 1960. *J. Brit. Astr. Assoc.*, **70**, 352.

Andrillat, Y. 1970. *Comptes Rendus*, **270**, 1066.

Andrillat, Y. 1973. *Mem. Soc. Roy. sci. Liege*,
6eme serie, **tome V**, 371.

Andrillat, Y. 1982a. in *IAU Colloquium No. 70,
The Nature of Symbiotic Stars*,
ed. M. Friedjung and R. Viotti (Dordrecht: Reidel), p. 47.

Andrillat, Y. 1982b. in *IAU Colloquium No. 70,
The Nature of Symbiotic Stars*,
eds. M. Friedjung and R. Viotti (Dordrecht: Reidel), p. 173.

Andrillat, Y., Ciatti, F. and Swings, J.P. 1982.
Astrophys. Space Sci., **83**, 423.

Andrillat, Y. and Houziaux, L. 1976. *Astr. Astrophys.*, **52**, 119.

Andrillat, Y. and Houziaux, L. 1982. in *IAU Colloquium No. 70,
The Nature of Symbiotic Stars*, ed. M. Friedjung and R. Viotti
(Dordrecht: Reidel), p. 57.

Antal, M. and Tremko, J. 1959. *BAC*, **10**, 171.

Arakelyan, M.A. and Ivanova, N.L. 1958. *Bull. Byurakan Obs.*, No. 24, 19.

Argelander, F.W.A. 1867. *Astr. Beob. Sternw. Bonn*, **6**, 142.

Arhipova, V.P. 1963. *Astr. Zh.*, **40**, 71.

Arhipova, V.P. 1977. *Perem. Zvedzy*, **20**, 345.

Arhipova, V.P. 1983. *Perem. Zvedzy*, **22**, 25.

Arhipova, V.P. and Dokuchaeva, O.D. 1962. *Astr. Zh.*, **39**, 613.

Arhipova, V.P. and Dokuchaeva, O.D. 1978. *Soviet Astr. Lett.*, **4**, 48.

Arhipova, V.P., Dokuchaeva, O.D. and Esipov, V.F. 1979.
Soviet Astr. J., **23**, 174.

Arhipova, V.P., Dokuchaeva, O.D. and Vorontsov-Velyaminov, B.A.
1961. *Astr. Tsirk.*, No. 223.

Arhipova, V.P. and Esipov, V.F. 1976. *Astr. Tsirk.*, No. 930, 5.

Arhipova, V.P. and Mandel, O.E. 1973. *Inf. Bull. Var. Stars*, No. 762.

Arhipova, V.P. and Mandel, O.E. 1975. in *IAU Symposium No. 67,
Variable Stars and Stellar Evolution*,
ed. V.E. Sherwood and L. Plaut (Dordrecht: Reidel), p. 391.

Arhipova, V.P., Noskova, R.I. and Gapbarov, A. 1974. *Astr. Zh.*, **51**, 532.

Ashbrook, J.A. 1946. *Bull. Harv. Coll. Obs.*, No. 918.

Aslanov, I.A. and Rustamov, Yu.S. 1970. *Tsirk. Shemakhin
Astrofiz. Obs.*, No. 6, 19.

Audouze, J., Bouchet, P., Fehrenbach, Ch. and Wosczyck, A. 1981.
Astr. Astrophys., **93**, 1.

Baade, W. 1943. *Ann. Rept. Dir. Mt. Wilson. Obs. 1942-43*, p. 17.

Babcock, H.W. 1950. *Pub. Astr. Soc. Pac.*, **62**, 277.

Babcock, H.W. and Cowling, T.G. 1953.
Mon. Not. Roy. Astr. Soc., **113**, 357.

Badalyan, G.S. 1958. *Bull. Byurakan Astrofiz. Obs.*, **25**, 49.

Bailey, J. 1975. *J. Brit. Astr. Assoc.*, **85**, 217.

Bailey, S.I. 1919a. *Bull. Harv. Coll. Obs.*, No. 693.

Bailey, S.I. 1919b. *Bull. Harv. Coll. Obs.*, No. 697.

Bailey, S.I. 1921. *Bull. Harv. Coll. Obs.*, No. 753.

Baratta, G.B., Altamore, A., Cassatella, A., Friedjung, M., Ponz, D.,
and Ricciardi, O. 1982. in
IAU Colloquium No. 70, The Nature of Symbiotic Stars,
ed. M. Friedjung and R. Viotti (Dordrecht: Reidel), p. 145.

Baratta, G.B., Altamore, A., Cassatella, A. and Viotti, R. 1979.
Mem. Soc. Astr. Ital., **50**, 221.

Baratta, G.B., Cassatella, A., Altamore, A., and Viotti, R. 1978.
Inf. Bull. Var. Stars, No. 1493.

Baratta, G.B., Cassatella, A., and Viotti, R. 1974.
Astrophys. J., **187**, 651.

Baratta, G.B., Mengoli, P.A., and Viotti, R. 1982.
Inf. Bull. Var. Stars, No. 2126.

Baratta, G.B. and Viotti, R. 1983. *Mem. Soc. Astr. Ital.,* **54**, 493.

Barbon, R., Mammano, A., and Rosino, L. 1969. in
Non-periodic Phenomena in Variable Stars,
ed. L. Detre (Budapest: Academic), p. 257.

Barker, T. 1978. *Astrophys. J.,* **219**, 914.

Barnard, E.E. 1907. *Astrophys. J.,* **25**, 279.

Barnard, E.E. 1921. *Astr. J.,* **32**, 48.

Barnes, T.G. III. 1973. *Astrophys. J. Suppl.,* **25**, 369.

Bartolini, C. 1965. *Coelum,* **33**, 89.

Barton, J.R., Phillips, B.A., and Allen, D.A. 1979.
Mon. Not. Roy. Astr. Soc., **187**, 813.

Bateson, F.M. 1971. *Circ. Roy. Astr. Soc. N. Z.,* No. 164.

Bateson, F.M., Jones, A.F., and Menzies, B. 1975.
Pub. Var. Star Sect., Roy. Astr. Soc. N. Z., No. 3, 40.

Bath, G.T. and Wallerstein, G. 1976. *Pub. Astr. Soc. Pac.,* **88**, 759.

Bauer, J. 1975. *BAV Rundbrief,* **24**, 45.

Bauer, J. 1977. *BAV Rundbrief,* **26**, 2.

Baxendell, J. 1866. *Mon. Not. Roy. Astr. Soc.,* **27**, 5.

Baxendell, J. 1868. *Mem. Manch. Lit. Phil. Soc.,* series 3, **3**, 279.

Beals, C.S. 1951. *Pub. DAO Victoria,* **9**, 1.

Becker, R.H., Boldt, E.A., Holt, S.S., Pravdo, S.H., Rothschild, R.E.,
Serlemitsos, P.J. and Swank, J.H. 1976.
Astrophys. J. Lett., **207**, L167.

Becker, R.H. and White, R. 1985. In *Radio Stars,* ed. R.H. Hjellming
and D.M. Gibson (Dordrecht: Reidel), p. 139.

Behrmann, C. 1866a. *Astr. Nachr.,* **67**, 111.

Behrmann, C. 1866b. *Astr. Nachr.,* **67**, 301.

Beliavsky, S. 1927. *Astr. Nachr.,* **230**, 153.

Belokon', E.T. and Shulov, O.S. 1974.
Trudy Leningrad Astr. Obs., **30**, 103.

Belti, A. and Rosino,L. 1952. *Mem. Soc. Astr. Ital.,* **23**, 115.

Belyakina, T.S. 1965a. *Izv. Krym. Astrofiz. Obs.,* **33**, 226.

Belyakina, T.S. 1965b. *Izv. Krym. Astrofiz. Obs.,* **34**, 100.

Belyakina, T.S. 1966. *Astrofizika,* **2**, 57.

Belyakina, T.S. 1968a. *Izv. Krym. Astrofiz. Obs.,* **38**, 171.

Belyakina, T.S. 1968b. *Astr. Zh.,* **45**, 139.

Belyakina, T.S. 1969. *Izv. Krym. Astrofiz. Obs.*, **40**, 39.

Belyakina, T.S. 1970a. *Izv. Krym. Astrofiz. Obs.*, **41-42**, 275.

Belyakina, T.S. 1970b. *Astrofizika*, **6**, 49.

Belyakina, T.S. 1974a. *Izv. Krym. Astrofiz. Obs.*, **50**, 103.

Belyakina, T.S. 1974b. *Inf. Bull. Var. Stars*, No. 863.

Belyakina, T.S. 1975. in
IAU Symposium No. 67, Variable Stars and Stellar Evolution,
ed. V.E. Sherwood and L. Plaut (Dordrecht: Reidel), p. 397.

Belyakina, T.S. 1976. *Inf. Bull. Var. Stars*, No. 1169.

Belyakina, T.S. 1979a. *Inf. Bull. Var. Stars*, No. 1602.

Belyakina, T.S. 1979b. *Izv. Krym. Astrofiz. Obs.*, **59**, 133.

Belyakina, T.S. 1980. *Inf. Bull. Var. Stars*, No. 1808.

Belyakina, T.S. 1981. *Inf. Bull. Var. Stars*, No. 1974.

Belyakina, T.S. 1983. *Inf. Bull. Var. Stars*, No. 2485.

Belyakina, T.S. 1984. *Izv. Krym. Astrofiz. Obs.*, **68**, 108.

Belyakina, T.S. 1985a. *Inf. Bull. Var. Stars*, No. 2697.

Belyakina, T.S. 1985b. *Inf. Bull. Var. Stars*, No. 2698.

Belyakina, T.S., Bondar, N.I., Chochol, D., Chuvaev, K.K., Efimov, Yu. S., Gershberg, R.E., Grygar, J., Ilric, L., Krasnobabtsev, V.I., Petrov, P.P., Piirola, V., Savanov, I.S., Shakhovskaya, N.I., Shakhovskoj, N.M., and Shenavrin, V.I., 1984. *Astr. Astrophys.*, **132**, L12.

Belyakina, T.S., Boyarchuk, A.A. and Gershberg, R.E. 1963.
Izv. Krym. Astrofiz. Obs., **30**, 25.

Belyakina, T.S. and Brodskaya, E.S. 1974.
Izv. Krym. Astrofiz. Obs., **49**, 58.

Belyakina, T.S., Efimov, Yu.S., Pavlenko, E.P.
and Shenavrin, V.I. 1982a. *Sov. Astr. J.*, **26**, 1.

Belyakina, T.S., Gershberg, R.E., Efimov, Yu.S., Krasnobabtsev, V.I., Pavlenko, E.P., Petrov, P.P., Chuvaev, K.K. and Shenavrin, V.I. 1982b.
Sov. Astr. J., **26**, 184.

Belyakina, T.S., Gershberg, R.E., Efimov, Yu.S., Krasnobabtsev, V.I., Pavlenko, E.P., Petrov, P.P., Chuvaev, K.K. and Shenavrin, V.I. 1982c.
in *IAU Colloquium No. 70, The Nature of Symbiotic Stars*,
ed. M. Friedjung and R. Viotti (Dordrecht: Reidel), p. 221.

Belyakina, T.S., Gershberg, R.E. and Shakhovskaya, N.I. 1978.
Soviet Astr. Lett., **4**, 219.

Belyakina, T.S., Gershberg, R.E. and Shakhovskaya, N.I. 1979.
Soviet Astr. Lett., **5**, 349.

Bensammar, S., Friedjung, M. and Assus, P. 1980.
Astr. Astrophys., **83**, 261.

Beriev, A.D. 1955. *Izv. Stalinabad Astrofiz. Obs.*, No. 15, 29.

Berman, L. 1932. *Pub. Astr. Soc. Pac.*, **44**, 318.

Bertaud, Ch. 1946. *Bull. Soc. Astr. France*, **60**, 74.

Bertaud, Ch. 1949. *Bull. Astr.*, **14**, 61.

Bertaud, Ch. 1958. *Bull. Soc. Astr. France*, **72**, 437.

Bertaud, Ch. 1960. *l'Astronomie*, **74**, 57.

Bertaud, Ch. and Weber, R. 1962. *l'Astronomie*, **76**, 41.

Bianchini, A. and Middleditch, J. 1976. *Inf. Bull. Var. Stars*, No. 1151.

Bidelman, W.P. 1954. *Astrophys. J. Suppl.*, **1**, 175.

Bidelman, W.P. and MacConnell, D.J. 1973. *Astr. J.*, **78**, 687.

Bird, F. 1866. *Astr. Reg.*, **4**, 230.

Blair, W.P., Stencel, R.E., Feibelman, W.A. and Michalitsianos, A.G. 1983. *Astrophys. J. Suppl.*, **53**, 573.

Blair, W.P., Stencel, R.E., Shaviv, G. and Feibelman, W.A. 1981. *Astr. Astrophys.*, **99**, 73.

Blanco, V.M. 1961. *Contr. Bosscha Obs.*, No. 13.

Bloch, M. 1934. *Bull. Assoc. Fran. Obs. Et. Var.*, **3**, 78.

Bloch, M. 1946. *Pub. Obs. Haute Prov.* Ser. B, No. 5.

Bloch, M. 1960. *J. Phys. Radium*, **21**, 118.

Bloch, M. 1961. *Comptes Rendus*, **253**, 801.

Bloch, M. 1964. *Ann. d'Astrophys.*, **27**, 292.

Bloch, M. 1965a. *Pub. Obs. Haute Prov.*, **7**, No. 21.

Bloch, M. 1965b. in *Coll. Int. sur les Novae, Novoides et Supernovae*, ed. M.J. Dufay (Paris: CNRS), p. 151.

Bloch, M. and Dufay, M.J. 1958. *Comptes Rendus*, **247**, 865.

Bloch, M., Dufay, M.J., Fehrenbach, Ch., and Tcheng, M.-L. 1946a. *Comptes Rendus*, **223**, 72.

Bloch, M., Dufay, M.J., Fehrenbach, Ch., and Tcheng, M.-L. 1946b. *Ann. d'Astrophys.*, **9**, 157.

Bloch, M., Ellsworth, J. and Liau, S.P. 1933a. *Comptes Rendus*, **197**,1095.

Bloch, M., Ellsworth, J. and Liau, S.P. 1933b. *Bull. Assoc. Fran. Obs. Et. Var.*, **2**, 125.

Bloch, M., Fehrenbach, Ch., and Tcheng, M.-L. 1946. *Comptes Rendus*, **223**, 196.

Bloch, M., Jousten, N. and Swings, J.P. 1969. *Bull. Soc. Roy. sci. Liege*, **38**, Nos. 5-6, 245.

Bloch, M. and Tcheng, M.-L. 1951a. *Pub. Obs. Haute Prov.* **2**, No. 27.

Bloch, M. and Tcheng, M.-L. 1951b. *Ann. d'Astrophys.*, **14**, 266.

Bloch, M. and Tcheng, M.-L. 1952a. *Comptes Rendus*, **234**, 810.

Bloch, M. and Tcheng, M.-L. 1952b. *Ann. d'Astrophys.*, **15** , 104.

Bloch, M. and Tcheng, M.-L. 1953. *Ann. d'Astrophys.*, **16**, 73.

Bloch, M. and Tcheng, M.-L. 1955. *Comptes Rendus*, **241**, 1105.

Bloch, M. and Tcheng, M.-L. 1958. *Mem. Soc. Roy. sci. Liege*, 4eme serie, **tome XX**, 458.

Bochonko, D.R. and Malville, J.M. 1968. *Pub. Astr. Soc. Pac.*, **80**, 177.

Boehm, C. and Hack, M. 1983. *Inf. Bull. Var. Stars*, No. 2398.

Bohme, S. 1938. *Astr. Nachr.*, **266**, 267.

Bohme, S. 1939. *Astr. Nachr.*, **268**, 71.

Bopp, B. 1977. *Inf. Bull. Var. Stars*, No. 1327.

Bopp, B. 1981. in *Proceedings of the North American Workshop on Symbiotic Stars*, ed. R.E. Stencel (Boulder: JILA), p. 11.

Bossen, H. 1972. *Inf. Bull. Var. Stars*, No. 722.

Bowers, P.F. and Kundu, M.R. 1979. *Astr. J.*, **84**, 791.

Boyarchuk, A.A. 1964a. *Perem. Zvedzy*, **15**, 36.

Boyarchuk, A.A. 1964b. *Perem. Zvedzy*, **15**, 48.

Boyarchuk, A.A. 1965. *Izv. Krym. Astrofiz. Obs.*, **33**, 186.

Boyarchuk, A.A. 1966a. *Izv. Krym. Astrofiz. Obs.*, **35**, 8.

Boyarchuk, A.A. 1966b. *Astrofizika*, **2**, 101.

Boyarchuk, A.A. 1966c. *Astr. Zh.*, **43**, 976.

Boyarchuk, A.A. 1967a. *Astrofizika*, **3**, 203.

Boyarchuk, A.A. 1967b. *Astr. Zh.*, **44**, 12.

Boyarchuk, A.A. 1968a. *Astrofizika*, **4**, 289.

Boyarchuk, A.A. 1968b. *Soviet Astr. J.*, **11**, 818.

Boyarchuk, A.A. 1968c. *Izv. Krym. Astrofiz. Obs.*, **38**, 155.

Boyarchuk, A.A. 1969a. *Izv. Krym. Astrofiz. Obs.*, **39**, 124.

Boyarchuk, A.A. 1969b. in *Non-Periodic Phenomena in Variable
Stars*, ed. L. Detre (Budapest: Academic Press), p. 395.

Boyarchuk, A.A. 1970. *Izv. Krym. Astrofiz. Obs.*, **41-42**, 264.

Boyarchuk, A.A. 1975. in
IAU Symposium No. 67, Variable Stars and Stellar Evolution,
ed. V.E. Sherwood and L. Plaut (Dordrecht: Reidel), p. 377.

Boyarchuk, A.A., Esipov, V.F. and Moroz, V.I. 1966.
Astr. Zh., **43**, 421.

Boyarchuk, A.A. and Gershberg, R.E. 1966.
Izv. Krym. Astrofiz. Obs., **35**, 3.

Brahde, R. 1949. *Ann. Stock. Obs.*, **15**, 1.

Bregman, J. D. 1982. *Bull. A.A.S.*, **14**, 982.

Bregman, J.D., Goebel, J.H. and Strecker, D.W. 1978. *Astrophys. J. (Letters)*, **223**, L45.

Brewster, M. 1975. *Inf. Bull. Var. Stars*, No. 1032.

Brewster, M. 1976. *J. American Assoc. Variable Star Obs.*,
4, No. 2, 97.

Brocka, B. 1979. *Pub. Astr. Soc. Pac.*, **91**, 519.

Brown, L.W., Feibelman, W.A., Hobbs, R.W. and McCracken, C.W. 1978.
Astrophys. Lett., **19**, 75.

Bruch, A. 1980. *Inf. Bull. Var. Stars*, No. 1805.

Bruck, H. 1935. *Publ. Potsdam Obs.*, **28**, 161.

Brugel, E.W. and Wallerstein, G. 1981. *Obs.*, **101**, 164.

Brugel, E.W., Cardelli, J.A., Szkody, P., and Wallerstein, G. 1984.
Pub. Astr. Soc. Pac., **96**, 78.

Bruhns, C. 1868. *Astr. Nachr.*, **72**, 365.

Brydon, H.B. 1933. *J. Roy. Astr. Soc. Can.*, **27**, 355.

Buchnicek, Z. 1946. *Rise Hvezd*, **27**, 81.

Burbidge, G.R. and Burbidge, E.M. 1954. *Astrophys. J.*, **120**, 76.

Burchi, R. 1980. *Inf. Bull. Var. Stars*, No. 1813.

Burchi, R., Chochol, D., DiPaolantonio, A., Mancuso, S., Milano, L.
and Vittone, A. 1983. *Astrophys. Space Sci.*, **89**, 387.

Burchi, R., DiPaolantonio, A., Mancuso, S., Milano, L. and Vittone, A.
1980. *Inf. Bull. Var. Stars*, No. 1871.

Burstein, D. and Heiles, C. 1982. *Astr. J.*, **87**, 1165.

Cameron, D. and Nassau, J.J. 1956. *Astrophys. J.*, **124**, 346.

Campbell, L. 1907. *Ann. Harv. Coll. Obs.*, **57**, 1.

Campbell, L. 1912. *Ann. Harv. Coll. Obs.*, **63**, ,1.

Campbell, L. 1918. *Ann. Harv. Coll. Obs.*, **79**, 1.

Campbell, L. 1933. *Bull. Harv. Coll. Obs.*, No. 893.

Campbell, L. 1940. *Pop. Astr.*, **48**, 506.

Campbell, L. 1942. *Harv. Repr.*, No. 231, 34.

Campbell, L. and Shapley, H. 1923. *Circ. Harv. Coll. Obs.*, No. 247.

Cannon, A.J. 1918. *Ann. Harv. Coll. Obs.*, **92**, 107.

Cannon, A.J. 1921. *Circ. Harv. Coll. Obs.*, No. 224.

Cannon, A.J. 1933. *Bull. Harv. Coll. Obs.*, No. 891.

Cannon, A.J. and Mayall, M.W. 1938. *Bull. Harv. Coll. Obs.*, No. 908.

Caputo, F. 1971a. *Pub. Astr. Soc. Pac.*, **83**, 62.

Caputo, F. 1971b. *Astrophys. Space Sci.*, **10**, 93.

Carlson, E.D. and Henize, K.G. 1974.
  Astrophys. J. (Letters), **188**, L47.

Cassatella, A. 1982.
  IAU Colloquium No. 70, The Nature of Symbiotic Stars,
  ed. M. Friedjung and R. Viotti (Dordrecht: Reidel), p. 157.

Cassatella, A., Patriarchi, P., Selvelli, P.L., Bianchi, L.,
  Cacciari, C., Heck, A., Perryman, M. and Wamsteker, W.
  1982a. in *Third European IUE Conference*, ed. E. Rolfe, A. Heck
  and B. Battrick (ESA SP-176), p. 229.

Cassatella, A., Patriarchi, P., Selvelli, P.L., Bianchi, L.,
  Cacciari, C., Heck, A., Perryman, M. and Wamsteker, W.
  1982b. in *Advances in Ultraviolet Astronomy*, ed. Y. Kondo, J.M. Mead
  and R.D. Chapman (NASA CP-2238), p. 482.

Catchpole, R.H., Robertson, B.S.C., Lloyd-Evans, T.H.H., Feast, M.W.,
  Glass, I.S. and Carter, B.S. 1979. *Circ. SAAO*, **1**, 61.

Cester, B. 1968. *Inf. Bull. Var. Stars*, No. 291.

Cester, B. 1969. *Astrophys. Space Sci.*, **3**, 198.

Chkhikvadze, I.N. 1970a. *Astr. Tsirk.*, No. 546, 7.

Chkhikvadze, I.N. 1970b. *Astr. Tsirk.*, No. 595, 1.

Chkhikvadze, I.N. 1981. *Astr. Tsirk.*, No. 1180.

Chochol, D., Grygar, J. and Hric, L. 1980. in *IAU Symposium No. 88*,
  Close Binary Stars: Observations and Interpretation,
  ed. M. Plavec, D.M. Popper and R.K. Ulrich (Dordrecht: Reidel), p. 547.

Chochol, D. and Hric, L. 1982. in
  IAU Colloquium No. 70, The Nature of Symbiotic Stars,
  ed. M. Friedjung and R. Viotti (Dordrecht: Reidel), p. 137.

Chochol, D., Hric, L. and Papousek, J. 1981.
  Inf. Bull. Var. Stars, No. 2059.

Chochol, D., Vittone, A.A., Milano, L., and Rusconi, L. 1984.
  Astr. Astrophys., **140**, 91.

Ciatti, F., D'Odorico, S. and Mammano, A. 1974.
  Astr. Astrophys., **34**, 181.

Ciatti, F. and Mammano, A. 1975a. *Astr. Astrophys.*, **38**, 435.

Ciatti, F. and Mammano, A. 1975b. in
  IAU Symposium No. 67, Variable Stars and Stellar Evolution,
  ed. V.E. Sherwood and L. Plaut (Dordrecht: Reidel), p. 409.

Ciatti, F. and Mammano, A. 1976. *Mem. Soc. Roy. sci. Liege*,
  6eme serie, **tome IX**, 379.

Ciatti, F., Mammano, A. and Rosino, L. 1971.
Veroff. Remeis Stern. Bamberg, **9**, 64.

Ciatti, F., Mammano, A. and Rosino, L. 1975. in
IAU Symposium No. 67, Variable Stars and Stellar Evolution,
ed. V.E. Sherwood and L. Plaut (Dordrecht: Reidel), p. 389.

Ciatti, F., Mammano, A. and Vittone, A. 1978a.
Astr. Astrophys., **61**, 459.

Ciatti, F., Mammano, A. and Vittone, A. 1978b.
Astr. As.rophys., **68**, 251.

Ciatti, F., Mammano, A. and Vittone, A. 1979.
Astr. Astrophys., **79**, 247.

Code, A.D. 1968. *Astrophys. J.*, **151**, L145.

Coe, M.J., Engel, A.R., Evans, A.J. and Quenby, J.J. 1981.
Astrophys. J., **243**, 155.

Cohen, M. and Barlow, M.J. 1974. *Astrophys. J.*, **193**, 401.

Cohen, M. and Barlow, M.J. 1980. *Astrophys. J.*, **238**, 585.

Colacevich, A. 1933. *Atti della Reale Accad. Naz. Lincei*, series 6,
18, 307.

Connelley, M. and Sandage, A. 1958. *Pub. Astr. Soc. Pac.*, **70**, 600.

Cordova, F.A. and Mason, K.O. 1984.
Mon. Not. Roy. Astr. Soc., **206**, 879.

Cordova, F.A., Mason, K.O. and Nelson, J.E. 1981.
Astrophys. J., **245**, 609.

Courbebaisse, M. 1866. *Comptes Rendus*, **62**, 1115.

Cousins, A.W.J. 1958. *Mon. Notes Astr. Soc. S. Africa*, **17**, 132.

Cousins, A.W.J. and Feast, M.W. 1954. *Obs.*, **74**, 88.

Cowley, A. and Stencel, R. 1973. *Astrophys. J.*, **184**, 687.

Cowley, C.R. 1956. *Pub. Astr. Soc. Pac.*, **68**, 537.

Cragg, T.A. 1959. *Pub. Astr. Soc. Pac.*, **71**, 47.

Crampton, D., Grygar, J., Kohoutek, L. and Viotti, R. 1970.
Astrophys. Lett., **6**, 5.

Davidsen, A., Malina, R. and Bowyer, S. 1976.
Astrophys. J., **203**, 448.

Davidsen, A., Malina, R. and Bowyer, S. 1977.
Astrophys. J., **211**, 866.

Davidson, K., Humphreys, R.M. and Merrill, K.M. 1978.
Astrophys. J., **220**, 239.

Davidson, M. 1946a. *Circ. Brit. Astr. Assoc.*, No. 259.

Davidson, M. 1946b. *Circ. Brit. Astr. Assoc.*, No. 260.

Dawes, W.R. 1866. *Astr. Reg.*, **4**, 164.

Dean, C.A. and van Citters, W. 1970. *Pub. Astr. Soc. Pac.*, **82**, 924.

Dermul, A. 1933. *Gaz. Astr.*, **20**, 135.

Dermul, A. 1958. *Gaz. Astr.*, **40**, 39.

de Roy, F. 1933. *Gaz. Astr.*, **20**, 152.

de Roy, F. 1934. *Gaz. Astr.*, **21**, 110.

Deuel, W. and Nussbaumer, H. 1983.
Astrophys. J. (Letters), **271**, L19.

Deutsch, A.J. 1948. *Pub. Astr. Soc. Pac.*, **60**, 120.

Deutsch, A.J., Lowen, L., Morris, S.C. and Wallerstein, G. 1974.
  Pub. Astr. Soc. Pac., **86**, 233.
Doan, N.H. 1970. *Astr. Astrophys.*, **8**, 307.
Doberck, W. 1924. *J. des Observ.*, **7**, 85.
Dokuchaeva, O.D. 1958. *Perem. Zvedzy*, **12**, 372.
Dokuchaeva, O.D. 1976. *Inf. Bull. Var. Stars*, No. 1189.
Dokuchaeva, O.D. and Balazs, B. 1976. *Astr. Tsirk.*, No. 929.
Dolidze, M.V. and Alanya, I.F. 1962.
  Bull. Abastumani Astr. Obs., No. 28, 113.
Dolidze, M.V. and Dzhimshelejshvili, G.N. 1966. *Izv. Abastumani*
  Astrofiz. Obs., No. 34, 27.
Dolidze, M.V. and Dzhimshelejshvili, G.N. 1969. *Izv. Abastumani*
  Astrofiz. Obs., No. 37, 35.
Dolidze, M.V. and Dzhimshelejshvili, G.N. 1970. *Communications*
  of the Academy of Sciences of Georgia SSR, **58**, 57.
Dolidze, M.V. and Pugach, A.F. 1962. *Izv. Abastumani Astrofiz.*
  Obs., No. 28, 121.
Dombrovoki, V.A. and Polyakora, T.A. 1974.
  Trudy Astr. Obs. Lenin., **30**, 89.
Doroshenko, V.T. and Magnitsky, A.K. 1982. *Astr. Tsirk.*, No. 1207.
Doroshenko, V.T. and Nikolov, N.S. 1967. *Astr. Zh.*, **44**, 570.
Doroshenko, V.T. and Rustamov, 1970. *Astr. Zh.*, **44**, 570.
Dossin, F. 1959. *Ann. d'Astrophys.*, **22**, 733.
Dossin, F. 1964. *Astr. J.*, **69**, 137.
Doty, J.A., Hoffman, J.A. and Lewin, W.H.G. 1981.
  Astrophys. J., **243**, 257.
Doxsey, D.E., Apparao, K.M.V., Bradt, H.V., Dower, R.G. and
  Jernigan, J.G. 1977. *Nature*, **270**, 586.
Drummond, J.D. 1980. *Astr. Astrophys.*, **88**, L11.
Dufay, J. and Bloch, M. 1964. *Ann. d'Astrophys.*, **27**, 462.
Dufay, J. and Bloch, M. 1965. in *Coll. Int. sur les Novae, Novoides et*
  Supernovae, ed. M. Dufay (Paris: CNRS), p. 53.
Dufay, J., Bloch, M., Bertaud, Ch. and Dufay, M. 1958. *Comptes*
  Rendus, **247**, 1316.
Dufay, J., Bloch, M., Bertaud, Ch. and Dufay, M. 1964. *Ann.*
  d'Astrophys., **27**, 555.
Dufay, J., Bloch, M., Bertaud, Ch. and Dufay, M. 1965a. *Pub. Obs. Haute*
  Prov., **7**, No. 20.
Dufay, J., Bloch, M., Bertaud, Ch. and Dufay, M. 1965b.
  in *Coll. Int. sur les Novae, Novoides et Supernovae*,
  ed. M. Dufay (Paris: CNRS), p. 18.
Dumortier, B. and Stram, E. 1970. *l'Astronomie*, **84**, 455.
Duschl, W.J. 1983. *Astr. Astrophys.*, **119**, 248.
Dzhimshelejshvili, G.N. 1968. *Astr. Tsirk.*, No. 505, 8.
Dzhimshelejshvili, G.N. 1970. *Astr. Tsirk.*, No. 557, 2.
Dzhimshelejshvili, G.N. 1971. *Astrofizika*, **7**, 363.
Edlen, B. 1969. *Mon. Not. Roy. Astr. Soc.*, **144**, 391.
Efimov, Yu. S. 1979. *Soviet Astr. Lett.*, **5**, 352.

Eggen, O.J. 1964. *Bull. Roy. Green. Obs.*, No. 84.

Eggen, O.J. and Rodgers, A.W. 1969. *Astrophys. J.*, **158**, L111.

Eiroa, C., Hefele, H. and Qian, Z.-U. 1982. in
IAU Colloquium No. 70, The Nature of Symbiotic Stars,
ed. M. Friedjung and R. Viotti (Dordrecht: Reidel), p. 43.

Elvey, C.T. 1941. *Astrophys. J.*, **94**, 140.

Elvey, C.T. and Babcock, H.W. 1943. *Astrophys. J.*, **97**, 412.

Engels, D. 1979. *Astr. Astrophys. Suppl.*, **36**, 337.

Erleksova, G.E. 1964. *Bull. Astr. Obs. Stalinabad*, No. 37, 43.

Ern, M. 1866. *Bull. Acad. Roy. des Sci. Lett. Beaux-Arts de Belgique*, 2eme serie, **21**, 535.

Ernilev, N.P. 1950. *Peremennye Zvedze*, **7**, 273.

Eskioglu, A.N. 1963. *Ann. d'Astrophys.*, **26**, 331.

Espin, T.E. 1893. *Astr. Nachr.*, **134**, 124.

Espin, T.E. 1900. *Astr. Nachr.*, **152**, 135.

d'Esterre, M. 1915. *Mon. Not. Roy. Astr. Soc.*, **75**, 292.

Faloma, R. 1983. *Astrophys. Space Sci.*, **91**, 63.

Faraggiana, R. 1967. *Inf. Bull. Var. Stars*, No. 219.

Faraggiana, R. 1980a. in *IAU Colloquium No. 88*,
Close Binary systems: Observations and Interpretation,
ed. M. Plavec, D.M. Popper and R.K. Ulrich (Dordrecht: Reidel), p. 549.

Faraggiana, R. 1980b. *Astr. Astrophys.*, **84**, 366.

Faraggiana, R. and Hack, M. 1969a. *15eme Coll. Astrophys. Liege*, p. 317.

Faraggiana, R. and Hack, M. 1969b. *Astrophys. Space Sci.*, **3**, 205.

Faraggiana, R. and Hack, M. 1971. *Astr. Astrophys.*, **15**, 55.

Faraggiana, R. and Hack, M. 1980. *Inf. Bull. Var. Stars*, No. 1861.

Faraggiana, R., Hack, M. and Selvelli, P.L. 1979. *Mem. Soc. Astr. Ital.*, **50**, 207.

Farquhar, E.J. 1866. *Am. Journ. Sci. Arts* (2), **42**, 79.

Feast, M.W. 1967. *Mon. Not. Roy. Astr. Soc.*, **135**, 287.

Feast, M.W. 1970. *Astrophys. J.*, **160**, L171.

Feast, M.W. 1971. *Obs.*, **91**, 112.

Feast, M.W. 1982. *Mon. Notes Astr. Soc. S. Africa*, **41**, 72.

Feast, M.W., Catchpole, R.M., Whitelock, P.A., Carter, B.S., and Roberts, G. 1983a. *Mon. Not. Roy. Astr. Soc.*, **203**, 373.

Feast, M.W. and Glass, I.S. 1974.
Mon. Not. Roy. Astr. Soc., **167**, 81.

Feast, M.W. and Glass, I.S. 1980. *Obs.*, **100**, 208.

Feast, M.W., Robertson, B.S.C. and Catchpole, R.M. 1977.
Mon. Not. Roy. Astr. Soc., **179**, 499.

Feast, M.W., Whitelock, P.A., Catchpole, R.M., Robertson, B.S.C. and Carter, B.S. 1983b. *Mon. Not. Roy. Astr. Soc.*, **202**, 951.

Fehrenbach, Ch. 1982. in
IAU Colloquium No. 70, The Nature of Symbiotic Stars,
ed. M. Friedjung and R. Viotti (Dordrecht: Reidel), p. 149.

Fehrenbach, Ch. and Huang, C.C. 1981. *Astr. Astrophys.*, **46**, 257.

Feibelman, W.A. 1982a. *Astrophys. J.*, **258**, 548.

Feibelman, W.A. 1982b. *Astrophys. J. (Letters)*, **263**, L69.

Feibelman, W.A. 1983. *Astrophys. J.*, **275**, 628.

Feibelman, W.A., Boggess, A., Hobbs, R.W. and McCracken, C.W. 1980.
  Astrophys. J., **241**, 725.

Feldman, P.A. 1977. *J. Roy. Astr. Soc. Canada*, **71**, 386.

Fernie, J.D. 1985. *Pub. Astr. Soc. Pac.*, **97**, 653.

Filin, A.I. 1959. *Astr. Tsirk.*, No. 200.

FitzGerald, M.P. 1973. *Nature Phys. Sci.*, **245**, 58.

FitzGerald, M.P. and Houk, N. 1970. *Astrophys. J.*, **159**, 963.

FitzGerald, M.P., Houk, N., McCuskey, S.W. and Hoffleit, D. 1966.
  Astrophys. J., **144**, 1135.

FitzGerald, M.P. and Pilavaki, A. 1974.
  Astrophys. J. Suppl., **28**, 147.

Flammarion, G.C. and Quenisset, F. 1934. *Comptes Rendus*, **198**, 154.

Fleming, W.P. 1907. *Ann. Harv. Coll. Obs.*, **47**, 1.

Fleming, W.P. 1908. *Circ. Harv. Coll. Obs.*, No. 143.

Fleming, W.P. 1912. *Ann. Harv. Coll. Obs.*, **56**, 165.

Flower, D.R., Nussbaumer, H. and Schild, H. 1979.
  Astr. Astrophys., **72**, L1.

Folkart, B., Pecker, J.-C. and Pottasch, S.R. 1964a.
  Ann. d'Astrophys., **27**, 249.

Folkart, B., Pecker, J.-C. and Pottasch, S.R. 1964b.
  Ann. d'Astrophys., **27**, 252.

Folkart, B., Pecker, J.-C. and Pottasch, S.R. 1965a. in
  Coll. Int. sur les Novae, Novoides et Supernovae,
  ed. M. Dufay (Paris: CNRS), p. 50.

Folkart, B., Pecker, J.-C. and Pottasch, S.R. 1965b. in
  Coll. Int. sur les Novae, Novoides et Supernovae,
  ed. M. Dufay (Paris: CNRS), p. 60.

Folkart, B., Pecker, J.-C., Pottasch, S.R. and Rountree-Lesh, J.
  1961. *Notes et Inf., Pub. Paris Obs.*, No. 1.

Frantsman, Yu. L. 1962. *Astr. Zh.*, **39**, 256.

Fried, J.W. 1980a. *Astr. Astrophys.*, **81**, 182.

Fried, J.W. 1980b. *Astr. Astrophys.*, **88**, 141.

Friedjung, M. 1966. *Mon. Not. Roy. Astr. Soc.*, **133**, 401.

Friedjung, M. 1967. *Astr. J.*, **72**, 797.

Friedjung, M. 1972. *Astrophys. Space Sci.*, **19**, 501.

Friedjung, M. 1974. *Astrophys. Space Sci.*, **29**, L5.

Friedjung, M. 1980. in *Close Binary Systems: Observations and
  Interpretation*, IAU Symposium No. 88, ed. M. Plavec, D.M. Popper
  and R.K. Ulrich (Dordrecht: Reidel), p. 543.

Friedjung, M., Ferrari-Toniolo, M., Persi, P., Altamore, A., Cassatella, A.,
  and Viotti, R. in *The Future of UV Astronomy Based on Six Years of IUE
  Research*, ed. J.M. Mead, R.D. Chapman, and Y. Kondo (NASA: CP-2349), p. 408.

Frogel, J.A., Persson, S.E., and Cohen, J.G. 1981.
  Astrophys. J., **246**, 842.

Galatola, A. 1973. *Bull. Am. Astr. Soc.*, **5**, 415.

Galkina, T.S. 1983. *Izv. Krym. Astrofiz. Obs.*, **67**, 33.

Gallagher, J.S., Holm, A.V., Anderson, C.M. and Webbink, R.F. 1979.
  Astrophys. J., **229**, 994.

Gaposchkin, S. 1945. *Ann. Harv. Coll. Obs.*, **115**, No. 2.

Gaposchkin, S. 1946. *Bull. Harv. Coll. Obs.*, No. 914.

Gaposchkin, S. 1952. *Ann. Harv. Coll. Obs.*, **118**, 155.

Gaposchkin, S. 1970. *Astr. J.*, **75**, 641.

Garstang, R.H. 1968. *Astrophys. Space Sci.*, **2**, 336.

Gauzit, J. 1955a. *Ann. d'Astrophys.*, **18**, 354.

Gauzit, J. 1955b. *Comptes Rendus*, **241**, 741.

Gauzit, J. 1955c. *Comptes Rendus*, **241**, 793.

Gehrz, R.D. and Woolf, N.J. 1971. *Astrophys. J.*, **165**, 285.

Geisel, S.L. 1970. *Astrophys. J.*, **161**, L105.

Geisel, S.L. and Kleinmann, D.E. and Low, F.J. 1970.
    Astrophys. J. (Letters), **161**, L101.

Gerasimovic, B.P. and Shapley, H. 1930. *Bull. Harv. Coll. Obs.*, No. 872.

Gershberg, R.E., Krasnobabtsev, V.I., Petrov, P.P. and Chuvaev, K.K.
    1982. *Sov. Astr. J.*, **26**, 3.

Gessner, H. 1959. *Mitt. Verand. Sterne*, No. 416.

Gessner, H. 1966. *Mitt. Verand. Sterne*, **4**, 37.

Gevorkian, G.H. 1981. *Inf. Bull. Var. Stars*, No. 2056.

Ghigo, F.D. and Cohen, N.L. 1981. *Astrophys. J.*, **245**, 988.

Gillett, F.C., Knacke, R.F. and Stein, W.A. 1971.
    Astrophys. J., **163**, L57.

Gillett, F.C., Merrill, K.M. and Stein, W.A. 1971.
    Astrophys. J., **164**, 87.

Gillett, F.C., Merrill, K.M. and Stein, W.A. 1972.
    Astrophys. J., **172**, 367.

Gilson, E., Petit, M. and Koeckelenbergh, A. 1963. *Ciel et Terre*, **79**, 312.

Glass, I.S. and Feast, M.W. 1973. *Nature Phys. Sci.*, **245**, 39.

Glass, I.S. and Webster, B.L. 1973.
    Mon. Not. Roy. Astr. Soc., **165**, 77.

Glebocki, R. 1970. *Postepy Astr.*, **18**, 1.

Gorbatskij, V.G. and Ivanova, L.N. 1973. *Bull. Lenin. Univ.*, No. 13, 144.

Gordon, C.P. 1968. *Pub. Astr. Soc. Pac.*, **80**, 597.

Gould, B.A. 1866a. *Astr. Nachr.*, **67**, 141.

Gould, B.A. 1866b. *Astr. Nachr.*, **68**, 183.

Gould, B.A. 1866c. *Am. J. Sci. Arts* (2), **42**, 80.

Graff, K. 1914. *Astr. Nachr.*, **197**, 233.

Grandjean, J. 1952. *Ann. d'Astrophys.*, **15**, 7.

Grant, G. and Aller, L.H. 1954. *Astr. J.*, **59**, 321.

Gratton, L. 1951. *Circ. Astr. Obs. La Plata*, No. 8, 17.

Gratton, L. 1952. *Trans. IAU*, **8**, 849.

Gratton, L. and Kruger, E.C. 1946. *Ric. Sci. Ricos.*, **16**, 299.

Gratton, L. and Kruger, E.C. 1949. *Mem. Soc. Astr. Ital.*, **20**, 197.

Grauer, A. and Bond, H. 1981. *Pub. Astr. Soc. Pac.*, **93**, 630.

Gravina, R. 1980. *Inf. Bull. Var. Stars*, No. 1759.

Gravina, R. 1981a. *Inf. Bull. Var. Stars*, No. 1945.

Gravina, R. 1981b. *Inf. Bull. Var. Stars*, No. 2041.

Greenstein, N.K. 1937. *Bull. Harv. Coll. Obs.*, No. 906.

Gregory, C.C.L. and Peachey, E.M. 1946.
    Mon. Not. Roy. Astr. Soc., **106**, 135.

Gregory, P.C., Kwok, S. and Seaquist, E.R. 1977.
  Astrophys. J., **211**, 429.
Gregory, P.C. and Seaquist, E.R. 1974. *Nature*, **247**, 532.
Grewing, M. and Habing, H.J. 1972. *Mitt. Astr. Ges.*, No. 31, 190.
Griffin, R. and Thackeray, A.D. 1958. *Obs.*, **78**, 245.
Groube, W. 1946. *Bull. Soc. Astr. France*, **60**, 139.
Grygar, J. and Chochol, D. 1982.
  IAU Colloquium No. 70, The Nature of Symbiotic Stars,
  ed. M. Friedjung and R. Viotti (Dordrecht: Reidel), p. 165.
Grygar, J., Hric, L., and Chochol, D. 1977.
  in *IAU Colloquium No. 42, Interaction of Variable Stars with Their Environment,*
  ed. R. Kippenhahn, J. Rahe, and W. Strohmeier,
  Veroff. Bamberg Remeis-Sternw., **11**, 383.
Grygar, J., Hric, L., Chochol, D. and Mammano, A. 1979. *Bull. Astr.*
  Inst. Czech., **30**, 8.
Guerrieri, E. 1934. *Mem. Soc. Astr. Ital.*, **8**, 97.
Gusev, E.B. 1969a. *Astr. Tsirk.*, No. 508, 6.
Gusev, E.B. 1969b. *Astr. Tsirk.*, No. 511, 6.
Gusev, E.B. 1969c. *Astr. Tsirk.*, No. 518, 2.
Gusev, E.B. 1970. *Perem. Zvedzy*, **17**, 312.
Gusev, E.B. 1971. *Astr. Tsirk.*, No. 667, 6.
Gusev, E.B. 1976. *Astr. Tsirk.*, No. 901, 2.
Gusev, E.B. 1977. *Astr. Tsirk.*, No. 944, 4.
Gusev, E.B. and Caramish, V.F. 1970. *Perem. Zvedzy*, **17**, 305.
Hachenberg, O. and Wellman, P. 1939. *Zeit. f. Astrophys.*, **17**, 246.
Hack, M. 1979. *Nature*, **279**, 305.
Hack, M. 1982. in *Advances in Ultraviolet Astronomy*, ed. Y. Kondo,
  J.M. Mead and R.D. Chapman (NASA CP-2238), p. 89.
Hack, M., Persic, M. and Selvelli, P. 1982. in
  Third European IUE Conference,
  ed. E. Rolfe, A. Heck and B. Battrick (ESA SP-176), p. 193.
Hack, M., Rusconi, L., Sedmak, G., Engin, S. and Yilmaz, N. 1982.
  Astr. Astrophys., **113**, 250.
Hack, M. and Selvelli, P.L. 1982a. *Astr. Astrophys.*, **107**, 200.
Hack, M. and Selvelli, P.L. 1982b. in
  IAU Colloquium No. 70, The Nature of Symbiotic Stars,
  ed. M. Friedjung and R. Viotti (Dordrecht: Reidel), p. 131.
Hang, H.J. and Mo, C.E. 1963. *Acta Astr. Sinica*, **11**, 27.
Hanley, C.M. 1943. *Ann. Harv. Coll. Obs.*, **109**, No. 3.
Haro, G. 1952. *Bol. Obs. Tonan. y Tacub.*, **1**, 93.
Harris, S. and Scott, P.F. 1976. *Mon. Not. Roy. Astr.*
  Soc., **175**, 371.
Hartwig, E. 1901. *Astr. Nachr.*, No. 3744.
Hartwig, E. 1910. *Veroff. Remeis-Sternw. Bamberg*, **1**, 1.
Hartwig, E., Kempf, P. and Muller, G. 1919.
  Astr. Nachr., **208**, 57.
Harvey, P.M. 1974. *Astrophys. J.*, **188**, 95.
Harwood, M. 1925. *Bull. Harv. Coll. Obs.*, No. 826.

Hawkins, F.J., Mason, K.O. and Sanford, P.W. 1973. *Nature Phys. Sci.* **241**, 109.

Heck, A., Beeckmans, F., Benvenuti, P., Cassatella, A., Clavel, J., Macchetto, F., Penston, M.V., Selvelli, P.L. and Stickland, D. 1978. *Mess.*, No. 15, p. 27.

Heck, A. and Manfroid, J. 1982. *Mess.*, No. 30, p. 6.

Heck, A. and Manfroid, J. 1985. *Astr. Astrophys.* **142**, 341.

Henize, K.G. 1952. *Astrophys. J.*, **115**, 133.

Henize, K.G. 1967. *Astrophys. J. Suppl.*, **14**, 125.

Henize, K.G. 1976. *Astrophys. J. Suppl.*, **30**, 491.

Henize, K.G. and Carlson, E.D. 1980. *Pub. Astr. Soc. Pac.*, **92**, 479.

Henize, K.G. and McLaughlin, D.B. 1951. *Astrophys. J.*, **114**, 163.

Henrichs, H.F. 1975. *Astr. Astrophys.*, **44**, 41.

Herbig, G.H. 1950. *Pub. Astr. Soc. Pac.*, **62**, 211.

Herbig, G.H. 1960. *Astrophys. J.*, **131**, 632.

Herbig, G.H. 1965. *Kleine Veroff. Remeis-Sternw. Bamberg*, **4**, 164.

Herbig, G.H. 1969. *Proc. Nat. Acad. Sci.*, **63**, 1045.

Herbig, G.H. and Hoffleit, D. 1975. *Astrophys. J.*, **202**, L41.

Hetzler, C. 1940. *Astrophys. J.*, **92**, 58.

Hicks, P.D. 1973. *Inf. Bull. Var. Stars*, No. 804.

Higgs, L.A. 1971. *Catalogue of Radio Observations of Planetary Nebulae and Related Optical Data*, Publication of the Astronomical Branch of the Radio and Electronics Division, NRC Canada, **1**, No. 1.

Himpel, K. 1938a. *Astr. Nachr.*, **267**, 55.

Himpel, K. 1938b. *Beob. Zirk.*, No. 13.

Himpel, K. 1940. *Astr. Nachr.*, **270**, 183.

Hind, J.R. 1866. *Astr. Reg.*, **4**, 229.

Hjellming, R.M. 1981. in *Proceedings of the North American Workshop on Symbiotic Stars*, ed. R.E. Stencel (Boulder: JILA), p. 15.

Hjellming, R.M. 1985. in *Radio Stars*, ed. R.M. Hjellming and O.M. Gibson (Dordrecht: Reidel), p. 97.

Hjellming, R.M. and Bignell, R.C. 1982. *Science*, **216**, 1279.

Hobbs, R.W. and Marionni, P. 1971. *Astrophys. J.*, **167**, 85.

Hobbs, R.W., Michalitsianos, A.G. and Kafatos, M. 1980. in *The Universe at Ultraviolet Wavelengths*, ed. R.D. Chapman (NASA CP-2171), p. 355.

Hoffleit, D. 1968a. *Irish Astr. J.*, **8**, 149.

Hoffleit, D. 1968b. *Inf. Bull. Var. Stars*, No. 254.

Hoffleit, D. 1970. *Inf. Bull. Var. Stars*, No. 469.

Hoffmeister, C. 1919. *Astr. Nachr.*, **208**, 239.

Hoffmeister, C. 1932. *Astr. Nachr.*, **247**, 281.

Hoffmeister, C. 1949. *Erg. Astr. Nachr.*, **12**, No. 1.

Hogg, F.S. 1932. *Pub. Astr. Soc. Pac.*, **44**, 328.

Hogg, F.S. 1934. *Pub. Am. Astr. Soc.*, **8**, 14.

Hollis, J.M., Kafatos, M., Michalitsianos, A.G. and McAlister, H.A. 1985. *Astrophys. J.*, **289**, 765.

Honda, M., Ishida, K., Naguchi, T., Norimoto, Y., Nakagiri, M., Sugano, T. and Yamashita, Y. 1979. *Tokyo Astr. Bull.*, No. 262.

Hopp, U. and Witzigmann, S. 1978. *Inf. Bull. Var. Stars*, No. 1494.
Hopp, U. and Witzigmann, S. 1981. *Inf. Bull. Var. Stars*, No. 2048.
Howarth, M. 1937. *Pop. Astr.*, **45**, 38.
Hric, L., Chochol, D. and Grygar, J. 1978.
    Inf. Bull. Var. Stars, No. 1525
Hric, L., Chochol, D. and Grygar, J. 1980.
    Inf. Bull. Var. Stars, No. 1835
Hron, J. and Maitzen, H.M. 1983. *Inf. Bull. Var. Stars*, No. 2273.
Huang, C.-C. 1982a. in *IAU Colloquium No. 70, The Nature of Symbiotic Stars*,
    ed. M. Friedjung and R. Viotti (Dordrecht: Reidel), p. 151.
Huang, C.-C. 1982b. in *IAU Colloquium No. 70, The Nature of Symbiotic Stars*,
    ed. M. Friedjung and R. Viotti (Dordrecht: Reidel), p. 185.
Huang, C.-C. 1984. *Astr. Astrophys.*, **135**, 410.
Hubscher, J. 1983. *BAV Rundbrief*, **32**, 130.
Huggins, W. 1866a. *Mon. Not. Roy. Astr. Soc.*, **26**, 275.
Huggins, W. 1866b. *Quart. J. Sci.*, **3**, 376.
Huggins, W. 1866c. *Astr. Nachr.*, **67**, 29.
Huggins, W. 1866d. *Astr. Nachr.*, **67**, 125.
Humason, M.L. 1938. *Astrophys. J.*, **88**, 228.
Huruhata, M. 1980. *Inf. Bull. Var. Stars*, No. 1896.
Hutchings, J.B. and Cowley, A.P. 1982.
    Pub. Astr. Soc. Pac., **94**, 107.
Hutchings, J.B., Cowley, A.P., Ake, T.B. and Imhoff, C.L. 1983.
    Astrophys. J., **275**, 271.
Hutchings, J.B., Cowley, A.P. and Redman, R.O. 1975.
    Astrophys. J., **201**, 404.
Hutchings, J.B. and Redman, R.O. 1972.
    Pub. Astr. Soc. Pac., **84**, 240.
Ianna, P.A. 1964. *Astrophys. J.*, **139**, 780.
Ichimura, K., Shimuzu, Y., Nakagiri, M. and Yamashita, Y. 1979.
    Tokyo Astr. Bull., *2nd ser.*, **No. 258**, 2953.
Iijima, T. 1981a. *Astr. Astrophys.*, **94**, 290.
Iijima, T. 1981b. in *Photometric and Spectroscopic Binary Systems*,
    eds. E.B. Carling and Z. Kopal, (Dordrecht: Reidel), p. 517.
Iijima, T. 1982. *Astr. Astrophys.*, **116**, 210.
Iijima, T. 1984. *Inf. Bull. Var. Stars*, No. 2491.
Iijima, T. and Mammano, A. 1981. *Astrophys. Space Sci.*, **79**, 55.
Iijima, T., Mammano, A. and Margoni, R. 1980. *Mem. Soc. Astr. Ital.*
    51, 683.
Iijima, T., Mammano, A. and Margoni, R. 1981.
    Astrophys. Space Sci., **75**, 237.
Iijima, T. and Ortolani, S. 1984. *Astr. Astrophys.* **136**, 1.
Ilovaisky, S.A. and Spinrad, H. 1966.
    Pub. Astr. Soc. Pac., **78**, 527.
Ilovaisky, S.A. and Wallerstein, G. 1968.
    Pub. Astr. Soc. Pac., **80**, 155.
Ipatov, A.P., Taranova, O.G., and Yudin, B.F. 1984.
    Astr. Astrophys., **135**, 325.

Ipatov, A.P., Taranova, O.G., and Yudin, B.F. 1985.
Astr. Astrophys., **142**, 85.

Ipatov, A.P. and Yudin, B.F. 1981. *Soviet Astr. Lett.*, **7**, 238.

Ipatov, A.P. and Yudin, B.F. 1983. *Soviet Astr. Lett.*, **9**, 222.

Isles, J.E. 1972. *Sky and Telescope*, **44**, 204.

Isles, J.E. 1974. *J. Brit. Astr. Assoc.*, **84**, 203.

Ivanova, N.L. 1960. *Soobshch. Byurakan Astrofiz. Obs.*, No. 28, 17.

Ivanova, N.L. 1963. *Soobshch. Byurakan Astrofiz. Obs.*, No. 34, 93.

Ivanova, N.L. 1969. *Astr. Tsirk.*, No. 508, 4.

Ivanova, N.L. and Khudyakova, T.N. 1980.
Soobshch. Byurakan Astrofiz. Obs., No. 52, 52.

Jacchia, L. 1934. *Coelum*, **4**, 63.

Jacchia, L. 1941. *Bull. Harv. Coll. Obs.*, No. 915.

Jacobsen, T.S. and Wallerstein, G. 1975.
Pub. Astr. Soc. Pac., **87**, 269.

Jakovkin, N. 1946. *Astr. Tsirk.*, No. 51.

Janssen, E.M. and Vyssotsky, A.N. 1943.
Pub. Astr. Soc. Pac., **55**, 244.

Jarzebowski, T. 1964. *Acta Astr.*, **14**, 77.

Johansson, S. 1983. *Mon. Not. Roy. Astr. Soc.*, **205**, 71p.

Johnson, H.M. 1980. *Astrophys. J.*, **237**, 840.

Johnson, H.M. 1981. *Astrophys. J.*, **244**, 551.

Johnson, H.M. 1982. *Astrophys. J.*, **253**, 224.

Johnson, H.M., Balick, B. and Thompson, A.R. 1979.
Astrophys. J., **233**, 919.

Johnson, H.M. and Golson, J.C. 1969.
Astrophys. J. (Letters), **155**, L91.

Joy, A.H. 1931. *Pub. Astr. Soc. Pac.*, **43**, 353.

Joy, A.H. 1942. *Astrophys. J.*, **96**, 344.

Joy, A.H. 1961. *Astrophys. J.*, **133**, 493.

Joy, A.H. and Swings, P. 1945. *Astrophys. J.*, **102**, 353.

Kafatos, M. 1983. *Obs.*, **103**, 51.

Kafatos, M., Hollis, J.M. and Michalitsianos, A.G. 1983.
Astrophys. J. (Letters), **267**, L103.

Kafatos, M. and Michalitsianos, A.G. 1982a. in
IAU Colloquium No. 70, The Nature of Symbiotic Stars,
ed. M. Friedjung and R. Viotti (Dordrecht: Reidel), p. 203.

Kafatos, M. and Michalitsianos, A.G. 1982b. in
Advances in Ultraviolet Astronomy,
ed. Y. Kondo, J.M. Mead and R.D. Chapman (NASA CP-2238), p. 452.

Kafatos, M. and Michalitsianos, A.G. 1982c. *Nature*, **298**, 540.

Kafatos, M., Michalitsianos, A.G., Allen, D.A., and Stencel, R.E. 1983.
Astrophys. J., **275**, 584.

Kafatos, M., Michalitsianos, A.G., and Feibelman, W.A. 1982.
Astrophys. J., **257**, 204.

Kafatos, M., Michalitsianos, A.G., and Hobbs, R.W. 1980a.
in *The Universe at Ultraviolet Wavelengths*,
ed. R.D. Chapman (NASA CP-2171), p. 349.

Kafatos, M., Michalitsianos, A.G., and Hobbs, R.W. 1980b.
Astrophys. J., **240**, 114.

Kaler, J.B. 1981. *Astrophys. J.*, **245**, 568.

Kaler, J.B. 1983. *Astrophys. J.*, **264**, 594.

Kaler, J.B. and Hickey, J.P. 1983. *Pub. Astr. Soc. Pac.*, **95**, 759.

Kaler, J.B., Kenyon, S.J., and Hickey, J.P. 1983.
Pub. Astr. Soc. Pac., **95**, 1006.

Kenyon, S.J. 1982. *Pub. Astr. Soc. Pac.*, **94**, 165.

Kenyon, S.J. 1983. Ph. D. thesis, Univ. Illinois.

Kenyon, S.J. and Bateson, F. M. 1984. *Pub. Astr. Soc. Pac.*, **96**, 321.

Kenyon, S.J. and Gallagher, J.S. 1983. *Astr. J.*, **88**, 666.

Kenyon, S.J., Michalitsianos, A.G., Lutz, J.H., and Kafatos, M. 1985.
Pub. Astr. Soc. Pac., **97**, 268.

Kenyon, S.J. and Truran, J.W. 1983. *Astrophys. J.*, **273**, 280.

Kenyon, S.J. and Webbink, R.F. 1984. *Astrophys. J.*, **279**, 252.

Kenyon, S.J., Webbink, R.F., Gallagher, J.S. and Truran, J.W. 1982.
Astr. Astrophys., **106**, 109.

Keyes, C.D. and Plavec, M.J. 1980a.
in *The Universe at Ultraviolet Wavelengths*,
ed. R.D. Chapman (NASA CP-2171), p. 443.

Keyes, C.D. and Plavec, M.J. 1980b.
in *IAU Symposium No. 88*,
Close Binary Stars: Observations and Interpretation,
ed. M.J. Plavec, D.M. Popper and R.K. Ulrich
(Dordrecht: Reidel), p. 365.

Kharadze, E.K. 1946. *Astr. Tsirk.*, No. 52.

Khozov, G.V. and Khudyakova, T.N. 1974.
Trudy Astr. Obs. Lenin., **30**, 48.

Khozov, G.V., Khudyakova, T.N. and Larionova, L.V. 1975.
Trudy Astr. Obs. Lenin., **31**, 123.

Khozov, G.V., Khudyakova, T.N., Larionova, L.V. and Larionov, V.M. 1977.
Trudy Astr. Obs. Lenin., **33**, 26.

Khozov, G.V., Khudyakova, T.N., Larionova, L.V. and Larionov, V.M. 1978.
Trudy Astr. Obs. Lenin., **34**, 68.

Khozov, G.V., Khudyakova, T.N. and Nikitin, S.N. 1976.
Trudy Astr. Obs. Lenin., **32**, 61.

Khudyakova, T.N. 1985. *Pis'ma Astr. Zh.*, **11**, 623.

Kindl, C., Marxer, N. and Nussbaumer, H. 1982.
Astr. Astrophys., **116**, 265.

Kindl, C. and Nussbaumer, H. 1982a.
in *IAU Colloquium No. 70, The Nature of Symbiotic Stars*,
ed. M. Friedjung and R. Viotti (Dordrecht: Reidel), p. 175.

Kindl, C. and Nussbaumer, H. 1982b.
in *IAU Colloquium No. 70, The Nature of Symbiotic Stars*,
ed. M. Friedjung and R. Viotti (Dordrecht: Reidel), p. 213.

Klutz, M. 1979. *Astr. Astrophys.*, **73**, 244.

Klutz, M., Simonetto, O. and Swings, J.P. 1978.
Astr. Astrophys., **66**, 283.

Klutz, M. and Swings, J.P. 1981. *Astr. Astrophys.*, **96**, 406.

Klyus, I.A. 1978. *Perem. Zvedzy*, **21**, 132.

Klyus, I.A. 1982. *Astr. Tsirk.*, No. 1226.

Knacke, R.F. 1972. *Astrophys. Lett.*, **11**, 201.

Knapp, G.R. 1985. *Astrophys. J.*, **293**, 268.

Knapp, G.R. and Morris, M. 1985. *Astrophys. J.*, **292**, 640.

Knight, N.F.H. 1946. *J. Brit. Astr. Assoc.*, **56**, 74.

Knipe, G.F.G. 1969. *Mon. Notes Astr. Soc. South Africa*, **28**, 54.

Knott, G. 1866. *Astr. Reg.*, **4**, 267.

de Kock, R.P. 1948. *Mon. Notes Astr. Soc. S. Africa*, **7**, 74.

Kohoutek, L. 1965. *Bull. Astr. Inst. Czech.*, **16**, 221.

Kohoutek, L. 1969. *Inf. Bull. Var. Stars*, No. 384.

Kohoutek, L. and Bossen, H. 1970. *Astrophys. Lett.*, **6**, 157.

Kohoutek, L. and Martin, W. 1981. *Astr. Astrophys.*, **94**, 365.

Kohoutek, L., Pekny, Z., and Perek, L. 1965.
Bull. Astr. Inst. Czech., **16**, 189.

Kolotilov, E.A. 1971. *Astr. Tsirk.*, No. 624, 2.

Kolotilov, E.A. 1975. *Astr. Tsirk.*, No. 866.

Kolotilov, E.A. 1983a. *Soviet Astr. J.*, **27**, 188.

Kolotilov, E.A. 1983b. *Soviet Astr. J.*, **27**, 432.

Kolotilov, E.A. and Belyakina, T.S. 1982. *Inf. Bull. Var. Stars*, No. 2077.

Kolotilov, E.A. and Shenavrin, V.P. 1976. *Astr. Tsirk.*, No. 918, 1.

Koo, J.-W.C. and Haymes, R.C. 1980.
Astrophys. J. (Letters), **239**, L57.

Kraft, R.E. 1958. *Astrophys. J.*, **127**, 625.

Kraft, R.E. 1964. *Astrophys. J.*, **139**, 457.

Kukarkin, B.V., Parenago, P.P., Efremov, Yu. N., and Kholopov, P.N. 1951.
Catalogue of Suspected Variable Stars,
(Moscow : Soviet Academy of Sciences).

Kurochkin, N.E. 1976. *Soviet Astr. Lett.*, **2**, 169.

Kurochkin, N.E. and Starikova, G.A. 1958. *Astr. Tsirk.*, No. 194.

Kwok, S. 1982a. in *IAU Colloquium No. 70, The Nature of Symbiotic Stars*,
ed. M. Friedjung and R. Viotti (Dordrecht: Reidel), p. 17.

Kwok, S. 1982b. in *IAU Colloquium No. 70, The Nature of Symbiotic Stars*,
ed. M. Friedjung and R. Viotti (Dordrecht: Reidel), p. 209.

Kwok, S., Bignell, N., and Purton, C.R. 1984.
Astrophys. J., **279**, 188.

Kwok, S. and Purton, C.R. 1979. *Astrophys. J.*, **229**, 187.

Kwok, S., Purton, C.R., and Keenan, D.W. 1981.
Astrophys. J., **250**, 232.

Lacchini, G.B. 1933. *Astr. Nachr.*, **248**, 365.

Lambert, D.L., Slovak, M.H., Shields, G.A., and Ferland, G.J. 1980.
in *The Universe at Ultraviolet Wavelengths*, ed. R.D. Chapman
(NASA CP-2171), p. 461.

Lambert, D.L. and Thackeray, A.D. 1969. *Astrophys. Space Sci.*, **5**, 283.

Lampland, C.O. 1922. *Pub. Astr. Soc. Pac.*, **34**, 218.

Landolt, A.U. 1975. *Pub. Astr. Soc. Pac.*, **87**, 265.

Larsson-Leander, G. 1947. *Ark. f. Mat., Astr. and Fys.*, **34**, 16.

Lawrence, G.M., Ostriker, J.P. and Hesser, J.E. 1967.
Astrophys. J. (Letters), **148**, L161.

Lee, S.G. 1973. *Inf. Bull. Var. Stars*, No. 813.

Lepine, J.R.D., LeSqueren, A.M., and Scalise, E. 1978.
Astrophys. J., **225**, 869.

Lepine, J.R.D. and Rien, N.Q. 1974. *Astr. Astrophys.*, **36**, 469.

LeVerrier, M. 1866. *Comptes Rendus*, **62**, 1108.

Lewin, W.H.G., Ricker, G.R., and McClintock, J.E. 1971.
Astrophys. J. (Letters), **169**, L17.

Liller, M.H. and Liller, W. 1979. *Astr. J.*, **84**, 1357.

Lindsay, E.M. 1932. *Bull. Harv. Coll. Obs.*, No. 888.

Lockwood, G.W. 1972. *Astrophys. J. Suppl.*, **24**, 375.

Lockyer, A.N. 1892. *Phil. Trans. Roy. Soc.*, **A182**, 408.

Longmore, A.J. and Allen, D.A. 1977. *Astrophys. Lett.*, **18**, 159.

Lopez, L.A. and Sahade, J. 1968. *Bol. Acos. Argent. Astr.*, No. 14, 68.

Lorenzetti, D., Saraceno, P., and Strafella, F. 1985. *Astrophys. J.*, **298**, 350.

Loreta, E. 1931. *Bull. Lyon Obs.*, **13**, 6.

Loreta, E. 1933. *Gaz. Astr.*, **20**, 153.

Loreta, E. 1934a. *Gaz. Astr.*, **21**, 41.

Loreta, E. 1934b. *Bull. Soc. Astr. Fran.*, **48**, 86.

Loreta, E. 1934c. *Mem. Soc. Astr. Ital.*, **8**, 235.

Low, F.J. 1970. *Sky Survey, Semi-annual technical report,*
15 March 1970, (U. Ariz.: AFCR-70-017).

Lukatskaya, F.I. 1973. *Peremennye. Zvedzy*, **19**, 253.

Lundmark, K. 1921a. *Astr. Nachr.*, **213**, 93.

Lundmark, K. 1921b. *Pub. Astr. Soc. Pac.*, **33**, 271.

Lundmark, K. 1922. *Pub. Astr. Soc. Pac.*, **34**, 225.

Luthardt, R. 1984. *Inf. Bull. Var. Stars*, No. 2495.

Luthardt, R. 1985. *Inf. Bull. Var. Stars*, No. 2789.

Lutz, J.H. 1977a. *Astr. Astrophys.*, **60**, 93.

Lutz, J.H. 1977b. *Pub. Astr. Soc. Pac.*, **89**, 10.

Lutz, J.H. 1984. *Astrophys. J.*, **279**, 714.

Lutz, J.H. and Kaler, J.B. 1983. *Pub. Astr. Soc. Pac.*, **95**, 739.

Lutz, J.H. and Lutz, T.E. 1981. in
Proceedings of the North American Workshop on Symbiotic Stars,
ed. R.E. Stencel (Boulder: JILA), p. 28.

Lutz, J.H., Lutz, T.E., Kaler, J.B., Osterbrock, D.E., and
Gregory, S.A. 1976. *Astrophys. J.*, **203**, 481.

Luud, L. 1970. *Eesti NSV Teaduste Akadeemia Toimetised,*
Fuusika-Matemaatika, **19**, 177.

Luud, L. 1973. *Pub. Tartu Astrofiz. Obs*, **40**, 163.

Luud, L. 1979. in *IAU Colloquium No. 53,*
White Dwarfs and Variable Degenerate Stars,
ed. H.M. Van Horn and V. Weidemann (Rochester: Univ. Rochester Press), p. 459.

Luud, L. 1980a. *Astrofizika*, **16**, 443.

Luud, L. 1980b. *Astr. Tsirk.*, No. 1098.

Luud, L. and Il'mas, M. 1965. *Pub. Tartu Astrofiz. Obs*, **35**, 176.

Luud, L. and Il'mas, M. 1966. *Astrofizika*, **2**, 205.

Luud, L., Ruusalepp, M., and Kuusk, T. 1971. *Pub. Tartu Astrofiz. Obs.*, **39**, 106.

Luud, L., Ruusalepp, M., and Vennik, J. 1977. *Pub. Tartu*
Astrofiz. Obs., **45**, 113.

Luud, L. and Tomov, T. 1984. *Astrofizika*, **20**, 419.

Luud, L., Tomov, T., Vennik, J., and Panov, K. 1982.
Soviet Astr. Lett., **8**, 257.

Luud, L., Vennik, J., and Pehk, M. 1978a.
Soviet Astr. Lett., **4**, 46.

Luud, L., Vennik, J., and Pehk, M. 1978b.
Pub. Tartu Astrofiz. Obs., **46**, 77.

Luyten, W.J. 1927. *Bull. Harv. Coll. Obs.*, No. 852.

Luyten, W.J. 1935. *Astr. Nach.*, **258**, 121.

Lynn, W.T. 1866a. *Astr. Reg.*, **4**, 218.

Lynn, W.T. 1866b. *Astr. Nachr.*, **68**, 105.

Lynn, W.T. 1866c. *Astr. Reg.*, **4**, 231.

Lynn, W.T. 1866d. *Astr. Reg.*, **4**, 310.

MacConnell, D.J. 1972. *Inf. Bull. Var. Stars*, No. 734.

MacConnell, D.J. 1978. *Astr. Astrophys. Suppl.*, **33**, 219.

McCall, A. and Hough, J.H. 1980. *Astr. Astrophys. Suppl.*, **42**, 141.

McCuskey, S.W. 1961. *Pub. Astr. Soc. Pac.*, **73**, 264.

McLaughlin, D.B. 1933. *Pub. Mich. Obs.*, **5**, 119.

McLaughlin, D.B. 1939. *Pop. Astr.*, **47**, 353.

McLaughlin, D.B. 1946. *Pub. Astr. Soc. Pac.*, **58**, 46.

McLaughlin, D.B. 1947. *Pub. Astr. Soc. Pac.*, **63**, 8.

McLaughlin, D.B. 1953. *Astrophys. J.*, **117**, 279.

Maehara, H. and Yamashita, Y. 1978.
Pub. Astr. Soc. Japan, **17**, 93.

Magalashvili, N.L. and Kumsishvili, I.I. 1958. *Astr. Tsirk.*, No. 199.

Mahra, H.S., Joshi, S.C., Srivastava, J.B., and Dhir, S.L. 1979.
Inf. Bull. Var. Stars, No. 1683.

Mammano, A. and Chincarini, G. 1966.
Mem. Soc. Astr. Ital., **37**, 421.

Mammano, A. and Ciatti, F. 1975. *Astr. Astrophys.*, **39**, 405.

Mammano, A. and Martini, A. 1969. in
Non-periodic Phenomena in Variable Stars,
ed. L. Detre (Budapest: Academic Press), p. 415.

Mammano, A. and Righini, G.M. 1973.
Mem. Soc. Astr. Ital., **44**, 23.

Mammano, A., Rosino, L., and Yildizdogdu, S. 1975. in *Variable Stars
and Stellar Evolution*, IAU Symposium No. 67, ed. V.E. Sherwood and
L. Plaut (Dordrecht: Reidel), p. 415.

Mammano, A. and Taffara, S. 1978.
Astr. Astrophys. Suppl., **34**, 211.

Marsh, K.A. 1976. *Astrophys. J.*, **203**, 552.

Marsh, K.A., Purton, C.R., and Feldman, P.A. 1976. *Astr. Astrophys.*, **49**, 211.

Martini, A. 1968. *Mem. Soc. Astr. Ital.*, **39**, 633.

Mason, K.O. 1977. *Mon. Not. Roy. Astr. Soc.*, **178**, 81P.

Mattei, J.A. 1972. *J. Am. Assoc. Var. Star Obs.*, **1**, 50.

Mattei, J.A. 1974. *J. Roy. Astr. Soc. Can.*, **68**, 221.

Mattei, J.A. 1976. *J. Roy. Astr. Soc. Can.*, **70**, 325.

Mattei, J.A. 1978. *J. Roy. Astr. Soc. Can.*, **72**, 61.

Mattei, J.A. and Allen, J. 1979. *J. Roy. Astr. Soc. Can.*, **73**, 173.

Mayall, M.W. 1937. *Ann. Harv. Coll. Obs.*, **105**, 491.
Mayall, M.W. 1940. *Bull. Harv. Coll. Obs.*, No. 913.
Mayall, M.W. 1949. *Astr. J.*, **54**, 191.
Mayall, M.W. 1952. *Bull. Harv. Coll. Obs.*, No. 920, 32.
Mayall, M.W. 1960. *J. Roy. Astr. Soc. Can.*, **54**, 43.
Mayall, M.W. 1969. *J. Roy. Astr. Soc. Can.*, **63**, 321.
Meinunger, L.V. 1966. *Mitt. Verand. Sterne*, **3**, 111.
Meinunger, L.V. 1970. *Mitt. Verand. Sterne*, **5**, 111.
Meinunger, L. 1979. *Inf. Bull. Var. Stars*, No. 1611.
Meinunger, L. 1981. *Inf. Bull. Var. Stars*, No. 2016.
Mendoza, E.E. 1972. *Bol. Obs. Tonan. Tacub.*, **6**, 211.
Menzies, J.W., Coulson, I.M., Caldwell, J.A.R., and Corben, P.M. 1982.
    Mon. Not. Roy. Astr. Soc., **200**, 463.
Merlin, P. 1972. *Comptes Rendus*, **275**, 45.
Merlin, P. 1973. *Astr. Astrophys.*, **23**, 363.
Merrill, K.M. and Stein, W.A. 1976.
    Pub. Astr. Soc. Pac., **88**, 285.
Merrill, P.W. 1916. *Pub. Michigan Obs.*, **2**, 71.
Merrill, P.W. 1919. *Pub. Astr. Soc. Pac.*, **31**, 305.
Merrill, P.W. 1920. *Pub. Astr. Soc. Pac.*, **32**, 247.
Merrill, P.W. 1921. *Astrophys. J.*, **53**, 375.
Merrill, P.W. 1922. *Pub. Astr. Soc. Pac.*, **34**, 134.
Merrill, P.W. 1923. *Astrophys. J.*, **58**, 215.
Merrill, P.W. 1926. *Pub. Astr. Soc. Pac.*, **38**, 175.
Merrill, P.W. 1927a. *Pub. Astr. Soc. Pac.*, **39**, 48.
Merrill, P.W. 1927b. *Astrophys. J.*, **65**, 286.
Merrill, P.W. 1929a. *Pub. Astr. Soc. Pac.*, **41**, 255.
Merrill, P.W. 1929b. *Astrophys. J.*, **69**, 330.
Merrill, P.W. 1932. *Astrophys. J.*, **75**, 413.
Merrill, P.W. 1933. *Astrophys. J.*, **77**, 44.
Merrill, P.W. 1934. *Pub. Astr. Soc. Pac.*, **46**, 296.
Merrill, P.W. 1935. *Astrophys. J.*, **81**, 312.
Merrill, P.W. 1936. *Astrophys. J.*, **83**, 272.
Merrill, P.W. 1940. *Spectra of Long Period Variable Stars*
    (Chicago: University of Chicago Press).
Merrill, P.W. 1941a. *Pub. Astr. Soc. Pac.*, **53**, 124.
Merrill, P.W. 1941b. *Pub. Am. Astr. Soc.*, **10**, 168.
Merrill, P.W. 1941c. *Astrophys. J.*, **93**, 40.
Merrill, P.W. 1941d. *Astrophys. J.*, **93**, 383.
Merrill, P.W. 1942. *Astrophys. J.*, **95**, 386.
Merrill, P.W. 1943. *Astrophys. J.*, **98**, 334.
Merrill, P.W. 1944. *Astrophys. J.*, **99**, 15.
Merrill, P.W. 1947. *Astrophys. J.*, **105**, 120.
Merrill, P.W. 1948. *Astrophys. J.*, **107**, 317.
Merrill, P.W. 1950a. *Astrophys. J.*, **111**, 484.
Merrill, P.W. 1950b. *Astrophys. J.*, **112**, 504.
Merrill, P.W. 1951a. *Astrophys. J.*, **113**, 605.
Merrill, P.W. 1951b. *Astrophys. J.*, **114**, 338.

Merrill, P.W. 1959. *Astrophys. J.*, **129**, 44.

Merrill, P.W. and Bowen, I.S. 1951. *Pub. Astr. Soc. Pac.*, **63**, 255.

Merrill, P.W. and Burwell, C.G. 1933. *Astrophys. J.*, **78**, 87.

Merrill, P.W. and Burwell, C.G. 1943. *Astrophys. J.*, **98**, 153.

Merrill, P.W. and Burwell, C.G. 1950. *Astrophys. J.*, **112**, 72.

Merrill, P.W. and Humason, M.L. 1932. *Pub. Astr. Soc. Pac.*, **44**, 56.

Merrill, P.W., Humason, M.L. and Burwell, C.G. 1932.
Astrophys. J., **76**, 156.

Mianes, P. 1965. *J. des Observ.*, **48**, 103.

Michalitsianos, A.G., Hobbs, R.W., and Kafatos, M. 1980. in
The Universe at Ultraviolet Wavelengths,
ed. R.D. Chapman (NASA CP-2171), p. 367.

Michalitsianos, A.G. and Kafatos, M. 1982a.
Astrophys. J. (Letters), **262**, L47.

Michalitsianos, A.G. and Kafatos, M. 1982b.
in *IAU Colloquium No. 70, The Nature of Symbiotic Stars*,
ed. M. Friedjung and R. Viotti (Dordrecht: Reidel), p. 191.

Michalitsianos, A.G. and Kafatos, M. 1984.
Mon. Not. Roy. Astr. Soc., **207**, 575.

Michalitsianos, A.G., Kafatos, M., and Hobbs, R.W. 1980.
Astrophys. J., **237**, 506.

Michalitsianos, A.G., Kafatos, M., Hobbs, R.W., and Maran, S.P. 1980.
Nature, **284**, 148.

Michalitsianos, A.G., Kafatos, M., Feibelman, W.A., and Hobbs, R.W.
1982a. *Astrophys. J.*, **253**, 735.

Michalitsianos, A.G., Kafatos, M., Feibelman, W.A., and Wallerstein, G. 1982b.
Astr. Astrophys., **109**, 136.

Michalitsianos, A.G., Kafatos, M., Stencel, R.E., and Boyarchuk, A.A. 1982c.
in *IAU Colloquium No. 70, The Nature of Symbiotic Stars*,
ed. M. Friedjung and R. Viotti (Dordrecht: Reidel), p. 141.

Mikolajewska, J. 1983. *Inf. Bull. Var. Stars*, No. 2355.

Mikolajewska, J. and Mikolajewski, M. 1980a.
Acta Astr., **30**, 347.

Mikolajewska, J. and Mikolajewski, M. 1980b.
Inf. Bull. Var. Stars, No. 1846.

Mikolajewska, J. and Mikolajewski, M. 1982. in
IAU Colloquium No. 70, The Nature of Symbiotic Stars,
ed. M. Friedjung and R. Viotti (Dordrecht: Reidel), p. 147.

Mikolajewska, J. and Mikolajewski, M. 1983.
Acta Astr., **33**, 403.

Mikolajewska, J., Mikolajewski, M. and Krelowski, J. 1983.
Inf. Bull. Var. Stars, No. 2356.

Miller, W.J. 1967. *Spec. Vat. Ric. Astr.*, **7**, 241.

Milne, D.K. 1979. *Astr. Astrophys. Suppl.*, **36**, 227.

Minkowski, R. 1943. *Pub. Astr. Soc. Pac.*, **55**, 101.

Minkowski, R. 1946. *Pub. Astr. Soc. Pac.*, **58**, 305.

Minkowski, R. 1948. *Pub. Astr. Soc. Pac.*, **60**, 386.

Mirzoyan, L.V. 1955. *Dokl. Akad. Nauk SSR*, **105**, 928.

Mirzoyan, L.V. 1956. *Izv. Byurakan Astrofiz. Obs.*, No. 19, 43.

Mirzoyan, L.V. 1958a. *Mem. Soc. Roy. Sci. Liege*, 4eme serie,
tome XX, 274.

Mirzoyan, L.V. 1958b. *Dokl. Akad. Nauk SSR*, **119**, 667.

Mirzoyan, L.V. and Bartaya, R.A. 1960. *Izv. Abastumani
Astrofiz. Obs.*, No. 25, 121.

Mitchell, W.E. 1956. *Astr. J.*, **61**, 187.

Mjalkovskij, M.I. 1977. *Peremennye Zvedzy Prilozhenie*, **3**, 71.

Mochnacki, S.W. and Starkman, G. 1985.
Pub. Astr. Soc. Pac., **97**, 151.

Model, A. and Lochel, K. 1942. **2**, 129.

Moore, J.H. 1919. *Pub. Astr. Soc. Pac.*, **31**, 909.

Morgan, W.W. and Deutsch, A.J. 1947. *Astrophys. J.*, **106**, 362.

Moroz, V.I. 1966. *Astr. Zh.*, **43**, 63.

Morrison, D. and Simon, T. 1973. *Astrophys. J.*, **186**, 193.

Mueller, B.E.A. and Nussbaumer, H. 1985.
Astr. Astrophys., **145**, 144.

Muller, G. and Kempf, P. 1907. *Publ. Potsdam Obs.*, **17**, 292.

Mumford, G.S. 1956. *Pub. Astr. Soc. Pac.*, **68**, 538.

Mumford, G.S. 1966. *Astrophys. J.*, **146**, 962.

Muratorio, G. and Friedjung, M. 1982a. in
IAU Colloquium No. 70, The Nature of Symbiotic Stars,
ed. M. Friedjung and R. Viotti (Dordrecht: Reidel), p. 83.

Muratorio, G. and Friedjung, M. 1982b. in
IAU Colloquium No. 70, The Nature of Symbiotic Stars,
ed. M. Friedjung and R. Viotti (Dordrecht: Reidel), p. 161.

Mustel, E.R., Boyarchuk, A.A., and Bartash, T.M. 1963. *Izv.
Krym. Astrofiz. Obs.*, **30**, 19.

Nakagiri, M. and Yamashita, Y. 1980.
Tokyo Astr. Bull., No. 263.

Nakagiri, M. and Yamashita, Y. 1982a.
Ann. Tokyo Obs., 2nd ser., **19**, 1.

Nakagiri, M. and Yamashita, Y. 1982b.
Ann. Tokyo Obs., 2nd ser., **19**, 8.

Nassau, J.J. and Blanco, V.M. 1954. *Astrophys. J.*, **120**, 129.

Nassau, J.J. and Cameron, D.M. 1954. *Astrophys. J.*, **119**, 75.

Netzer, H., Leibowitz, E.M., and Ferland, G. 1983.
in *IAU Colloquium No. 72, Cataclysmic Variables and Related Objects*,
ed. M. Livio and G. Shaviv (Dordrecht: Reidel), p. 35.

Newell, R.T. and Hjellming, R.M. 1981.
in *Proceedings of the North American Workshop on Symbiotic Stars*,
ed. R.E. Stencel (Boulder: JILA), p. 16.

Neugebauer, G. and Leighton, R.B. 1969. in
Two Micron Sky Survey, A Preliminary Catalog, NASA SP-3047.

Nicholson, S.B. and Pettit, E. 1922.
Pub. Astr. Soc. Pac., **34**, 290.

Nikitin, S.N. and Khudyakova, T.N. 1979.
Soviet Astr. Lett, **5**, 327.

Noguchi, K. Maihara, T., Okuda, H. and Sato, S. 1977.
Pub. Astr. Soc. Japan, **29**, 511.
Noskova, R.I., Saveljeva, M.V., Arhipova, V.P., Goranskij, V.P.
and Dokuchaeva, O.D. 1979. *Peremennye Zvezdy Prilozhenie*, **3**, 755.
Nussbaumer, H. 1982a. in
IAU Colloquium No. 70, The Nature of Symbiotic Stars,
ed. M. Friedjung and R. Viotti (Dordrecht: Reidel), p. 85.
Nussbaumer, H. 1982b. in *Third European IUE Conference*, ed.
E. Rolfe, A. Heck and B. Battrick (ESA SP-176), p. 201.
Nussbaumer, H. and Schild, H. 1981. *Astr. Astrophys.*, **101**, 118.
Nussbaumer, H. and Schmutz, W. 1983. *Astr. Astrophys.*, **126**, 59.
O'Dell, C.R. 1963. *Astrophys. J.*, **138**, 1018.
O'Dell, C.R. 1966. *Astrophys. J.*, **145**, 487.
O'Dell, C.R. 1967. *Astrophys. J.*, **149**, 373.
Oliversen, N.G. and Anderson, C.M. 1982a. in
IAU Colloquium No. 70, The Nature of Symbiotic Stars,
ed. M. Friedjung and R. Viotti (Dordrecht: Reidel), p. 71.
Oliversen, N.G. and Anderson, C.M. 1982b. in
IAU Colloquium No. 70, The Nature of Symbiotic Stars,
ed. M. Friedjung and R. Viotti (Dordrecht: Reidel), p. 177.
Oliversen, N.G. and Anderson, C.M. 1983. *Astrophys. J.*, **268**, 250.
Oliversen, N.G., Anderson, C.M., Cassinelli, J.P., Sanders, W.T. and
Kaler, J.B.
1980. *Bull. Amer. Astr. Soc.*, **12**, 819.
Oliversen, N.G., Anderson, C.M., and Nordsieck, K.H. 1982. in
IAU Colloquium No. 70, The Nature of Symbiotic Stars,
ed. M. Friedjung and R. Viotti (Dordrecht: Reidel), p. 153.
Oliversen, N.G., Anderson, C.M., Stencel, R.E., and Slovak, M.H. 1985.
Astrophys. J., **295**, 620.
Oskanian, A.V. 1983. *Inf. Bull. Var. Stars*, No. 2349.
Osterbrock, D.E. 1974. *Astrophysics of Gaseous Nebulae*
(San Francisco: W.H. Freeman and Company).
Paczynski, B. 1965. *Acta Astr.*, **15**, 197.
Pagel, B.E.J. 1958. *Mem. Soc. Roy. Sci. Liege*,
4eme serie, **tome XX**, 177.
Palmer, L.H. and Africano, J.L. 1982. *Inf. Bull. Var. Stars*, No. 2069.
Parenago, P.P. 1933. *Peremennye Zvezdy*, **4**, 275.
Parenago, P.P. 1936. *Peremennye Zvezdy*, **5**, 139.
Parkhurst, H.M. 1890. *Ann. Harv. Coll. Obs.*, **29**, 89.
Payne, C.H. 1928. *Bull. Harv. Coll. Obs.*, No. 861.
Payne, C.H. 1930. *Bull. Harv. Coll. Obs.*, No. 878.
Payne, C.H. 1930. *The Stars of High Luminosity*
(New York: McGraw-Hill), p. 67.
Payne-Gaposchkin, C. 1946. *Astrophys. J.*, **104**, 362.
Payne-Gaposchkin, C. 1952. *Ann. Harv. Coll. Obs.*, **115**, No. 22.
Payne-Gaposchkin, C. 1954. *Ann. Harv. Coll. Obs.*, 113, No. 4.
Payne-Gaposchkin, C. 1957. *The Galactic Novae* (Amsterdam: North Holland).
Payne-Gaposchkin, C. 1958. *Mem. Soc. Roy. Sci. Liege*, 4eme serie,
tome XX, 483.

Payne-Gaposchkin, C. and Boyd, C. 1946. *Astrophys. J.*, **104**, 357.
Payne-Gaposchkin, C. and Gaposchkin, S. 1938. *Variable Stars*
(Cambridge: Harvard University Press).
Payne-Gaposchkin, C. and Wright, F.W. 1946. *Astrophys. J.*, **104**, 75.
Pearce, J.A. 1933. *J. Roy. Astr. Soc. Can.*, **27**, 337.
Peltier, L. 1933. *Gaz. Astr.*, **20**, 135.
Penston, M.V. and Allen, D.A. 1985.
Mon. Not. Roy. Astr. Soc., **212**, 939.
Penston, M.V., Benvenuti, P., Cassatella, A., Heck, A., Selvelli, P.,
Macchetto, F., Ponz, D., Jordan, C., Cramer, N., Rufener, F., and
Manfroid, J. 1980. in *The Universe at Ultraviolet Wavelengths*,
ed. R.D. Chapman (NASA CP-2171), p. 469.
Penston, M.V., Benvenuti, P., Cassatella, A., Heck, A., Selvelli, P.,
Macchetto, F., Ponz, D., Jordan, C., Cramer, N., Rufener, F., and
Manfroid, J. 1983. *Mon. Not. Roy. Astr. Soc.*, **202**, 833.
Perek, L. 1960. *Bull. Astr. Inst. Czech.*, **11**, 256.
Perek, L. 1962. *Bull. Astr. Inst. Czech.*, **13**, 67.
Perek, L. 1963. *Bull. Astr. Inst. Czech.*, **14**, 201.
Perek, L. and Kohoutek, L. 1967. *Catalog of Galactic Planetary*
Nebulae (Prague: Czech. Acad. Sci.).
Persic, M., Hack, M., and Selvelli, P.L. 1984.
Astr. Astrophys., **140**, 317.
Persson, S.E. and Frogel, J.A. 1973. *Astrophys. J.*, **182**, 502.
Peterson, A.W. 1977. *Pub. Astr. Soc. Pac.*, **89**, 129.
Petit, M. 1960. *J. des Observ.*, **43**, 79.
Pettit, E. 1946a. *Pub. Astr. Soc. Pac.*, **58**, 153.
Pettit, E. 1946b. *Pub. Astr. Soc. Pac.*, **58**, 213.
Pettit, E. 1946c. *Pub. Astr. Soc. Pac.*, **58**, 255.
Pettit, E. 1946d. *Pub. Astr. Soc. Pac.*, **58**, 359.
Pettit, E. 1950. *Pub. Astr. Soc. Pac.*, **62**, 65.
Pettit, E. and Nicholson, S.B. 1933. *Astrophys. J.*, **78**, 320.
Pickering, E.C. 1897. *Circ. Harv. Coll. Obs.*, No. 17.
Pickering, E.C. 1901a. *Astrophys. J.*, **13**, 226.
Pickering, E.C. 1901b. *Circ. Harv. Coll. Obs.*, No. 54.
Pickering, E.C. 1904. *Circ. Harv. Coll. Obs.*, No. 76.
Pickering, E.C. 1905. *Circ. Harv. Coll. Obs.*, No. 99.
Pickering, E.C. 1910. *Circ. Harv. Coll. Obs.*, No. 158.
Pickering, E.C. 1911a. *Circ. Harv. Coll. Obs.*, No. 167.
Pickering, E.C. 1911b. *Circ. Harv. Coll. Obs.*, No. 168.
Pickering, E.C. 1914a. *Circ. Harv. Coll. Obs.*, No. 182.
Pickering, E.C. 1914b. *Circ. Harv. Coll. Obs.*, No. 184.
Piirola, V. 1982. in *IAU Colloquium No. 70, The Nature of Symbiotic Stars*,
ed. M. Friedjung and R. Viotti (Dordrecht: Reidel), p. 139.
Plakidis, S. 1933. *Bull. Assoc. Fran. Obs. Et. Var.*, **2**, 144.
Plaskett, H.H. 1923. *Pop. Astr.*, **31**, 658.
Plaskett, H.H. 1927. *Pub. Am. Astr. Soc.*, **5**, 77.
Plaskett, H.H. 1928. *Pub. DAO Victoria*, **4**, 119.
Plaut, L. 1932. *Astr. Nachr.*, **244**, 289.

Ponz, D., Cassatella, A., and Viotti, R. 1982.
in *IAU Colloquium No. 70, The Nature of Symbiotic Stars*,
ed. M. Friedjung and R. Viotti (Dordrecht: Reidel), p. 217.

Pottasch, S.R. 1967. *Bull. Astr. Inst. Neth.*, **19**, 227.

Pottasch, S.R. and Varsavsky, C.M. 1960.
Ann. d'Astrophys., **23**, 516.

Prager, R. 1934. *Geschichte und Literatur des Lichtwechels
der Veranderlichen Sterne* (Berlin: Ferd. Dummlers
Verlasbuchhandlung).

Prager, R. 1940. *Bull. Harv. Coll. Obs.*, No. 912.

Price, S.D. and Walker, R.G. 1976. *Four Color IR Survey*, AFGL-TR-76-0208.

Price, S.D. and Walker, R.G. 1983.
Revised Four Color IR Survey, AFGL-TR-83-0161.

Piirola, V. 1983. in *IAU Colloquium No. 72, Cataclysmic Variables and Related
Objects*, ed. M. Livio and G. Shaviv (Dordrecht: Reidel), p. 211.

Pucinskas, A. 1964. *Bull. Vilniaus Univ. Astr. Obs.*, No. 10, 28.

Pucinskas, A. 1965. *Bull. Vilniaus Univ. Astr. Obs.*, No. 14, 45.

Pucinskas, A. 1970. *Bull. Vilniaus Univ. Astr. Obs.*, No. 27, 24.

Pucinskas, A. 1972. *Bull. Vilniaus Univ. Astr. Obs.*, No. 33, 50.

Puetter, R.C., Russell, R.W., Soifer, B.T., and Willner, S.P. 1978.
Astrophys. J. (Letters), **233**, L93.

Purgathofer, A. and Schnell, A. 1982. *Inf. Bull. Var. Stars*, No. 2071.

Purgathofer, A. and Schnell, A. 1983. *Inf. Bull. Var. Stars*, No. 2264.

Purton, C.R., Allen, D.A., Feldman, F.A., and Wright, A.E. 1977.
Mon. Not. Roy. Astr. Soc., **180**, 97p.

Purton, C.R. and Feldman, P.A. 1978. in
IAU Symposium No. 76, Planetary Nebulae: Observations and Theory,
ed. Y. Terzian (Dordrecht: Reidel), p.325.

Purton, C.R., Feldman, P.A. and Marsh, K.A. 1973.
Nature Phys. Sci., **245**, 5.

Purton, C.R., Feldman, P.A., Marsh, K.A., Allen, D.A., and Wright, A.E. 1982.
Mon. Not. Roy. Astr. Soc., **198**, 321.

Purton, C.R., Kwok, S., and Feldman, P.A. 1983. *Astr. J.*, **88**, 1825.

Quester, W. 1972. *BAV Rundbrief*, **21**, 25.

Raassen, A.J.J. and Hansen, J.E. 1981. *Astrophys. J.*, **243**, 217.

Raassen, A.J.J. 1985. *Astrophys. J.*, **292**, 696.

Rathman, G. 1961. *Mitt. Verand. Sterne*, No. 537.

Razmazde, N.A. 1960a. *Astr Zh.*, **37**, 342.

Razmazde, N.A. 1960b. *Astr Zh.*, **37**, 1005.

Reinmuth, K. 1926. *Astr. Nachr.*, **226**, 193.

Richer, N. 1971. *Astrophys. J.*, **167**, 521.

Richter, N. 1972. *Inf. Bull. Var. Stars*, No. 708.

Ricker, G.R., McClintock, J.E., Gerassimenko, M., and Lewin, W.H.G. 1973.
Astrophys. J., **184**, 237.

Ricker, G.R., Gerassimenko, M., McClintock, J.E., Ryckman, S.G.,
and Lewin, W.H.G. 1976. *Astrophys. J.*, **207**, 333.

Rigollet, R. 1947. *l'Astronomie*, **61**, 247.

Rigollet, R. 1948. *l'Astronomie*, **62**, 297.

Rives, E. 1946. *B.S.A.F.*, **60**, 20.

Robinson, E.L. 1975. *Astr. J.*, **80**, 515.

Robinson, L.J. 1964. *Inf. Bull. Var. Stars*, No. 73.

Robinson, L.J. 1969. *Peremennye Zvedzy*, **16**, 507.

Roche, P.F., Allen, D.A., and Aitken, D.K. 1983.
   Mon. Not. Roy. Astr. Soc., **104**, 1009.

Rodriguez, M.H. 1985. *Inf. Bull. Var. Stars*, No. 2753.

Roman, N.G. 1953. *Astrophys. J.*, **117**, 467.

Romano, G. 1960. *Mem. Soc. Astr. Ital.*, **31**, 11.

Romano, G. 1966. *Mem. Soc. Astr. Ital.*, **37**, 535.

Rosino, L., Taffara, S., and Pinto, G. 1960. *Mem. Soc. Astr.*
   Ital., **31**, 251.

Ross, F.E. 1926. *Astr. J.*, **36**, 122.

Roumens, M. 1934. *Bull. Soc. Astr. Fran.*, **48**, 139.

Rybski, P.M. 1973. *Pub. Astr. Soc. Pac.*, **85**, 653.

Sahade, J. 1949. *Astrophys. J.*, **109**, 541.

Sahade, J. 1960. *Kleine Veroff. Remeis-Sternwarte Bamberg*, **4**, 140.

Sahade, J. 1976.
   Mem. Soc. Roy. Sci. Liege, 6eme serie, **tome IX**, 303.

Sahade, J. and Brandi, E. 1980. in
   The Universe at Ultraviolet Wavelengths,
   ed. R.D. Chapman (NASA CP-2171), p. 451.

Sahade, J., Brandi, E., and Fontenla, J.M. 1981. *Rev. Mex.*
   Astr. Astrophys., **6**, 201.

Sahade, J., Brandi, E., and Fontenla, J.M. 1984.
   Astr. Astrophys. Suppl., **56**, 17.

Sanduleak, N. and Stephenson, C.B. 1972. *Pub. Astr. Soc.*
   Pac., **84**, 816.

Sanduleak, N. and Stephenson, C.B. 1973. *Astrophys. J.*, **185**, 899.

Sanford, R.F. 1935. *Astrophys. J.*, **82**, 202.

Sanford, R.F. 1944a. *Astrophys. J.*, **99**, 145.

Sanford, R.F. 1944b. *Pub. Astr. Soc. Pac.*, **56**, 122.

Sanford, R.F. 1947a. *Pub. Astr. Soc. Pac.*, **59**, 331.

Sanford, R.F. 1947b. *Pub. Astr. Soc. Pac.*, **59**, 333.

Sanford, R.F. 1949a. *Pub. Astr. Soc. Pac.*, **61**, 261.

Sanford, R.F. 1949b. *Astrophys. J.*, **109**, 81.

Sanford, R.F. 1950. *Astrophys. J.*, **111**, 270.

Sayer, A.R. 1938. *Ann. Harv. Coll. Obs.*, **105**, 21.

Schaifers, K. 1960. *Mitt. Verand. Sterne*, No. 431.

Schaifers, K. 1961. *Berlin Sternwarte - Mitt. Veran.*
   Sterne, No. 506-507.

Schmidt, J.F.J. 1866a. *Astr. Nachr.*, **67**, 87.

Schmidt, J.F.J. 1866b. *Astr. Nachr.*, **67**, 257.

Schmidt, J.F.J. 1866c. *Astr. Nachr.*, **68**, 69.

Schmidt, J.F.J. 1884. *Astr. Nachr.*, **108**, 131.

Schonfeld, E. 1867. *Astr. Nachr.*, **69**, 257.

Schwartz, P.R. and Spencer, J.H. 1977. *Mon. Not. Roy.*
   Astr. Soc., **180**, 297.

Schweitzer, E. 1976. *Bull. Assoc. Francaise Obs. Etoiles Var.*, **20**, 24.

Seaquist, E.R. 1977. *Astrophys. J.*, **211**, 547.
Seaquist, E.R. and Gregory, P.C. 1973.
  Nature Phys. Sci., **245**, 85.
Seaquist, E.R., Taylor, A.R., and Button, S. 1984.
  Astrophys. J., **284**, 202.
Seidel, Th. 1956. *Mitt. Verand. Sterne*, No. 238, 1.
Serkowski, K. 1970. *Astrophys. J.*, **160**, 1083.
Shaganian, B.L. 1979. *Peremennye Zvedzy Prilozhenie*, **3**, 735.
Shanin, G.I. 1978. *Soviet Astr. Lett.*, **4**, 100.
Shao, C.Y. and Liller, W. 1971.
  Bull. Am. Astr. Soc., **3**, 443.
Shapley, H. 1919. *Pub. Astr. Soc. Pac.*, **32**, 226.
Shapley, H. 1922a. *Bull. Harv. Coll. Obs.*, No. 762.
Shapley, H. 1922b. *Bull. Harv. Coll. Obs.*, No. 778.
Shapley, H. 1923. *Bull. Harv. Coll. Obs.*, No. 789.
Shapley, H. 1924a. *Bull. Harv. Coll. Obs.*, No. 809.
Shapley, H. 1924b. *Bull. Harv. Coll. Obs.*, No. 810.
Shapley, H. 1940a. *Harv. Announ. Card*, No. 535.
Shapley, H. 1940b. *Harv. Announ. Card*, No. 545.
Shapley, H. 1949a. *Harv. Announ. Card*, No. 1019.
Shapley, H. 1949b. *Harv. Announ. Card*, No. 1020.
Shapley, H. 1950. *Harv. Announ. Card*, No. 1075.
Sharov, A.S. 1954. *Peremennye Zvedzy*, **10**, 55.
Sharov, A.S. 1960. *Peremennye Zvedzy*, **13**, 54.
Shawl, S.J. 1975. *Astr. J.*, **80**, 602.
Simon, T., Morrison, D. and Cruikshank, D.P. 1972.
  Astrophys. J., **177**, L17.
Slovak, M.H. 1978. *Astr. Astrophys.*, **70**, L75.
Slovak, M.H. 1980. *Bull. Am. Astr. Soc.*, **12**, 868.
Slovak, M.H. 1982a. Ph. D. thesis, U. Texas.
Slovak, M.H. 1982b. *Astrophys. J.*, **262**, 282.
Slovak, M.H. 1982c. in *Advances in Ultraviolet Astronomy*, ed.
  Y. Kondo, J.M. Mead and R.D. Chapman (NASA CP-2238), p. 448.
Slovak, M.H. 1982d. *J. Am. Assoc. Var. Star Obs.*, **11**, 67.
Slovak, M.H. and Africano, J. 1978.
  Mon. Not. Roy. Astr. Soc., **185**, 591.
Slovak, M.H. and Lambert, D.L. 1982. in
  IAU Colloquium No. 70, The Nature of Symbiotic Stars,
  ed. M. Friedjung and R. Viotti (Dordrecht: Reidel), p. 103.
Smak, J. 1964. *Astrophys. J. Suppl.*, **9**, 141.
Smak, J. 1966. *Acta Astr.*, **16**, 109.
Smith, S.E. 1979. *Astr. J.*, **84**, 795.
Smith, S.E. 1980. *Astrophys. J.*, **237**, 831.
Smith, S.E. 1981. *Astrophys. J. (Letters)*, **243**, L95.
Smith, S.E. and Bopp, B.W. 1981.
  Mon. Not. Roy. Astr. Soc., **195**, 733.
Sol, H. 1983. *Messenger*, No. 32, p. 37.
Solf, J. 1983. *Astrophys. J. (Letters)*, **266**, L113.
Solf, J. 1984. *Astr. Astrophys.*, **139**, 296.

Soloviev, A.V. 1946a. *Peremennye Zvedzy*, **6**, 68.

Soloviev, A.V. 1946b. *Astr. Tsirk.*, No. 48.

Soloviev, A.V. 1949. *Astr. Tsirk.*, No. 88.

Sopka, R.J., Herbig, G., Kafatos, M., and Michalitsianos, A.G. 1982.
    Astrophys. J. (Letters), **258**, L35.

Spergel, D.N., Giuliani, J.L., Jr., and Knapp, G.R. 1983.
    Astrophys. J., **275**, 330.

Spiesman, W.J. 1984. *Mon. Not. Roy. Astr. Soc.*, **206**, 77.

Splittgerber, E. 1975a. *Mitt. Verand. Sterne*, **6**, 193.

Splittgerber, E. 1975b. *Mitt. Verand. Sterne*, **7**, 39.

Splittgerber, E. 1976. *Mitt. Verand. Sterne*, **7**, 114.

Starikova, G.A. 1960. *Astr. Tsirk.*, No. 208.

Stauffer, J.R. 1984. *Astrophys. J.*, **280**, 695.

Steavenson, W.H. 1946. *Mon. Not. Roy. Astr. Soc.*, **106**, 280.

Stebbins, J., Huffer, C.M., and Whitford, A.E. 1940. *Astrophys. J.*, **91**, 20.

Stebbins, J. and Kron, G.E. 1956. *Astrophys. J.*, **123**, 440.

Stebbins, J. and Whitford, A.E. 1945. *Astrophys. J.*, **102**, 318.

Stein, W.A., Gaustad, J.E., Gillett, F.C., and Knacke, R.F. 1969.
    Astrophys. J., **155**, L3.

Stencel, R.E. 1982. in *Advances in Ultraviolet Astronomy*, ed. Y. Kondo,
    J.M. Mead and R.D. Chapman (NASA CP-2238), p. 219.

Stencel, R.E. 1984. *Astrophys. J. (Letters)*, **281**, L75.

Stencel, R.E., Michalitsianos, A.G., and Kafatos, M. 1982. in
    Advances in Ultraviolet Astronomy,
        ed. Y. Kondo, J.M. Mead and R.D. Chapman (NASA CP-2238), p. 509.

Stencel, R.E., Michalitsianos, A.M., Kafatos, M., and
    Boyarchuk, A.A. 1980. in *The Universe at Ultraviolet Wavelengths*,
        ed. R.D. Chapman (NASA CP-2171), p. 459.

Stencel, R.E., Michalitsianos, A.M., Kafatos, M., and
    Boyarchuk, A.A. 1982. *Astrophys. J. (Letters)*, **253**, L77.

Stencel, R.E. and Sahade, J. 1980. *Astrophys. J.*, **238**, 929.

Stephenson, C.B. and Sanduleak, N. 1971.
    Pub. Warner and Swasey Obs., **1**, No. 1.

Sterne, T.E. and Campbell, L. 1937.
    Ann. Harv. Coll. Obs., **105**, 470.

Stienon, F.M. 1973. *Bull. Am. Astr. Soc.*, **5**, 17.

Stienon, F.M., Chartrand, M.R., and Shao, C.Y. 1974.
    Astrophys. J., **79**, 47.

Stone, E.J. 1866. *Mon. Not. Roy. Astr. Soc.*, **27**, 57.

Stover, R.J. and Sivertsen, S. 1977.
    Astrophys. J. Lett., **214**, L33.

Strafella, F. 1980. *Astrophys. J.*, **243**, 583.

Strickman, M.S., Johnson, W.N., and Kurfess, J.D. 1980. *Astrophys. J.*
    (Letters), **240**, L21.

Struve, O. 1944. *Astrophys. J.*, **99**, 209.

Struve, O. and Elvey, C.T. 1939.
    Pub. Astr. Soc. Pac., **51**, 297.

Svatos, J. and Solc, M. 1981.
    Astrophys. Space Sci., **78**, 503.

Svovpoulos, S.N. 1966. *Pub. Astr. Soc. Pac.*, **78**, 157.

Swings, J.P. 1968. *Bull. Astr. Inst. Czech.*, **19**, 274.

Swings, J.P. 1973. *Astrophys. Lett.*, **15**, 71.

Swings, J.P. 1981. *Astr. Astrophys. Suppl.*, **43**, 331.

Swings, J.P. and Allen, D.A. 1972. *Pub. Astr. Soc. Pac.*, **84**, 523.

Swings, J.P. and Andrillat, Y. 1981. *Astr. Astrophys.*, **103**, L1.

Swings, J.P. and Klutz, M. 1976. *Astr. Astrophys.*, **46**, 303.

Swings, J.P. and Swings, P. 1967. *Astrophys. Lett.*, **1**, 54.

Swings, P. 1970. in *Spectroscopic Astrophysics*,
ed. G. Herbig (Berkeley: U. California), p. 125.

Swings, P., Elvey, C.T., and Struve, O. 1940.
Pub. Astr. Soc. Pac., **52**, 199..

Swings, P. and Struve, O. 1940a. *Astrophys. J.*, **91**, 546.

Swings, P. and Struve, O. 1940b. *Astrophys. J.*, **92**, 295.

Swings, P. and Struve, O. 1941a. *Astrophys. J.*, **93**, 356.

Swings, P. and Struve, O. 1941b. *Astrophys. J.*, **94**, 291.

Swings, P. and Struve, O. 1942a. *Astrophys. J.*, **95**, 152.

Swings, P. and Struve, O. 1942b. *Astrophys. J.*, **96**, 254.

Swings, P. and Struve, O. 1942c. *Astrophys. J.*, **96**, 468.

Swings, P. and Struve, O. 1943a. *Astrophys. J.*, **97**, 194.

Swings, P. and Struve, O. 1943b. *Astrophys. J.*, **98**, 91.

Swings, P. and Struve, O. 1945. *Astrophys. J.*, **101**, 224.

Swope, H.H. 1928. *Bull. Harv. Coll. Obs.*, No. 862.

Swope, H.H. 1938. *Ann. Harv. Coll. Obs.*, **90**, No. 8.

Swope, H.H. 1941. *Astrophys. J.*, **94**, 140.

Swope, H.H. 1943. *Ann. Harv. Coll. Obs.*, **109**, No. 9.

Szkody, P. 1977. *Astrophys. J.*, **217**, 140.

Szkody, P., Michalsky, J. and Stockes, G.M. 1982.
Pub. Astr. Soc. Pac., **94**, 137.

Taffara, S. 1949a. *Atti Ist. Sci. Lett. Arti*, **58**, 95.

Taffara, S. 1949b. *Contr. Asiago Obs.*, No. 13.

Tamura, S. 1977. *Astrophys. Lett.*, **19**, 57.

Tamura, S. 1978. in *IAU Symposium No. 76,
Planetary Nebulae: Observations and Theory*,
ed. Y. Terzian (Dordrecht: Reidel), p.357.

Tamura, S. 1981a. *Astrophys. Lett.*, **22**, 165.

Tamura, S. 1981b. *Pub. Astr Soc. Japan*, **33**, 701.

Tamura, S. 1983. *Pub. Astr. Soc. Japan*, **35**, 317.

Taranova, O.G. and Shenavrin, V.I. 1981. *Astr. Tsirk.*, No. 1185.

Taranova, O.G. and Shenavrin, V.I. 1983. *Peremennye Zvedzy*, **21**, 817.

Taranova, O.G. and Yudin, B.F. 1979. *Astr. Tsirk.*, No. 1034.

Taranova, O.G. and Yudin, B.F. 1980. *Soviet Astr. Lett.*, **6**, 273.

Taranova, O.G. and Yudin, B.F. 1981a. *Astr. Tsirk.*, No. 1188.

Taranova, O.G. and Yudin, B.F. 1981b. *Soviet Astr. J.*, **25**, 598.

Taranova, O.G. and Yudin, B.F. 1981c. *Soviet Astr. J.*, **25**, 710.

Taranova, O.G. and Yudin, B.F. 1982a. *Soviet Astr. J.*, **26**, 57.

Taranova, O.G. and Yudin, B.F. 1982b. *Soviet Astr. Lett.*, **8**, 46.

Taranova, O.G. and Yudin, B.F. 1982c. *Soviet Astr. Lett.*, **8**, 389.

Taranova, O.G. and Yudin, B.F. 1983a. *Astr. Astrophys.*, **117**, 209.

Taranova, O.G. and Yudin, B.F. 1983b. *Sov. Astr. Lett.*, **9**, 19.

Taranova, O.G. and Yudin, B.F. 1983c. *Sov. Astr. Lett.*, **9**, 322.

Taranova, O.G. and Yudin, B.F. 1983d. *Astr. Tsirk.*, No. 1256.

Taranova, O.G. and Yudin, B.F. 1984. *Astr. Tsirk.*, No. 1325.

Taranova, O.G., Yudin, B.F. and Shenavrin, V.I. 1982. *Astr. Tsirk.*, No. 1219.

Taylor, A.R. and Seaquist, E.R. 1984. *Astrophys. J.*, **286**, 263.

Tcheng, M.-L. 1949a. *Comptes Rendus*, **229**, 348.

Tcheng, M.-L. 1949b. *Ann. d'Astrophys.*, **12**, 264.

Tcheng, M.-L. 1950. *Ann. d'Astrophys.*, **13**, 51.

Tcheng, M.-L. and Bloch, M. 1952. *Ann. d'Astrophys.*, **15**, 104.

Tcheng, M.-L. and Bloch, M. 1954. *Ann. d'Astrophys.*, **17**, 6.

Tcheng, M.-L. and Bloch, M. 1955. *Ann. d'Astrophys.*, **18**, 450.

Tcheng, M.-L. and Bloch, M. 1956. *Vistas in Astr.*, **2**, 1412.

Tcheng, M.-L. and Bloch, M. 1957. *Ann. d'Astrophys.*, **20**, 86.

Tempesti, P. 1975. *Inf. Bull. Var. Stars*, No. 974.

Tempesti, P. 1976. *Inf. Bull. Var. Stars*, No. 1094.

Tempesti, P. 1977. *Inf. Bull. Var. Stars*, No. 1291.

Terzian, Y. and Dickey, J. 1973. *Astr. J.*, **78**, 875.

Thackeray, A.D. 1950. *Mon. Not. Roy. Astr. Soc.*, **110**, 45.

Thackeray, A.D. 1954a. *Obs.*, **74**, 90.

Thackeray, A.D. 1954b. *Obs.*, **74**, 257.

Thackeray, A.D. 1954c. *Obs.*, **74**, 258.

Thackeray, A.D. 1959. *Mon. Not. Roy. Astr. Soc.*, **119**, 629.

Thackeray, A.D. 1969. in *Les Transitions Interdites dans Les Spectres des Astres*, 15eme Coll. Astrophys. de Liege, p. 327.

Thackeray, A.D. 1977. *Mem. Roy. Astr. Soc.*, **83**, 1.

Thackeray, A.D. and Hutchings, J.B. 1974. *Mon. Not. Roy. Astr. Soc.*, **167**, 319.

Thackeray, A.D. and Webster, B.L. 1974. *Mon. Not. Roy. Astr. Soc.*, **168**, 101.

The, P.-S. 1962. *Contr. Bosscha Obs.*, No. 17.

The, P.-S. 1964. *Contr. Bosscha Obs.*, No. 26.

Thomas, R.M., Davison, P.J.N., Clancy, M.C., and Buselli, G. 1975. *Mon. Not. Roy. Astr. Soc.*, **170**, 569.

Thompson, G.T., Nandy, K., Jamar, C., Monfils, A., Houziaux, L., Carnochan, D.J., and Wilson, R. 1978. *Catalogue of Stellar UV Fluxes*, (Science Research Council).

Thronson, H.A. and Harvey, P.M. 1981. *Astrophys. J.*, **248**, 584.

Tifft, W.G. and Greenstein, J.L. 1958a. *Mem. Soc. Roy. Sci. Liege*, 4eme serie, **tome XX**, 449.

Tifft, W.G. and Greenstein, J.L. 1958b. *Astrophys. J.*, **127**, 160.

Tolbert, C.R., Pecker, J.-C., and Pottasch, S.R. 1967. *Bull. Astr. Inst. Neth.*, **19**, 17.

Tomov, T. 1984. *Inf. Bull. Var. Stars*, No. 2610.

Tomov, T. 1986. *Inf. Bull. Var. Stars*, No. 2866.

Tomov, T. and Luud, L. 1984. *Astrofizika*, **20**, 99.

Torres-Peimbert, S., Recillas-Cruz, E., and Peimbert, M. 1980. *Rev. Mex. Astr. Astrofiz.*, **5**, 51.

Townley, S.D. 1906. *Bull. Lick Obs.*, **4**, 38.

Townley, S.D., Cannon, A.J., and Campbell, L. 1928. *Ann. Harv. Coll. Obs.*, **79**, 161.

Tsesevich, B.R. 1956. *Astr. Tsirk.*, No. 167.

Uitterdijk, J. 1934. *Bull. Astr. Soc. Neth.*, 256.

Valentiner, W. 1900. *Veroff. Gross. Sternwarte Heidelberg*, **1**, 1.

Vandekerkhove, E. 1934. *Univ. de Brux. Inst. Astr.*, 2nd series, No. 22.

Vandekerkhove, E. 1946. *Ciel et Terre*, **22**, 124.

van Genderen, A.M. 1977. *Astr. Astrophys.*, **58**, 439.

van Genderen, A.M. 1980. *Astr. Astrophys.*, **88**, 77.

van Genderen, A.M. 1982. *Astr. Astrophys.*, **105**, 250.

van Genderen, A.M. 1983. *Astr. Astrophys.*, **119**, 265.

van Genderen, A.M., Glass, I.S. and Feast, M.W. 1974.
Mon. Not. Roy. Astr. Soc., **167**, 283.

van Maanen, A. 1936. *Astrophys. J.*, **84**, 409.

van Maanen, A. 1938. *Astr. J.*, **47**, 23.

Velghe, A.G. 1957. *Astrophys. J.*, **126**, 302.

Vennik, J., Tuvikene, T., and Luud, L. 1984.
Pub. Tartu Astrofiz. Obs., **50**, 170.

Viaro, M. 1934. *Mem. Soc. Astr. Ital.*, **8**, 95.

Vidal, N.V. and Rodgers, A.W. 1974. *Pub. Astr. Soc. Pac.*, **86**, 26.

Viotti, R., Altamore, A., Baratta, G.B., Cassatella, A., Freidjung, M. 1984.
Astrophys. J., **283**, 226.

Viotti, R., Altamore, A., Baratta, G.B., Cassatella, A., Freidjung, M.,
Giangrande, A., Ponz, D., and Ricciardi, O. 1982a.
in *Advances in Ultraviolet Astronomy*,
ed. Y. Kondo, J.M. Mead and R.D. Chapman (NASA CP-2238), p. 446.

Viotti, R., Giangrande, A., Ricciardi, O., and Cassatella, A. 1982b.
in *The Nature of Symbiotic Stars*, IAU Colloquium No. 70, ed.
M. Friedjung and R. Viotti (Dordrecht: Reidel), p. 125.

Viotti, R., Ricciardi, O., Ponz, D., Giangrande, A., Friedjung, M.,
Cassatella, A., Baratta, G.B., and Altamore, A. 1983. *Astr. Astrophys.*,
119, 285.

Vorontsov-Velyaminov, B.A. 1946. *Astr. Zh.*, **23**, 129.

Wackerling, L.R. 1970. *Mem. Roy. Astr. Soc.*, **73**, 153.

Walker, A.R. 1977a. *Mon. Not. Roy. Astr. Soc.*, **178**, 245.

Walker, A.R. 1977b. *Mon. Not. Roy. Astr. Soc.*, **179**, 587.

Walker, A.R. 1983. *Mon. Not. Roy. Astr. Soc.*, **203**, 25.

Walker, G.A.H., Morris, S.C., and Younger, P.F. 1969.
Astrophys. J., **156**, 117.

Walker, M.F. 1957. in *IAU Symposium No. 3, Non-stable Stars*,
ed. G. Herbig (London: Cambridge Univ.), p. 46.

Wallerstein, G. 1958. *Pub. Astr. Soc. Pac.*, **70**, 537.

Wallerstein, G. 1963. *Pub. Astr. Soc. Pac.*, **75**, 26.

Wallerstein, G. 1968. *Obs.*, **88**, 111.

Wallerstein, G. 1969. *Pub. Astr. Soc. Pac.*, **81**, 672.

Wallerstein, G. 1978. *Pub. Astr. Soc. Pac.*, **90**, 36.

Wallerstein, G. 1983. *Pub. Astr. Soc. Pac.*, **95**, 135.

Wallerstein, G. and Cassinelli, J.P. 1968.
Pub. Astr. Soc. Pac., **80**, 589.

Wallerstein, G. and Greenstein, J.L. 1980.
Pub. Astr. Soc. Pac., **92**, 275.

Wallerstein, G., Willson, L.A., Salzer, J., and Brugel, E. 1984.
    Astr. Astrophys., **133**, 137.
Walraven, J.H. and Walraven, Th. 1970.
    Inf. Bull. South. Hemis., No. 17, 31.
Wdowiak, T.J. 1977. *Pub. Astr. Soc. Pac.*, **89**, 569.
Weaver, W.B. 1972. *Pub. Astr. Soc. Pac.*, **84**, 854.
Webbink, R.F. 1976. *Nature*, **262**, 271.
Weber, R. 1958. *Bull. Soc. Astr. Fran.*, **72**, 332.
Weber, R. 1963. *J. des Observ.*, **46**, 175.
Webster, B.L. 1966. *Pub. Astr. Soc. Pac.*, **78**, 136.
Webster, B.L. 1973. *Mon. Not. Roy. Astr. Soc.*, **164**, 381.
Webster, B.L. 1974. in
    IAU Symposium No. 59, Stellar Instability and Evolution,
    ed. P. Ledoux, A. Noels, and A.W. Rodgers (Dordrecht: Reidel), p. 123.
Webster, B.L. and Allen, D.A. 1975.
    Mon. Not. Roy. Astr. Soc., **171**, 171.
Wellman, P. 1940. *Zeit. f. Astrophys.*, **19**, 16.
Welty, D.E. 1983. *Pub. Astr. Soc. Pac.*, **95**, 217.
Wendker, H.J. 1978 *Abhand. Hamb. Stern.*, **X**, No. 1.
Wendker, H.J., Baars, J.W.M., and Altenhoff, W.J. 1973.
    Nature Phys. Sci., **245**, 118.
Wendell, O.C. 1900. *Ann. Harv. Coll. Obs.*, **37**, 267.
Wenzel, W. 1955a. *Mitt. Verand. Sterne*, No. 203.
Wenzel, W. 1955b. *Mitt. Verand. Sterne*, No. 205.
Wenzel, W. 1956. *Mitt. Verand. Sterne*, No. 227.
Wenzel, W. 1976. *Inf. Bull. Var. Stars*, No. 1222.
Westgate, C. 1933. *Astrophys. J.*, **78**, 372.
White, N.E., Mason, K.E., Huckle, H.E., Charles, P.A., and
    Sanford, P.W. 1976. *Astrophys. J. (Letters)*, **209**, L119.
Whitelock, P.A. and Catchpole, R.M. 1982. in
    IAU Colloquium No. 70, The Nature of Symbiotic Stars,
    ed. M. Friedjung and R. Viotti (Dordrecht: Reidel), p. 207
Whitelock, P.A. and Catchpole, R.M. 1983. *Inf. Bull. Var. Stars*, No. 2296.
Whitelock, P.A., Catchpole, R.M., and Feast, M.W. 1982. in
    IAU Colloquium No. 70, The Nature of Symbiotic Stars,
    ed. M. Friedjung and R. Viotti (Dordrecht: Reidel), p. 215.
Whitelock, P.A., Catchpole, R.M., Feast, M.W., Roberts, G., and
    Carter, B.S. 1983a. *Mon. Not. Roy. Astr. Soc.*, 203, 363.
Whitelock, P.A., Feast, M.W., Catchpole, R.M., Carter, B.S., and
    Roberts, G. 1983b. *Mon. Not. Roy. Astr. Soc.*, **203**, 351.
Whitelock, P.A., Feast, M.W., Roberts, G., Carter, B.S., and
    Catchpole, R.M. 1983c. *Mon. Not. Roy. Astr. Soc.*, **205**, 1207.
Whitelock, P.A., Menzies, J.W., Evans, T.L., and Kilkenny, D. 1984.
    Mon. Not. Roy. Astr. Soc., **208**, 161.
Wilde, K. 1965. *Pub. Astr. Soc. Pac.*, **77**, 208.
Williams, G. 1983. *Astrophys. J. Suppl.*, **53**, 523.
Willson, L.A., Garnavich, P., and Mattei, J.A. 1981.
    Inf. Bull. Var. Stars, No. 1961.

Willson, L.A., Wallerstein, G., Brugel, E., and Stencel, R.E. 1984. *Astr. Astrophys.*, **133**, 154.

Wilson, R.E. 1923. *Astr. J.*, **34**, 183.

Wilson, R.E. 1942. *Astrophys. J.*, **96**, 371.

Wilson, R.E. 1943. *Pub. Astr. Soc. Pac.*, **55**, 282.

Wilson, R.E. 1945. *Pub. Astr. Soc. Pac.*, **57**, 309.

Wilson, R.E. 1950. *Pub. Astr. Soc. Pac.*, 62, 14.

Wilson, R.E. and Merrill, P.W. 1942. *Astrophys. J.*, **95**, 248.

Wilson, O.C. and Williams, E.G. 1934. *Astrophys. J.*, **80**, 344.

Wing, R.E. 1962. *Mon. Not. Roy. Astr. Soc.*, **125**, 189.

Wing, R.E. and Carpenter, K.G. 1980. in *The Universe at Ultraviolet Wavelengths*, ed. R.D. Chapman (NASA CP-2171), p. 341.

Wolf, W. 1919. *Astr. Nachr.*, **208**, 363.

Woodsworth, A.W. and Hughes, V.A. 1977. *Astr. Astrophys.*, **58**, 105.

Woolf, N.J. 1969. *Astrophys. J.*, **157**, L37.

Woolf, N.J. 1973. *Astrophys. J.*, **185**, 229.

Wray, J.D. 1966. Ph. D. thesis, U. Northwestern.

Wright, A.E. and Allen, D.A. 1978. *Mon. Not. Roy. Astr. Soc.*, **184**, 893.

Wright, F.W. 1946. *Bull. Harv. Coll. Obs.*, No. 918.

Wright, W.H. 1919. *Pub. Astr. Soc. Pac.*, **31**, 309.

Wright, W.H. and Neubauer, F.J. 1933. *Pub. Astr. Soc. Pac.*, **45**, 252.

Wu, C.-C., Ake, T.B., Boggess, A., Bohlin, R.C., Imhoff, C.L., Holm, A.V., Levay, Z.G., Panek, R.J., Schiffer, III, F.H., and Turnrose, B.E. 1983. *The IUE Ultraviolet Spectral Atlas, IUE NASA Newsletter No. 22.*

Yamamoto, I. 1924a. *Bull. Harv. Coll. Obs.*, No. 797.

Yamashita, Y. 1972. *Ann. Tokyo Astr. Obs.*, **3**, 169.

Yamashita, Y. and Maehara, H. 1979. *Pub. Astr. Soc. Japan*, **31**, 307.

Yamashita, Y., Maehara, H., and Norimoto, Y. 1982. *Pub. Astr. Soc. Jap.*, **34**, 269.

Yamashita, Y., Norimoto, Y., and Yoo, K.H. 1983. *Pub. Astr. Soc. Japan*, **35**, 521.

Yildizdogdu, F.S. 1981. *Publ. Istanbul Univ. Obs.*, No. 117.

Yoo, K.H. 1984. *Ann. Tokyo Obs. 2nd ser.*, **20**, 75.

Yoo, K.H. and Yamashita, Y. 1982. *Ann. Tokyo Obs. 2nd ser.*, **19**, 38.

Yoo, K.H. and Yamashita, Y. 1984. *Pub. Astr. Soc. Japan*, **36**, 567.

Yudin, B.F. 1981a. *Astr. Tsirk.*, No. 1160, p. 4.

Yudin, B.F. 1981b. *Astr. Tsirk.*, No. 1160, p. 7.

Yudin, B.F. 1981c. *Astr. Tsirk.*, No. 1192.

Yudin, B.F. 1982a. *Astr. Tsirk.*, No. 1220.

Yudin, B.F. 1982b. *Astr. Tsirk.*, No. 1221.

Yudin, B.F. 1982c. *Soviet Astr. J.*, **26**, 187.

Yudin, B.F. 1986. *Astr. Zh.*, **63**, 137.

Zakarov, G.P. 1951. *Peremennye Zvedzy*, **8**, 369.

Zipoy, D.M. 1975. *Astrophys. J.*, **201**, 397.

Zirin, H. 1976. *Nature*, **259**, 466.

Zongli, L. 1984. *Inf. Bull. Var. Stars*, No. 2576.

Zongli, L. and Xiangliang, H. 1983. *Inf. Bull. Var. Stars*, No. 2446.

Zongli, L., Xiangliang, H., and Bao, M. 1983.
  Inf. Bull. Var. Stars, No. 2291.

Zuckerman, B. 1979. *Astrophys. J.*, **230**, 442.

Subject index

absorption bands
  H_2O 5, 41–44, 65, 105, 157–160
  CN 41–42
  CO 5, 41–44, 105, 157–160
  OH 65
  CaCl 41
  SiO 65–66
  TiO 2, 5, 11, 31, 40–42, 44–45, 67, 94,
    98–99, 105–108, 157–160
  VO 105
accretion 15–17, 22–23, 82–90, 97,
  100–102, 112–113
Algol binaries 117

binary models 11, 13–23, 71–73
birthrates 116–119
bright spot 30

carbon stars 8
Chandrasekhar limit 95, 125
CNO cycle 73–75
common envelope evolution 118–121,
  123–125

disk instabilities 84–87
dust 32–34, 56–57, 77, 122–125

Eddington limit 15–16, 86, 88, 95, 100, 123
emission lines
  HI 1–3, 5, 13–16, 19–21, 28, 31, 40–41,
    52, 54–56, 58, 61–63, 65, 89, 94–95,
    98–99, 104–105, 107–108, 130–131,
    157–160
  He I 2–3, 5, 20–21, 52, 54, 61, 94–95,
    105, 130, 157–160
  He II 1–3, 12–14, 19–21, 28, 40–41,
    51–52, 54–56, 58, 61, 64, 89, 94–95,
    98–99, 104–105, 107–108, 130–131,
    157–160

C III 1
C III] 40–41, 52, 58
C IV 13, 40–41, 52, 65, 157–160
[N II] 105
N III 1–2, 94–95, 106
N III] 58
N IV] 58
N V 13, 41, 52
[O III] 1–2, 5, 21, 54, 58–59, 94–95,
  105
[Ne III] 2, 21, 58, 94–95
Ne IV] 58
[Ne V] 54, 107, 131
Na I 41, 98
Mg I 40, 192
Mg II 52, 57–58
Si IV 41, 105
[S III] 105
[A V] 105
Ca I 40–41
Ca II 31, 41, 98–99
Ti II 2, 31, 57, 98
Fe I 40–41
Fe II 2, 26, 57–58, 94–95
[Fe II] 26, 57, 94–95
[Fe VII] 54, 105, 107–108, 131, 157–160
[Fe X] 54, 94–95
[Fe XIV] 94, 96
Ba II 99
La II 99
$\lambda 6830$ 54, 107–108, 157–160

flickering 29–31

helium shell flashes 71–72, 82–83
hydrogen shell flashes 73–81, 89–90, 95–96,
  100–102, 103–112

infrared photometry 31–34

Individual stars